ECONOMIC AND FINANCIAL ANALYSIS OF CAPITAL INVESTMENTS

ECONOMIC AND FINANCIAL ANALYSIS OF CAPITAL INVESTMENTS

G.T. Stevens, Jr.
The University of Texas at Arlington

JOHN WILEY & SONS
New York
Chichester
Brisbane
Toronto

Library of Congress Cataloging in Publication Data:

Stevens, Jr. Gladstone Taylor, 1930–
Economic and financial analysis of capital investments.

Includes index.
1. Capital Investments—Evaluation. I. Title.
HG4028.C4S715 658.1′527 78-27851
ISBN 0-471-04851-8

Printed in the United States of America

10 9 8 7 6 5 4 3 2 1

PREFACE

This book is intended to be used as a general text in the area of economic analysis. To a large extent, it deals with the techniques for evaluating and comparing capital expenditures. Although much has been written about these techniques, it is felt that this book provides additional depth and insight into these techniques as a result of its detailed tax considerations and explicit consideration of financing capital expenditures by combinations of debt and equity. Furthermore, it shows the direct relationship between the definition of cash flow and cost of capital. Much of this is accomplished through a technique used by public utility companies for the evaluation of capital expenditures: the generation of minimum annual revenue requirements. This technique has long been used by public utilities but, for some reason, is not usually included in most discussions regarding the techniques for evaluating capital expenditures. This is unfortunate since it provides an excellent vehicle by which to show, numerically, the relationship between cash flow and cost of capital. In addition, it allows the inclusion of detailed tax considerations into cost comparisons and replacement analysis. Also, the technique provides a means of determining the short-term (yearly) effect of a particular capital expenditure on the gross income requirements of the company.

The basic approach taken in this book is to show and explain the "traditional approaches" and techniques for evaluating capital expenditures and then extends them to include, explicitly, taxes and debt financing. This is accomplished through the extensive use of detailed numerical examples followed by a discussion of the theoretical aspects. In these discussions a definite attempt has been made to correlate some of the vocabulary used in the areas of finance, accounting, and engineering economy. To some readers of this text, this might be a problem. However, it is worth the effort since, in practice, the vocabulary and terminology in these areas is often used interchangeably.

In general, this book is limited to point estimates of the variables involved in the analysis of capital expenditures. However, Chapter 12 does provide an introduction to the probabilistic case. It is believed that this chapter provides a foundation on which to consider more advanced topics.

I take this opportunity to thank my wife Jane and our children Bob and Bartie for their patience and encouragement. Also, I am grateful to Sandra Brewster who typed the final manuscript.

G.T. Stevens, Jr.

CONTENTS

ECONOMIC AND FINANCIAL ANALYSIS OF CAPITAL INVESTMENTS

CHAPTER 1
INTRODUCTION

Economic analysis is studied in most business and engineering colleges. Usually, in the business college, it is called managerial accounting, economics, or finance and in the engineering college, engineering economy. Whatever the name, the study of the methods and procedures for making sound economic decisions is most important. For most business and engineering college graduates are faced, at some point in their professional careers, with the responsibility of making economic decisions.

NATURE AND SCOPE OF ECONOMIC ANALYSIS

This book discusses economic analysis from the standpoint of making economic decisions at the company level. At this level, there are three major economic decisions: (1) the investment decision, (2) the financing decision, and (3) the dividend decision. All of these decisions are interrelated.

The Investment Decision

The investment decision is concerned with the evaluation of contemplated capital investments (expenditures) and the allocation of capital between competing investments. For example, the investment decision is concerned with answering such questions as:

1. Should a company build a new plant?
2. Should a company expand its production facilities?
3. Should an existing piece of equipment be replaced?
4. Which alternative should be chosen if investment funds are limited or the alternatives are mutually exclusive?

The answers to these questions must be based on a sound economic evaluation.

The discussion and presentation of various methods and techniques for evaluating prospective capital investments constitutes a major portion of this text.

At the onset, it must be understood what is meant by the word "decision." The methods and procedures presented in this text do not, by themselves, provide decisions. They provide guidelines for making decisions. In the final analysis, some person or group of persons must, ultimately, make the decision due to the uncertainties that exist in any economic evaluation. For, by its basic nature, an economic evaluation requires the estimation of many variables, usually over a considerable period of time. Consequently, there is always an element of risk in any economic evaluation. This risk must be understood and accepted by the decision maker and economic analyst. After all, one of the basic ideas in a capitalistic economy is that people receive "profits" as a result of their willingness to accept risk.

The Financing Decision

In order for a company to make capital investments it must have a source of funds. In general, these funds are obtained from two sources: debt and equity sources. Funds from these sources "cost" money and these costs are different. The total cost is a result of the individual costs and the amounts of money from each source. Therefore, the financing decision is concerned with the determination of the best mix of debt and equity funds for a particular company.

The Dividend Decision

Basically, the dividend decision is concerned with the amount of dividends paid to stockholders and the dependability of these dividends over time. The amount of dividends paid to stockholders has an effect on the amount of money available for equity financing and the cost of equity funds. Consequently, a balance must be made between dividend payments and investment needs.

Objective of the Company

In this book the basic long-range objective of the company is to maximize the stockholder's wealth. That is, to maximize the market's value of a share of stock. This value is an index of a company's progress and reflects the market's attitude toward investment, financing, and dividend decisions made by a company's management. This objective is equivalent to maximizing the net present value of the company and is much more inclusive than the frequently stated objective of profit maximization. For example, if profit maximization is the objective, no dividends would even be paid because total profits could always be increased by investing the dividends that would normally be paid to stockholders in some other ventures—or, at the very least, putting them in a savings account (or certificates of deposit). Such a policy would, of course, decrease the market value of the company's stock.

The objective of maximizing the stockholder's wealth is often subject to constraints. These constraints can be self-imposed due to short-term considerations such as liquidity requirements or a result of social mores or governmental restrictions. Even with constraints, the objective of maximizing the stockholders wealth serves as a sound basis on which to base investment, financing, and dividend decisions.

AN OVERVIEW OF THE BOOK

As previously stated this book is primarily concerned with the evaluation and comparison of capital investments. The techniques and methods used for these evaluations and comparisons have received considerable attention in the literature. However, this book is somewhat different as a result of its detail from the standpoint of tax regulations, the inclusion of such financial considerations as cost of capital and combinations of debt and equity financing and their relationship to the evaluation of capital expenditures, and the presentation of the methodology for generating minimum annual revenue requirements.

The approach taken in this book is primarily quantitative. This does not imply that all variables can be determined quantitatively. There are always qualitative factors that a decision maker must consider. However, even when economic decisions are largely qualitative, the quantitative approach can often lead to a reduction of the decision area.

Chapters 2, 3, and 4 are fundamental since they provide the foundation for the discussion of various topics in later chapters. Chapter 5 is a general discussion of such important topics as internal rate of return, net present value, and payback period. The discussion in this chapter is limited to a simple definition of cash flow that does not explicitly include debt financing. More involved definitions of cash flow are provided in Chapter 7. To some, it might appear that Chapter 5 is out of context. However, it is felt that by putting this chapter at this point in the book, a "sense of direction" is provided as well as a foundation, from a vocabulary standpoint, for later topics. Chapter 6 discusses cost of capital from different sources (bonds, retained earnings, etc.). It is considered fundamental since cost of capital is a basis for establishing the minimum attractive rate of return.

Chapter 7 is considered important from several standpoints. First, it shows the methodology for generating minimum annual revenue requirements—a method that can be used (and is by public utility companies) for evaluating capital expenditures, as well as those discussed in Chapter 5. Second, since minimum annual revenue requirements are annual costs, Chapter 7 provides a basis for the discussion of cost comparisons and replacement analysis. Finally, Chapter 7 shows the numerical relationships between cost of capital and definitions of cash flow.

Chapter 8 discusses before and after tax breakeven models, as well as the concept of profit from the standpoint of the accountant and economist and the relationship between profit and cash flow.

Chapters 9, 10 and 11 are, respectively, discussions of capital budgeting, cost comparisons, and replacement. These chapters include detailed tax considerations. Finally, Chapter 12 concludes with some probabilistic cash flow models.

CHAPTER 2
INTEREST AND INTEREST FACTORS

This chapter is concerned with interest, interest factors, and the concept of equivalence. Basically, interest is a payment for the use of money. It has been a factor in economic exchange for thousands of years, and it is probably safe to say that it will always be a factor in economic exchange. The ethics of interest has been a point of contention between philosophers through the ages; the fact is, however, interest is here to stay. Consequently, the considerations of interest and interest rates is required in any text on economic analysis and is fundamental in the evaluation of capital investments.

Interest can be considered from two viewpoints. If interest is paid, it can be considered as a cost. On the other hand, if interest is received, it can be considered as a return. Since money can earn a return over a period of time, interest rates are often considered as an expression of the *time value of money*.

Note that there is a difference between interest and interest rates, although these terms are often used synonymously. Strictly speaking, interest is the amount of money paid or received. That is, the units of interest are dollars. While, interest rates are usually expressed as a percentage.

SIMPLE INTEREST

When interest is paid (or received) only on the principal, this is called simple interest or, sometimes, "add-on" interest. Simple interest is calculated using the relationship

$$I = Pni \tag{2-1}$$

Where, P is the principal, n is the number of interest periods, and i is the interest rate. The final amount, F, due on a simple interest loan is the principal plus the interest:

$$F = P + I = P + Pni \tag{2-2}$$

EXAMPLE 2-1

If $2000 is borrowed for four years at 8% per year simple interest, how much interest is paid and what is the amount due at the end of four years?

The total interest is determined using Eq. (2-1)

$$I = (2000)(4)(0.08) = \$640$$

and the total amount due at the end of four years is

$$F = 2000 + 640 = \$2640$$

As an extension, simple interest may be based on a period of time that involves fractions of a year. In these cases, it is common to consider a year as being 12 months of 30 days each. This is called an annum. For example, if the loan in Example 2-1 is made for 100 days and the interest is 8% per annum, the amount due at the end of the 100 days is

$$F = 2000 + 2000\left(\frac{100}{360}\right)(0.08) = \$2044.44$$

COMPOUND INTEREST

When interest is paid on the principal plus accumulated interest, it is called compound interest.

EXAMPLE 2-2

Suppose in the first part of Example 2-1 the interest rate is compounded annually, what is the amount due at the end of four years?

The amount due in four years is $2720.97. The calculations to obtain this amount are shown in Table 2-1. The total interest paid is

$$I = 2720.97 - 2000 = \$720.97$$

Note that there is a significant dfference between the simple interest in Example 2-1 and the compound interest in Example 2-2.

DISCRETE INTEREST FACTORS

Calculations such as those shown in Table 2-1 are not difficult. However, they can be tedious when the number of years involved is large. Consequently,

Table 2-1
CALCULATIONS FOR EXAMPLE 2-2

End of Year	Principal	Interest	Amount Due
0	$2000	—	—
1	—	2000(0.08) = $160.00	2000 + 160.00 = $2160.00
2	—	2160(0.08) = 172.80	2160 + 172.80 = 2332.80
3	—	2332.80(0.08) = 186.62	2332.80 + 186.62 = 2519.42
4	—	2519.42(0.08) = 201.55	2519.42 + 201.55 = 2720.97

interest factors are defined and evaluated that greatly reduce the amount of work required in interest calculations. These interest factors are discussed in this section. Initially, this discussion is limited to interest periods of one year and discrete interest rates. Other periods of time and continuous interest rates are considered in later sections. This section also uses the end-of-period convention. That is, all cash transactions are considered to occur at the end of an interest period. The nomenclature that is used in this discussion of interest factors follows:

i = interest rate per interest period
n = number of interest periods
P = a single amount at the present time ($t = 0$)
This is often called the principal or initial amount.
F = a future amount at the end of n interest periods
A = a series of consecutive equal amounts made at end of each interest period

Single-Payment Compound-Amount Factor

If Table 2-1 is considered, it can be seen that the amount due at the end of each year, in general, is

$$F = P(1 + i)^n \tag{2-3}$$

For example, at the end of three years the amount due, F, is

$$F = 2000(1 + 0.08)^3 = \$2519.42$$

The factor, $(1 + i)^n$, in Eq. (2-3) is known as the single-payment compound-amount factor and is designated in this text as $(FP\ i,n)$. That is,

$$(FP\ i,n) = (1 + i)^n \tag{2-4}$$

This factor can be used in the formulation

$$F = P(FP\ i,n) \tag{2-5}$$

and gives, for the data in Example 2-2,

$$F = 2000(FP\ 8,4) = 2000(1.360) = \$2720$$

The value 1.360 is obtained from the interest tables given in Appendix A. The difference in this answer (\$2720) and the answer given in Table 2-1 (\$2720.97) is due to round-off error.

Single-Payment Present-Worth Factor

This factor is just the reciprocal of the single-payment compound-amount factor. That is,

$$(PF\ i,n) = \frac{1}{(1 + i)^n} \tag{2-6}$$

which is used in the formulation

$$P = F(PF\ i,n) \tag{2-7}$$

EXAMPLE 2-3

Suppose that it is known that \$5000 is needed five years from now to pay a certain debt. If it is possible to put money into a savings account that earns 6% compounded annually, how much money should be put into the account in order to have the \$5000 at the end of the five years?

The solution to this problem is

$$P = 5000(PF\ 6,5) = 5000(0.7473) = \$3736.50$$

Where the value 0.7473 is also obtained from Appendix A. Of course, this problem could also be solved using Eq. (2-5). However for more involved problems it is desirable to express interest factors in terms that require only multiplication.

Equal-Payment-Series Compound-Amount Factor

Frequently, in the analysis of economic alternatives, a series of equal costs (receipts) occur at the end of successive interest periods as shown in Table 2-2. The future amount occurring at the end of n years could be determined in the following manner:

$$F = A(1 + i)^{n-1} + A(I + i)^{n-2} + \ldots + A(1 + i)^0 \tag{2-8}$$

However, this can lead to tedious calculations. Therefore, another interest factor can be obtained that simplifies this type of calculation. Multiplying Eq. (2-8) by (1 + i) gives

$$F(1 + i) = A(1 + i)^n + A(1 + i)^{n-1} + \ldots + A(1 + i) \tag{2-9}$$

Table 2-2
SERIES OF YEAR-END PAYMENTS

End of Year	Payment
0	—
1	A
2	A
.	.
.	.
.	.
n	A

Subtracting Eq. (2-8) from Eq. (2-9) gives

$$F(1 + i) - F = -A + A(1 + i)^n \tag{2-10}$$

and rearranging Eq. (2-10) results in

$$F = A\left[\frac{(1 + i)^n - 1}{i}\right] \tag{2-11}$$

The term in the parentheses in Eq. (2-11) is called the equal-payment-series compound-amount factor and designated $(FA\ i,n)$. That is,

$$(FA\ i,n) = \frac{(1 + i)^n - 1}{i} \tag{2-12}$$

and

$$F = A(FA\ i,n) \tag{2-13}$$

EXAMPLE 2-4

How much money will be in account at the end of 10 years if 10 equal annual deposits of \$1000 are made and the interest rate is 7%? Assume the first deposit is made 1 year from now.

The solution to this problem is obtained using Eq. (2-13) as shown below:

$$F = 1000(FA\ 7,10) = 1000(13.186) = \$13,186$$

A Comment: To some, the idea of starting the deposits one year from now might not seem practical. That is, why not start the deposits now? Also, why make the deposit in the last year of the series since no interest is earned? The answer is simply that some convention must be used when establishing equal-payment interest factors and the convention shown in Table 2-2 is the choice of this text as it is in most all texts that define equal-payment factors. This convention poses no limitations since, with a slight extension, the problem of beginning-of-the-year equal payments can be solved using end-of-the-year equal-payment factors. Suppose, in this example, the first deposit is made now. This results in the situation depicted in Table 2-3. The amount in the account is

$$F = 1000(FA\ 7,10)(FP\ 7,1)$$
$$F = 1000(13.816)(1.07) = \$14,783.12$$

Equal-Payment-Series Sinking-Fund Factor

This factor is the reciprocal of the equal-payment-series compound-amount factor:

$$(AF\ i,n) = \frac{i}{(1 + i)^n - 1} \tag{2-19}$$

Eq. (2-19) is used in the formulation

$$A = F(AF\ i,n) \tag{2-20}$$

Table 2-3
BEGINNING-OF-THE-YEAR DEPOSITS FOR EXAMPLE 2-4

End of the Year	Deposit
0	$1000
1	1000
2	1000
3	1000
.	.
.	.
.	.
9	1000
10	$F = ?$

Equal-Payment-Series Present-Worth Factor

This interest factor is the combination of the $(FA\ i,n)$ and $(PF\ i,n)$ factors. It is defined as

$$(PA\ i,n) = (FA\ i,n)(PF\ i,n) = \frac{(1 + i)^n - 1}{i(1 + i)^n} \qquad (2\text{-}21)$$

and used in the formulation

$$P = A(PA\ i,n) \qquad (2\text{-}22)$$

EXAMPLE 2-5

A man anticipates that he must have $3000 per year for the next five years in order to pay certain debts. He has decided to deposit into an account now an amount that will just meet his debt obligations over the next five years. If the account pays 6%, how much must the man deposit?

The solution to this problem is to apply Eq. (2-22) and is shown below:

$$P = 3000(PA\ 6,5) = 3000(4.2124) = \$12,637.20$$

Equal-Payment Series Capital-Recovery Factor

This factor is the reciprocal of the equal-payment series present-worth factor:

$$(AP\ i,n) = \frac{i(1 + i)^n}{(1 + i)^n - 1} \qquad (2\text{-}23)$$

This factor is used in the formulation,

$$A = P(AP\ i,n) \qquad (2\text{-}24)$$

EFFECTIVE AND NOMINAL INTEREST

Up to this point, the discussion of interest rates has been limited to periods of one year. Often, however, interest rates and payments (or receipts) use periods of time other than a year.

In order for an interest rate to be completely defined, two time periods must be specified (or at least understood). For example,

6% per year compounded per year

and

4% per year compounded semiannually

are completely defined since the time basis of the interest rate and compounding periods are specified. Many times an interest rate is specified as

4% compounded semiannually

which can be misleading to those who are not familiar with interest rates. Just as specifying an interest rate of

$1\frac{1}{2}$% per month

can be misleading. These considerations lead to definitions of an effective *annual* interest rate, effective interest rate, and a nominal interest rate.

Effective Annual Interest Rates

With an effective *annual* interest rate, both time periods are expressed on an annual basis. Thus, 6% per year compounded per year is an effective *annual* interest rate. In this text if no time units are specified it should be assumed that the interest rate is an effective annual rate.

Effective Interest Rates

With an effective interest rate, both time periods are the same. Thus, 2% per month compounded per month is an effective interest rate. Note that with this definition an effective *annual* interest rate is always an effective interest rate. However, an effective interest is not, necessarily, an effective *annual* interest rate. For example, the 2% figure previously mentioned is an effective interest rate but not an effective *annual* interest rate.

Nominal Interest Rates

When an interest rate is expressed, for example, as 6% per year compounded semiannually, it is called a nominal interest rate. With a nominal interest rate the time basis is always per year. However, the compounding period may be different. With this definition, an effective annual interest rate could also be considered a nominal interest rate. However, a nominal interest rate is not necessarily, an effective annual rate. Quite often an interest rate is specified, for example, as 6% compounded semiannually. In this case, the phrase "per year" is understood.

It is a simple matter to convert nominal rates to effective rates or vice versa as shown below:

$$\text{Nominal rate} = (c)\,(\text{effective rate}) \tag{2-25}$$

Where, c is the number of compounding periods per year. For example, 6% per year compounded semiannually is equivalent to 3% semiannually compounded semiannually ($c = 2$); 18% per year compounded monthly is equivalent to 1½% per month compounded monthly ($c = 12$); and 12% per year compounded quarterly is equivalent to 3% per quarter compounded quarterly ($c = 4$).

In order to convert a nominal interest rate, r, to an effective annual interest rate the following relationship is used.

$$\text{effective annual interest rate} = \left(1 + \frac{r}{c}\right)^c - 1 \tag{2-26}$$

Where, r is the nominal interest rate and c is the same as previously defined. As an example, 18% per year compounded monthly is equivalent to an effective annual rate of

$$\left(1 + \frac{0.18}{12}\right)^{12} - 1 = 0.196 = 19.6\%$$

EXAMPLE 2-6

A woman borrows \$2000 and is told that she can pay back the loan in 24 monthly payments of \$94.15. Determine the effective interest rate, nominal interest rate, and the effective annual interest rate.

The effective interest rate is determined in the following manner.

$$2000 = 94.15(PA\ i,24)$$
$$(PA\ i,24) = 21.243$$
$$i = 1\% \text{ per month compounded per month}$$

The nominal rate is

$$(1)(12) = 12\% \text{ per year compounded monthly}$$

which is often abbreviated to 12% compounded monthly; the "per year" phrase being understood. The annual effective rate is

$$\left(1 + \frac{0.12}{12}\right)^{12} - 1 = 0.127 = 12.7\%$$

A Comment: Sometimes interest rates on automobile loans are "quoted" at values considerably lower than the effective annual rates. When this is done the monthly payments are usually calculated on the basis of simple interest. Fortunately, as a result of the fair-disclosure acts, the final loan papers, if studied carefully, will show the interest rate that is actually being paid. Most of the time this interest rate is the nominal rate rather than the effective annual rate. For example, an automobile loan of \$2000 for 24 months at 6.5% means that the monthly payment is

$$\frac{2000 + 2000(0.065)(2)}{24} = \$94.17$$

In this case, the nominal interest rate and annual effective interest rate are essentially the same as previously calculated in this example since the monthly payments are approximately equal, that is, 12% and 12.7%.

INTEREST PERIODS OTHER THAN A YEAR

Sometimes the compounding period and the timing of the payments (receipts, withdrawals, etc.) are not the same. In these cases, the basic approach is to convert the given interest rate to a nominal rate with compounding periods that are the same as the payments. This conversion can be accomplished by using Eq. (2-26).

EXAMPLE 2-7

A man deposits \$1000 each quarter into an account that pays 6% compounded semiannually. How much money is in the account at the end of three years?

Since the compounding period and deposit periods are not the same, it is necessary to convert this 6% interest rate to a nominal rate that is the same as the deposits, that is, quarterly. The first step is to convert the interest rate to annual effective rate using Eq. (2-26).

$$\left(1 + \frac{0.06}{2}\right)^2 - 1 = 0.0609 = 6.09\%$$

The next step is to convert this value (6.09%) to a nominal rate compounded, in this case, quarterly. This is accomplished using Eq. (2-26) as shown below:

$$0.0609 = \left(1 + \frac{r}{4}\right)^4 - 1$$

$$(1.0609)^{0.25} = 1 + \frac{r}{4}$$

$$r = (0.01489)(4) = 0.0596 = 5.96\% \text{ compounded quarterly}$$

Therefore,

$$\begin{aligned} F &= A(FA\ i,n) \\ &= 1000\left(FA\ \frac{5.96}{4},\ 4(3)\right) \\ &= 1000(FA\ 1.49,\ 12) \\ &= 1000\left[\frac{(1 + 0.0149)^{12} - 1}{0.0149}\right] \\ &= 1000(13.0336) = \$13{,}033.60 \end{aligned}$$

There are approximate approaches to this problem that will give almost the same answer. However, this solution is theoretically correct.

In cases where the deposits and compounding period are the same or only a single amount is involved, the calculations are more direct as shown in the next example.

EXAMPLE 2-8

A present deposit of \$1000 will amount to the following amount after five years at 6% compounded monthly.

$$F = 1000\left(FP\ \frac{6}{12}, 12(5)\right)$$

$$F = 1000(FP\ 0.5,60) = 1000(1.349) = \$1349$$

The future amount of a series of quarterly payments of \$1000 at 12% compounded quarterly for five years is

$$F = 1000\left(FA\ \frac{12}{4}, 5(4)\right)$$

$$F = 1000(FA\ 3,20) = 1000(26.870) = \$26,870$$

CONTINUOUS INTEREST FACTORS

Interest rates are sometimes compounded continuously. When interest rates are compounded continuously, this implies a nominal rate. For example, 6% compounded continuously implies 6% *per year* compounded continuously. Furthermore, it implies interest factors different than those previously defined.

Single-Payment Compounded-Amount Factor

In Example 2-8, the future amount is given as

$$F = P\left(FP\ \frac{r}{c}, cn\right) \tag{2-27}$$

or

$$F = P\left(1 + \frac{r}{c}\right)^{cn} \tag{2-28}$$

Eq. (2-28) can also be written as

$$F = P\left[\left(1 + \frac{r}{c}\right)^{c/r}\right]^{rn} \tag{2-29}$$

Now, continuous compounding implies that c approaches infinity. Consequently, the limit of the term in the brackets as $c \to \infty$ is $e = 2.7182$ and Eq. (2-29) can be written as

$$F = Pe^{rn} \tag{2-30}$$

The factor e^{rn} is called the single-payment compound-amount factor and designated $(FP\ r,n)$. It is used in the formulation

$$F = P(FP\ r,n) \tag{2-31}$$

EXAMPLE 2-9

If \$1000 is deposited into an account that pays 6% compounded continuously, the final amount at the end of 10 years is

$$F = 1000(FP\ 6,10) = 1000(1.822) = \$1822$$

The value 1.822 is obtained from Appendix A.

Single-Payment Present-Worth Factor

This factor is the inverse of the previous factor, that is,

$$(PF\ r,n) = \frac{1}{e^{rn}} \tag{2-32}$$

This factor is used in the formulation

$$P = F(PF\ r,n) \tag{2-33}$$

Equal-Payment-Series Compound-Amount Factor

If each equal payment, A, is considered individually, the future amount is

$$F = A[1 + e^{r} + e^{2r} + \ldots + e^{r(n-1)}] \tag{2-34}$$

The terms in brackets is a geometric progression with a common ratio of e^r. Therefore, the sum of n terms is

$$\frac{e^{rn} - 1}{e^{r} - 1} \tag{2-35}$$

Eq. (2-35) is called the equal-payment compound-amount factor and designated $(FA\ r,n)$. It is used in the formulation

$$F = A(FA\ r,n) \tag{2-36}$$

EXAMPLE 2-10

If \$1000 is deposited each quarter into an account that pays 8% compounded continuously, the amount in the account after three years is

$$\begin{aligned} F &= 1000\left(FA\ \frac{8}{4}, 3(4)\right) \\ &= 1000(FA\ 2,12) = 1000(13.427) = \$13,427 \end{aligned}$$

A Comment: It should be noted that at this point the deposits (\$1000 in this example) are still discrete and assumed to occur at the end of each period. Also, the time period of the interest rate is per year.

Equal-Payment-Series Sinking-Fund Factor

This factor is the inverse of the equal-payment-series compound-amount factor and is designated $(AF\ r,n)$. That is,

$$(AF\ r,n) = \frac{e^r - 1}{e^{rn} - 1} \tag{2-37}$$

This factor is used in the formulation

$$A = F(AF\ r,n) \tag{2-38}$$

Equal-Payment-Series Present-Worth Factor

This factor combines the $(FA\ r,n)$ and $(PF\ r,n)$ factors and is designated $(PA\ r,n)$. Mathematically, it is equal to

$$(FA\ r,n)(PF\ r,n) = \left(\frac{e^{rn} - 1}{e^r - 1}\right)\left(\frac{1}{e^{rn}}\right) = \frac{1 - e^{-rn}}{e^r - 1} \tag{2-39}$$

and used in the formulation

$$P = A(PA\ r,n) \tag{2-40}$$

Equal-Payment-Series Capital-Recovery Factor

This factor is the inverse of Eq. (2-39) and is designated $(AP\ r,n)$. Mathematically, it is equal to

$$(AP\ r,n) = \frac{e^r - 1}{1 - e^{-rn}} \tag{2-41}$$

and used in the formulation

$$A = P(AP\ r,n) \tag{2-42}$$

EXAMPLE 2-11

A man deposits \$2000 into an account that pays 9% compounded continuously. He wishes to know what equal monthly withdrawal is possible over the next three years such that the last withdrawal exhausts the account.

The solution is

$$A = 2000\left(AP\ \frac{9}{12}, 3(12)\right)$$
$$A = 2000(AP\ 0.75,36)$$

Unfortunately, the numerical value for the interest factor is not available in Appendix A. Therefore, it must be determined using Eq. (2-41).

$$(AP\ 0.75,36) = \frac{e^{0.0075} - 1}{1 - e^{-(0.0075)(36)}} = 0.03181$$

Consequently, the monthly withdrawal is

$$A = 2000(0.03181) = \$63.63$$

CONVERSION OF A CONTINUOUS RATE TO AN EQUIVALENT ANNUAL RATE

A continuous interest rate can be converted to an effective annual rate using the equation

$$\text{effective annual rate} = e^r - 1 \tag{2-43}$$

For example, the effective annual rate for 6% compounded continuously is 6.18% and for 10% compounded continuously is 10.51%.

CONTINUOUS PAYMENTS AND FUNDS-FLOW FACTORS

In previous discussions, payments (or receipts) are assumed to occur at the end of a time period. That is, they are assumed to be discrete. In some instances, it is desirable to consider that the cash transactions are uniform throughout a particular year. When cash transactions are considered uniform, it is often called fund flow. Funds-flow calculations imply continuous interest by their basic nature.

Problems that involve only single payments (F and P) can be solved using the same formulations given by Eqs. (2-21) and (2-23). When equal annual amounts are involved, new interest factors are necessary. The approach that is used in arriving at these funds-flow interest factors is first to determine a factor that will convert a funds-flow amount to an equivalent discrete amount. This, then, allows the use of the continuous interest factors given in the previous section.

Funds-Flow Conversion Factor

It is shown in the previous section on continuous interest that

$$F = A\left(\frac{e^{rn} - 1}{e^r - 1}\right)$$

Now if this equation is considered for only one year and the number of interest periods per year is c, it can be rewritten as

$$F = \frac{A}{c}\left[\frac{e^{r/c(c)} - 1}{e^{r/c} - 1}\right] \tag{2-44}$$

Taking the following limit of Eq. (2-44) gives

$$F = \lim_{c\to\infty}\left[\frac{A}{c}\left(\frac{e^r - 1}{e^{r/c} - 1}\right)\right] = \bar{A}\left(\frac{e^r - 1}{r}\right) \tag{2-45}$$

The term in parentheses in Eq. (2-45) is called the funds-flow conversion factor and converts a flow of funds, $\bar{A}$, to an equivalent annual discrete amount A. In order to make this conversion more specific, Eq. (2-45) is usually written as

$$A = \bar{A}(A\bar{A}\ r) \tag{2-46}$$

Values for the $(A\bar{A}\ r)$ factor are available in Appendix A. With Eq. (2-46) it is possible to define the following formulations:

$$F = \bar{A}(FA\ r,n)(A\bar{A}\ r) \tag{2-47}$$

$$\bar{A} = \frac{F(AF\ r,n)}{(A\bar{A}\ r)} \tag{2-48}$$

$$P = \bar{A}(PA\ r,n)(A\bar{A}\ r) \tag{2-49}$$

$$\bar{A} = \frac{P(AP\ r,n)}{(A\bar{A}\ r)} \tag{2-50}$$

If the mathematical expressions for the continuous interest factors are multiplied or divided by the funds-flow conversion factor as indicated in Eqs. (2-47) through (2-50), the actual funds flow factors are obtained. These results are shown in Table 2-4. However, it should be realized that numerical values for the factors shown in Table 2-4 are not *directly* obtainable in this text.

EXAMPLE 2-12

If the flow of funds is \$1000 semiannually, determine the present amount for five years if the interest rate is 6% compounded continuously. The solution to this problem is obtained using Eq. (2-49).

$$P = 1000\left(PA\ \frac{6}{2}, 2(5)\right)\left(A\bar{A}\ \frac{6}{2}\right)$$

$$P = 1000(PA\ 3,10)(A\bar{A}\ 3)$$

$$P = 1000(8.5105)(1.01515) = \$8639.40$$

Table 2-4
FUNDS-FLOW FACTORS

Factor	Mathematical Expression
$(F\bar{A}\ r,n)$	$\dfrac{e^{rn}-1}{r}$
$(\bar{A}F\ r,n)$	$\dfrac{r}{e^{rn}-1}$
$(P\bar{A}\ r,n)$	$\dfrac{e^{rn}-1}{re^{rn}}$
$(\bar{A}P\ r,n)$	$\dfrac{re^{rn}}{e^{rn}-1}$

It is interesting to note that this same answer can be obtained in the following manner:

$$P = 2(1000)(PA\ 6{,}5)(A\bar{A}\ 6)$$
$$P = 2000(4.1914)(1.030608) = \$8639.40$$

This implies that a solution to funds-flow problems can be obtained by correcting the funds flow to an annual amount and then evaluate the factors on the basis of the annual interest rate (with continuous compounding, of course). In general, this cannot be done with the discrete case. The reason it is possible with funds flow is due to the funds flow-factors. For example, using the data in this example, the relationship between the interest factors is

$$\begin{aligned}(PA\ 3{,}10)(A\bar{A}\ 3) &= \left(\frac{1 - e^{-0.03(10)}}{e^{0.03} - 1}\right)\left(\frac{e^{0.03} - 1}{0.03}\right)\\ &= \frac{1 - e^{-0.03(10)}}{.03} = \frac{1 - e^{-(0.06/2)(2)(5)}}{\frac{0.06}{2}}\\ &= 2\,\frac{1 - e^{-0.06(5)}}{0.06} = 2(PA\ 6{,}5)(A\bar{A}\ 6)\end{aligned} \tag{2-51}$$

EQUIVALENCE

The evaluation of economic alternatives often requires a comparison of receipts and disbursements of money occurring at different times. In order to make such comparisons, the receipts and disbursement for each alternative must be placed on an equivalent basis that accounts for the time value of money. This can be accomplished using the interest factors and is demonstrated in the next example.

EXAMPLE 2-13

Suppose the monetary values given in Table 2-5 represents costs. If the time value of money is 10%, which alternative should be chosen? In order to determine which alternative should be chosen, equivalent amounts must be determined. Equivalent amounts usually are determined on the basis of a present amount, future amount, or an annual amount.

The present equivalent amounts for the alternatives given in Table 2-5 are

$$P_x = 5000(PA\ 10{,}5) + 10{,}000(PA\ 10{,}5)(PF\ 10{,}5)$$
$$P_x = 5000(3.7908) + 10{,}000(3.7098)(0.6209) = \$42{,}491.08$$
$$P_y = 7000(PF\ 10{,}8) + 9500(PF\ 10{,}2)(PF\ 10{,}8)$$
$$P_y = 7000(5.3349) + 9500(1.7355)(0.4665) = \$45{,}035.60$$

Comparison of these two values indicates that alternative x has the least equivalent cost.

The future equivalent amount can be determined either using the previous calculations or directly from the values in Table 2-5. Both procedures are shown below:

$$F_x = 42{,}491.08(PF\ 10{,}10) = 42{,}491.08(2.594) = \$110{,}221.86$$
$$F_x = 500(FA\ 10{,}5)(FP\ 10{,}5) + 10000(FA\ 10{,}5)$$
$$F_x = 5000(6.105)(1.611) + 10000(6.105) = \$110{,}225.78$$
$$F_y = 45035.60(FP\ 10{,}10) = 45035.60(2.594) = \$116{,}822.35$$
$$F_y = 7000(FA\ 10{,}8)(FP\ 10{,}2) + 9500(FA\ 10{,}2)$$
$$F_y = 7000(11.436)(1.21) + 9500(2.100) = \$116{,}812.92$$

Table 2-5
DATA FOR EXAMPLE 2-13

End of Year	Alternative x	Alternative y
0	—	—
1	$ 5,000	$7,000
2	5,000	7,000
3	5,000	7,000
4	5,000	7,000
5	5,000	7,000
6	10,000	7,000
7	10,000	7,000
8	10,000	7,000
9	10,000	9,500
10	10,000	9,500

Theoretically, the answers for each method should be the same; differences are due to round-off error.

An equivalent annual amount is determined in the following manner.

$$A_x = (42{,}491.08)(AP\ 10{,}10)$$
$$A_x = (42{,}491.08)(0.1628) = \$6917.55$$
$$A_y = (45{,}035.60)(AP\ 10{,}10)$$
$$A_y = (45{,}035.60)(0.1628) = \$7221.80$$

These values could have also been determined using the equivalent future amounts (F_x and F_y) and the value for the (AF 10,10) interest factor given in Appendix A. It should be realized in determining an equivalent annual amount that it is first required to determine the equivalent present or future amount. Also, all three methods of computing equivalent amounts should give the same ranking of the alternatives.

A Comment: Sometimes the meaning of equivalence is misunderstood. Equivalent amounts do not represent actual monetary transactions. They are simply amounts that have the same monetary effect, taking into account interest, as the actual cash transactions. Of course, the equivalent amounts might be accepted in place of a series of disbursements or receipts if both parties are in agreement on the interest rate used in determining the equivalent amount.

EXAMPLE 2-14

A football coach has three years remaining on his contract at a yearly salary of $30,000. The alumni group is unhappy with his performance and wishes to pay off his contract now. If the football coach agrees that a 5% discount rate is satisfactory, how much must the alumni group pay him?

The answer to this question is

$$P = 30000(PA\ 5,3) = 30000(2.7233) = \$81,699$$

It should be noted that from the coach's standpoint the lower the interest rate the better. However, from the alumni's standpoint the higher the interest rate the better.

ANOTHER INTEREST FACTOR

There is another interest factor that must be considered because it is particularly useful in equivalence calculations. This interest factor provides a method for determining an equivalent annual amount for a constantly decreasing or increasing annual series. A decreasing series is shown in Table 2-6*a* and generalized in

Table 2-6
A CONSTANTLY DECREASING SERIES
(*a*) A NUMERICAL EXAMPLE OF A CONSTANTLY DECREASING SERIES

End of Year	Cash Transaction
0	—
1	\$1000
2	900
3	800
4	700
5	600

(*b*) A GENERALIZED CONSTANTLY DECREASING SERIES

End of Year	Cash Transaction
0	—
1	A_1
2	$A_1 - g$
.	.
.	.
.	.
$n - 1$	$A_1 - (n-2)g$
n	$A_1 - (n-1)g$

Table 2-6*b*. Where g is the annual change (gradient). Now the cash transactions in a particular year, A_i, consist of two parts, that is,

$$A_i = A_1 - A_2 \tag{2-52}$$

Where A_1 is a constantly yearly amount and A_2 is a multiple value of g as shown in Table 2-7. Each column of g in Table 2-7 is a constant amount. Consequently, the future amount of all of these columns is

$$F = g(FA\ i,n - 1 + FA\ i,n - 2\ \ldots\ FA\ i,2 + FA\ i,1) \tag{2-53}$$

$$F = g\left[\frac{(1+i)^{n-1} - 1}{i} + \frac{(1+i)^{n-2} - 1}{i} + \ldots + \frac{(1+i) - 1}{i}\right] \tag{2-54}$$

$$F = \frac{g}{i}\left[(1+i)^{n-1} + (1+i)^{n-2} + \ldots + (1+i) + 1\right] - \frac{ng}{i} \tag{2-55}$$

Since the terms in the brackets in Eq. (2-55) define the equal-payment-series compound amount factor [see Eq. (2-8)], Eq. (2-55) can be written as

$$F = \frac{g}{i}\left[\frac{(1+i)^n - 1}{i}\right] - \frac{ng}{i} \tag{2-56}$$

Converting Eq. (2-56) to an equivalent annual amount gives

$$A_2 = \left\{\frac{g}{i}\left[\frac{(1+i)^n - 1}{i}\right] - \frac{ng}{i}\right\}\left[\frac{i}{(1+i)^n - 1}\right] \tag{2-57}$$

Table 2-7
GRADIENT SERIES

End of Year	A_2	Expansion of A_2
0	—	—
1	0	0
2	g	g
3	$2g$	$g + g$
.	.	.
.	.	.
.	.	.
$n - 1$	$(n - 2)g$	$g + g + \ldots + g$
n	$(n - 1)g$	$g + g + \ldots + g + g$

$$A_2 = g\left[\frac{1}{i} - \frac{n}{i}\left(\frac{i}{(1+i)^n - 1}\right)\right] \tag{2-58}$$

$$A_2 = g\left[\frac{1}{i} - \frac{n}{i}(AF\ i,n)\right] \tag{2-59}$$

The term in the brackets in Eq. (2-59) is called the gradient factor and designated $(AG\ i,n)$. It is used in the formulation

$$A = A_1 \pm g(AG\ i,n) \tag{2-60}$$

In Eq. (2-60) the minus sign is used for a decreasing series and the plus sign for an increasing series. Gradient factor values for both the discrete and continuous case are given in Appendix A.

EXAMPLE 2-15

Determine an equivalent present amount for the values given in Table 2-6a using an interest rate of 10%. One method then can be used to find an equivalent present amount is to simply discount all the cash transactions back to the present as shown below:

$$P = 1000(PF\ 10,1) + 900(PF\ 10,2) + \ldots + 600(PF\ 10,5).$$

However, a more direct method is to use the gradient factor as shown below:

$$P = [1000 - 100(AG\ 10,5)]\ (PA\ 10,5)$$
$$P = [1000 - 100(1.8101)]\ (3.7908) = \$3104.63$$

Problems

2-1. If a man borrows \$2000 at 6% simple interest, how much will he owe at the end of four years? How much interest will be paid?

2-2. A woman deposits \$1000 into an account that pays 5% simple interest. After three years, she deposits the amount in this account into another account that pays 6% compounded annually for an additional five years. How much money will the woman have after the eight-year period?

2-3. Determine the final accumulated amount resulting from a series of equal annual deposits of \$2000 for 10 years if the interest rate is 8% compounded per year.

2-4. A man desires to have \$100,000 in an account at the end of 10 years. If the account pays 6% compounded annually, what series of equal annual deposits must be made assuming the first deposit is made 1 year from now?

2-5. Five years ago, a company borrowed \$200,000 at 9% compounded per year that it agreed to pay back in 15 equal annual installments. The company now wishes to cancel this debt by paying the remainder of the load in a single payment. What should be the amount of this single payment?

2-6. If a deposit of \$1000 is made now, \$800 3 years from now, \$800 5 years from now, and \$1000 7 years from now into an account that pays 6%, how much money will be in the account at the end of 12 years?

2-7. From the interest tables, determine values for the following interest factors:
(a) (*PA* 6.2,30)
(*b*) (*PA* 6,37)
(c) (*PA* 6.2,38)

2-8. If a woman borrows \$1000 and pays back at the end of five years \$1492, what was the rate of interest paid?

2-9. If a man loans \$13,500 and is paid back the loan as shown below, what rate of interest did the man earn on his loan?

End of Year	Payment
0	—
1	\$2000
2	2000
3	2000
4	2000
5	2000
6	3000
7	3000
8	3000

2-10. What is the minimum number of equal annual withdrawals of \$2000 that can be made if \$10,000 is deposited into an account that pays 7%?

2-11. How many years would it take a sum of money to double if the interest rate is 7%?

2-12. A woman desires to have \$10,000 at the end of eight years. What beginning of the year equal annual deposit must she make if the interest rate is 5%?

2-13. A man is considering placing some money in a savings account. If the interest rates quoted on various accounts are those given below, which account should he choose?
(a) 8% per year compounded per year
(b) 7.8% per year compounded semiannually
(c) 7.6% per year compounded quarterly
(d) 7.4% compounded continuously

2-14. If a credit company charges 1¼% on the unpaid monthly balance, what is the effective annual rate?

2-15. What is the final amount in an account after five years of \$5000 semiannual payments if the interest rate is
(a) 12% compounded semiannually
(b) 12% compounded continuously

2-16. If a present single deposit of \$20,000 is made, how much will be in the account at the end of five years using the following interest rates?
(a) 6% per year compounded per year
(b) 6% per year compounded semiannually
(c) 6% compounded continuously

2-17. Determine an equivalent present amount of equal quarterly payments of $600 for five years if the interest rate is
(a) 12% compounded quarterly
(b) 12% compounded continuously
(c) 12% compounded yearly

2-18. Determine an equivalent annual amount for the following end of the year series if the interest rate is
(a) 8% compounded annually
(b) 8% compounded continuously

Year	Amount	Year	Amount
0	$10,000	5	$4,000
1	2,000	6	4,500
2	2,500	7	5,000
3	3,000		
4	3,500		

2-19. For a four-year equal annual amount of $1000, determine and equivalent semiannual series for four years using an interest rate of 6% compounded per year.

2-20. A woman wishes to accumulate $10,000 in three years by making an equal monthly deposit into an account that pays 8% annually. What must the monthly deposit be if she makes the first deposit one month from now?

2-21. A man deposits $100 a month for two years into an account that pays 6% per year compounded per year. How much money will he have in the account?

2-22. Determine an equivalent annual amount for the series of payments shown below using 10%.

End of Year	Payment
0	—
1	$1000
2	0
3	1500
4	5000
5	5000
6	5000
7	5000
8	1000
9	1500
10	2000

2-23. For the payments given below, determine an equivalent annual amount assuming $i = 8\%$.

End of Year	Payment
0	—
1	$2000
2	2200
3	2400
4	2600
5	2800
6	3000
7	3200
8	3400
9	3600
10	3800

2-24. What is the equal quarterly funds-flow series for a present amount of $10,000 over five years at an interest rate of 8% compounded continuously?

2-25. Determine the future amount for the following funds-flow series.
(a) $1000 per year in seven years at 10% compounded continuously
(b) $1000 semiannually in five years at 10% compounded continuously

CHAPTER 3
DEPRECIATION AND DEPRECIATION MODELS

This chapter discusses depreciation and depreciation models. Depreciation may be considered as the lessening in value of a capital asset through wear, deterioration, or obsolescence. Whatever the reason for this reduction in value, it is a cost and must be taken into account in economic analysis. From an accountant's viewpoint, depreciation is a process of allocation not of valuation. That is, it is a method for distributing the initial cost of a capital asset over its estimated life. With this viewpoint, depreciation can be considered as a *systematic* method for the *recovery of the capital* invested in an asset. Also, of course, depreciation is important from the standpoint of taxes since it is a deduction in computing taxable income. In order to provide a systematic method for the recovery of capital and the computation of depreciation amounts for tax purposes certain depreciation models are used. These models are discussed in this chapter. In the discussion of these models the following nomenclature is used.

P = initial investment (capital investment, first cost, or capital expenditure)
L = salvage value (terminal value, final value, or final worth); this value occurs at the end of the depreciation period
n = depreciation period (usually in years); this is sometimes called the estimated life or the useful life of an asset
D_t = depreciation charge during the year t; it can also be considered as an estimate of the capital recovered during year t
B_t = book value at the end of year t; it also can be considered as an estimate of the unrecovered capital at the end of year t

Two useful expressions that relate book values and depreciation amounts are

$$B_t = P - \sum_{x=1}^{t} D_x \tag{3-1}$$

$$D_t = B_{t-1} - B_t \tag{3-2}$$

Eqs. (3-1) and (3-2) are true regardless of the depreciation model.

STRAIGHT-LINE DEPRECIATION

Straight-line depreciation is a constant amount for each year. That is, the book value decreases at a constant rate. The depreciation amount in any year is

$$D_t = \frac{P - L}{n} \tag{3-3}$$

and the book value at the end of year t is

$$B_t = P - \frac{t}{n}(P - L) \tag{3-4}$$

EXAMPLE 3-1

Using straight-line depreciation, determine the depreciation amounts and book values for each year if an asset has an initial cost of \$94,000, a salvage value of \$10,000, and a depreciation life of six years.

Using Eq. (3-3), the depreciation amount for each year is

$$D_t = \frac{94000 - 10000}{6} = \$14,000$$

The book values are determined using either Eqs. (3-1) or (3-4) and are given in Table 3-1.

SUM OF THE YEARS-DIGITS DEPRECIATION

The sum of the years-digits depreciation (SYD) model is often classified as an accelerated depreciation model because the depreciation amounts decrease with time and, consequently, the book values decrease at a decreasing rate. The depreciation amount in the year t is given by the relationship

$$D_t = \frac{n - t + 1}{n(n + 1)/2}(P - L) \tag{3-5}$$

Table 3-1
STRAIGHT-LINE DEPRECIATION FOR EXAMPLE 3-1

End of Year	Depreciation Amount	Book Value
0	—	\$94,000
1	\$14,000	80,000
2	14,000	66,000
3	14,000	52,000
4	14,000	38,000
5	14,000	24,000
6	14,000	10,000

and the book value is

$$B_t = (P - L)\left(\frac{n-t}{n}\right)\left(\frac{n-t+1}{n+1}\right) + L \tag{3-6}$$

EXAMPLE 3-2

Determine the depreciation amounts and book values for each year using SYD method for the capital asset described in Example 3-1.

The sum of the digits is

$$\frac{n(n+1)}{2} = \frac{6(7)}{2} = 21$$

and the depreciation amounts for years one and two are

$$D_1 = \frac{6}{21}(94000 - 10000) = \$24{,}000$$

$$D_2 = \frac{5}{21}(94000 - 10000) = \$20{,}000$$

The remaining depreciation amounts and book values are given in Table 3-2.

DECLINING-BALANCE DEPRECIATION

Declining-balance depreciation is another accelerated depreciation model. It is sometimes referred to as the fixed-percentage depreciation model. Basically, the depreciation amount for a particular year is determined by multiplying the book value at the beginning of the year by a fixed percentage. Mathematically, this can be expressed as

$$D_t = r(B_{t-1}) \tag{3-7}$$

where r is the fixed percentage. The book value at the end of a particular year is

$$B_t = (1 - r)^t P \tag{3-8}$$

Table 3-2

SUM OF THE YEARS-DIGITS DEPRECIATION FOR EXAMPLE 3-2

End of Year	Depreciation Amount	Book Value
0	—	\$94,000
1	\$24,000	70,000
2	20,000	50,000
3	16,000	34,000
4	12,000	22,000
5	8,000	14,000
6	4,000	10,000

If Eq. (3-8) is substituted into Eq. (3-7), the result is

$$D_t = r(1 - r)^{t-1} P \qquad (3\text{-}9)$$

which can also be used to calculate the depreciation amount.

The fixed percentage, r, can be determined by one of the following three methods.

1. Assume a value for r; when this is done, it is often necessary to adjust the depreciation amounts in the later years in order to arrive at the estimated salvage value.
2. If there is a salvage value, a rate can be defined by the equation

$$r = 1 - \sqrt[n]{\frac{L}{P}} \qquad (3\text{-}10)$$

   that will give the salvage value in the last year. Equation (3-10) is of course, not usable when the salvage value is zero. In these cases, a value for r must be estimated and adjustments made in the depreciation amounts in the latter years in order to attain a salvage value of zero. A common adjustment is to switch to straight-line depreciation in the latter years.
3. As a result of certain tax laws, it is common to set r equal to $2/n$. When this is done, it is called a double-declining-balance depreciation model.

EXAMPLE 3-3

Determine the yearly depreciation amounts and book values for an asset that costs \$94,000, has a salvage of \$10,000, and a life of six years using declining balance and the following rates: (a) 25%, (b) Eq. (3-10), (c) double-declining balance.

For part (a), the depreciation amounts for years one and two, using Eq. (3-9), are

$$D_1 = (0.25)(0.75)^0(94{,}000) = \$23{,}500$$
$$D_2 = (0.25)(0.75)^1(94{,}000) = \$17{,}625$$

The remaining depreciation amounts and book values are given in Table 3-3. It should be

Table 3-3
DECLINING-BALANCE DEPRECIATION WITH $r = 25\%$

End of Year	Depreciation Amount	Book Value
0	—	\$94,000
1	\$23,500	70,500
2	17,625	52,875
3	13,219	39,656
4	9,914	29,742
5	7,435	22,307
6	12,307	10,000

noted that the depreciation amount in year six has been adjusted in order to obtain the specified salvage value of $10,000. If Eq. (3-9) is used to calculate the depreciation amount in the sixth year, the result is

$$D_6 = (0.25)(0.75)^5(94{,}000) = \$5577$$

Which means that the book value, if this value is used, at the end of the sixth year is

$$B_6 = 22{,}307 - 5{,}577 = \$16{,}730$$

This book value is not the salvage value. Consequently, the depreciation amount in the sixth year is increased to $12,307 in order to give a salvage value of $10,000.

Another method for adjusting the depreciation amounts in order to obtain a specified salvage value is to switch to straight-line depreciation. Because of tax considerations (more specifically, the present worth of taxes), the optimum point at which to switch to straight-line depreciation is the year in which the declining balance depreciation amount is less than the yearly straight-line depreciation amount for the remaining life of the asset. The yearly straight-line depreciation amount for the remaining life of the asset is calculated using the relationship

$$D_x = \frac{BV_{x-1} - L}{n - x + 1} \tag{3-11}$$

where D_x is the straight-line depreciation amount for years x through n. For example, if the straight-line conversion is started in year two, the depreciation amounts for years two through six are

$$D_{2-6} = \frac{70{,}500 - 10{,}000}{5} = \$12{,}100$$

and similarly in year five

$$D_{5-6} = \frac{29{,}742 - 10{,}000}{2} = \$9{,}871$$

Other values and the final depreciation schedule are given in Table 3-4. In this example, the switch to straight-line depreciation occurs in year five.

For Part (b), the constant percentage, using Eq. (3-10), is

$$r = 1 - \sqrt[6]{\frac{10{,}000}{94{,}000}} = 0.3117$$

Using this value and Eqs. (3-2) and (3-9), gives the depreciation amounts and book values shown in Table 3-5.

For double-declining balance, the constant percentage is

$$r = \frac{2}{6} = \frac{1}{3}$$

The depreciation amounts and book values are shown in Table 3-6. It should be noted that the depreciation amount in year six is not obtained using Eq. (3-9) since this would result in a depreciation amount of

$$D_6 = \frac{1}{3}\left(\frac{2}{3}\right)^5(94{,}000) = \$4126$$

Table 3-4
DECLINING-BALANCE DEPRECIATION WITH r = 25% AND SWITCHOVER TO STRAIGHT LINE

End of Year	Depreciation Comparison: Declining Balance	Depreciation Comparison: Straight Line	Final Depreciation	Book Value
0	—	—	—	$94,000
1	$23,500	$14,000	$23,500	70,500
2	17,625	12,100	17,625	52,875
3	13,219	10,719	13,219	39,656
4	9,914	9,885	9,914	29,742
5	7,435	9,871	9,871	19,871
6	5,577	—	9,871	10,000

Table 3-5
DECLINING-BALANCE DEPRECIATION WITH r = 0.3117

End of Year	Depreciation Amount	Book Value
0	—	$94,000
1	$29,300	64,700
2	20,167	44,533
3	13,881	30,652
4	9,554	21,098
5	6,576	14,522
6	4,522	10,000

Table 3-6
DOUBLE DECLINING-BALANCE DEPRECIATION FOR EXAMPLE 3-3

End of Year	Depreciation Amount	Book Value
0	—	$94,000
1	$31,333	62,667
2	20,889	41,778
3	13,926	27,852
4	9,287	18,568
5	6,190	12,378
6	2,378	10,000

and a book value in the sixth year of

$$12{,}378 = 4126 = \$8252$$

This is not possible since the salvage value is $10,000. Consequently, the depreciation amount in the sixth year is taken as

$$BV_5 - L = 12{,}378 - 10{,}000 = \$2378$$

SINKING-FUND DEPRECIATION

Sinking-fund depreciation is, in effect, an annuity fund that just recovers the depreciable amount (first cost minus salvage). It assumes the value of an asset decreases at an increasing rate. The depreciation charge in any year consists of the sinking-fund deposit plus interest on the accumulated fund. Consequently the depreciation charge in any year is

$$D_t = (P - L)(AF\ i,n)(FP\ i,(t - 1)) \tag{3-12}$$

and the book value is

$$B_t = P-(P - L)(AF\ i,n)(FA\ i,t) \tag{3-13}$$

EXAMPLE 3-4

Determine the yearly depreciation amounts and book values for the asset described in Example 3-3 using sinking-fund depreciation and an interest rate of 7%.

The required values are given in Table 3-7. Sample calculations for the first two years are given below:

Using Eq. (3-12), the depreciation amount in years one and two are

$$D_1 = (94000 - 10000)(AF\ 7,6)(FP\ 7,0) = \$11{,}743$$
$$D_2 = (94000 - 10000)(AF\ 7,6)(FP\ 7,1) = \$12{,}565$$

Table 3-7

SINKING-FUND DEPRECIATION FOR EXAMPLE 3-4

End of Year	Depreciation Amount	Book Value
0	—	$94,000
1	$11,743	82,257
2	12,565	69,692
3	13,445	56,247
4	14,384	41,863
5	15,392	26,471
6	16,471	10,000

and the book values are

$$B_1 = 94{,}000 - 11{,}743 = \$82{,}257$$
$$B_2 = 82{,}257 - 12{,}565 = \$69{,}692$$

DEPRECIATION BASED ON USAGE

In the previously discussed depreciation models the depreciation amount is based on time. There are other depreciation models that are based on usage. When the depreciation charge is based on usage, the depreciation per unit of usage (miles, pounds, gallons, etc.) is defined as

$$d = \frac{P - L}{\text{lifetime usage}} \tag{3-14}$$

and the depreciation charge in a year is d multiplied by the annual usage. Book values are determined using Eq. (3-1). In usage models the depreciation charges cease when the salvage value is reached.

EXAMPLE 3-5

A large piece of earth-moving equipment costs \$50,000 and has a salvage value of \$5000. It is estimated that its lifetime usage is 5,000,000 cubic yards of earth. Determine the depreciation charge if it moves 1,000,000 cubic yards in a year.

The depreciation charge is

$$D = \frac{50{,}000 - 5000}{5{,}000{,}000}(1{,}000{,}000) = \$9000$$

From the standpoint of taxes, depreciation charges are calculated after the fact, that is, when the actual yearly usage is known. From the standpoint of estimating capital recovery, the yearly usage must be estimated. This procedure is shown in Table 3-8. It should be noted that the depreciation charge in year four is adjusted in order to obtain the salvage value.

Table 3-8

DEPRECIATION BASED ON USAGE FOR EXAMPLE 3-5

End of Year	Estimated Usage (Cubic Yards)	Depreciation Amount	Book Value
0	—	—	\$50,000
1	1,000,000	\$ 9,000	41,000
2	1,000,000	9,000	32,000
3	2,000,000	18,000	14,000
4	1,000,000	9,000	5,000

DEPLETION

Depletion is the consumption of natural resources and can be calculated by two methods: cost depletion and percentage depletion. Cost depletion is very similar to usage depreciation while percentage depletion is based on gross revenues (incomes). It is much more instructive to discuss depletion in relation to tax considerations. Consequently, a detailed discussion of depletion is deferred until Chapter 4.

CAPITAL RECOVERY WITH A RETURN

As previously mentioned, a provision for recovering the capital involved in an investment is required in most all economic analysis. If it is assumed that the depreciation amount represents the capital recovery, a rather simple formula can be obtained that includes capital recovery and a return on the unrecovered capital.

In Table 3-9, the yearly amounts of capital recovery (depreciation) plus a return (10%) on the unrecovered are given using the data from Example 3-1 and Table 3-1. If an equivalent annual amount of the capital recovery plus return is determined, the result is

$$[23{,}400(PF\ 10{,}1) + 22{,}000(PF\ 10{,}2) + \dots + 16{,}400(\mathrm{PF}\ 10{,}6)]\,(AP\ 10{,}6) = \$20{,}286$$

This same value can be obtained using the formula

$$\begin{aligned} &(P - L)(AP\ i{,}n) + L(i) \\ &(94{,}000 - 10{,}000)(AP\ 10{,}6) + 10{,}000(0.1) = \$20{,}286 \end{aligned} \tag{3-15}$$

Consequently, Eq. (3-15) provides a means of calculating an equivalent annual

Table 3-9
CAPITAL RECOVERY AND RETURN (10%) USING STRAIGHT-LINE DEPRECIATION

End of Year	Capital Recovery (Depreciation)	Capital Unrecovered (Book Value)	Return on Unrecovered Capital	Capital Recovery Plus Return
0	—	$94,000	—	—
1	$14,000	80,000	($94,000)(0.10) = $9,400	$23,400
2	14,000	66,000	(80,000)(0.10) = 8,000	22,000
3	14,000	52,000	(66,000)(0.10) = 6,600	20,600
4	14,000	38,000	(52,000)(0.10) = 5,200	19,200
5	14,000	24,000	(38,000)(0.10) = 3,800	17,800
6	14,000	10,000	(24,000)(0.10) = 2,400	16,400

amount of capital recovery and a return. Furthermore, Eq. (3-15) is not dependent upon the depreciation model. That is, the same result is obtained using any of the depreciation models given in this chapter. As an additional example, the same procedure is used for the data given in Example 3-5 and Table 3-8. Using a return of 10%, the yearly capital recovery and return amounts are given in Table 3-10. The equivalent annual amount using the values in Table 3-10 is

$$[14{,}000(PF\ 10{,}1) + 13{,}100(PF\ 10{,}2) + 21{,}200(PF\ 10{,}3) + 10{,}400(PF\ 10{,}4)](AP\ 10{,}4) = \$14{,}698$$

and from Eq. (3-15) is

$$(50{,}000 - 5000)(AP\ 10{,}4) + 5000(0.10) = \$14{,}698$$

Again the values are the same.

A Comment: It is noted that the equivalent annual amount of capital recovery with return can also be calculated using the formula

$$(P - L)(AF\ i{,}n) + P(i) \tag{3-16}$$

For example,

$$(84{,}000)(AF\ 10{,}6) + 94{,}000\ (0.10) = \$20{,}286$$

which is the same result previously obtained. Sometimes Eq. (3-16) is referred to as interest plus amortization. The problem with Eq. (3-16) is that it is often misinterpreted. That is, it is not unusual to hear such comments as, "there is no recovery of capital" because of the $P(i)$ term. Or, "sinking fund depreciation is not used." These comments indicate a misunderstanding of Eq. (3-16). Consequently, it is recommended that Eq. (3-15) be used to calculate capital recovery with a return.

Table 3-10

CAPITAL RECOVERY AND RETURN (10%) WITH USAGE DEPRECIATION

End of Year	Capital Recovery (Depreciation)	Capital Unrecovered (Book Value)	Return on Unrecovered Capital	Capital Recovery Plus Return
0	—	$50,000	—	—
1	$ 9,000	41,000	$5,000	$14,000
2	9,000	32,000	4,100	13,100
3	18,000	14,000	3,200	21,200
4	9,000	5,000	1,400	10,400

Problems

3-1. An asset costs \$100,000, has a salvage value of \$10,000, and an estimated life of 10 years. Determine the depreciation schedule (amounts and book values) for this asset using straight-line depreciation.

3-2. Repeat Problem 3-1 only using sum of the years-digits depreciation.

3-3. Repeat Problem 3-1 only using sinking-fund depreciation and an interest rate of 6%.

3-4. Repeat Problem 3-1 only using declining-balance depreciation and the following rates.

(a) 10%

(b) Eq. (3-10)

(c) $2/n$

3-5. For the data given in Tables 3-2 and 3-4, determine the yearly amounts of capital recovery plus a return of 10%, then calculate an equivalent annual amount for these values and check your answers using Eq. (3-15).

3-6. An asset originally cost \$80,000 and now has a book value of \$58,000. What is the accumulated depreciation amount that has been taken on this asset?

3-7. A particular asset is estimated to provide 40,000 hours of service during its life. This asset has a cost of \$80,000 and salvage value of \$10,000. What are the depreciation charges and book value for each year if the usage in the first year is 5000 hours, 10,000 hours in the second year, and 15,000 hours in the third year?

3-8. An asset is purchased that costs \$200,000. It is estimated to have a life of 15 years and a zero salvage value. Determine the depreciation amount for the eighth year using:

(a) Straight-line depreciation.

(b) Sum of the years-digits depreciation.

(c) Double declining-balance depreciation.

(d) Sinking-fund depreciation using 10%.

3-9. What are the two major uses of depreciation models?

3-10. An asset is purchased that costs \$100,000. It is estimated to have a life of eight years. Determine the depreciation schedule using double declining balance and a salvage value of (a) \$15,000 and (b) zero.

3-11. An asset has a first cost of \$40,000. The estimated life of the asset is six years and the salvage value, at that time, is \$5000. Determine the accumulated depreciation after three years using

(a) Straight-line depreciation.

(b) Sum of the years-digits depreciation.

(c) Double declining-balance depreciation.

3-12. An asset has a first cost of \$80,000 and life of eight years. Its estimated salvage at that time is \$5000. Assuming a return on unrecovered capital of 15%, determine:

(a) The capital recovery and return for the fifth year using sum of the years-digits depreciation.

(b) An equivalent annual amount of capital recovery and return.

CHAPTER 4
SOME TAX CONSIDERATIONS

This chapter presents some tax considerations that have a direct influence on some of the topics and calculations in subsequent chapters. It is not meant to be an all inclusive discussion of taxes but, rather, a foundation on which to include (or, at least, approximate) some of the more pertinent tax effects in economic studies.

There are considerable similarities between the basic tax regulations for individuals and corporations. However, at the same time, there are considerable differences. This chapter emphasizes the tax regulations as related to corporations and not individuals.

Tax regulations change from year to year. Consequently, the current tax regulations should be consulted to determine the applicability of some of the topics and values discussed and used in this chapter.

FEDERAL TAXES

Income Taxes

Federal income taxes are assessed on the net earnings (tabable income) of corporations and are made up of two components: a normal tax and a surtax. Basically, federal income taxes are given by the following equation.

$$\begin{aligned}\text{federal taxes} &= (\text{taxable income})(\text{normal tax rate}) \\ &\quad + (\text{taxable income} - 25{,}000)(\text{surtax}) \qquad (4\text{-}1)\end{aligned}$$

At present, the normal tax rate is 22% and the surtax 26%. Thus, the effective tax rate on an increment of taxable income can be between 22% and 48%. For example, the tax on a taxable income of \$20,000 is

$$\text{tax} = 20{,}000(0.22) = \$4400$$

Now if the taxable income increases to \$30,000 (an increment of \$10,000), the tax is

$$\text{tax} = (30{,}000)(0.22) + (30{,}000 - 25{,}000)(126) = \$7900$$

meaning a tax rate of the \$10,000 increment

$$\frac{7900 - 4400}{30{,}000 - 20{,}000} = 0.35 = 35\%$$

If the taxable income is above \$25,000 then the tax rate on an increment of taxable income is 48%. For example, if a corporation presently has a taxable income of \$30,000, its taxes are \$7900 (previous calculation). If the taxable income is increased to \$50,000, its taxes are

$$7900 + (50{,}000 - 30{,}000)(0.48) = \$17{,}500$$

Since most economic studies are done on the basis of incremental costs and incomes, the tax rate on the increments are most often used.

For fiscal years 1976 and 1977, the computation of corporate income tax is different than Eq. (4-1). In these years, the normal tax rate is 20% of the first \$25,000 of a company's taxable income, 22% of the all taxable income over \$25,000, and a 26% surtax for all taxable income in excess of \$50,000. This means that a company with a taxable income of \$25000 or less has a tax liability of

$$\text{tax} = (0.20)(\text{taxable income})$$

For a company with a taxable income greater than \$25,000 but equal to or less than \$50,000, the tax liability is

$$\text{taxes} = (0.20)(25{,}000) + (.22)(\text{taxable income} - 25{,}000)$$

A company's tax liability with a taxable income greater than \$50,000 is

$$\text{taxes} = (0.20)(25{,}000) + (0.22)(25{,}000) + (0.48)(\text{taxable income} - 50{,}000)$$

The computation of taxes using these rates is, supposedly, temporary. After 1977, tax rates are to revert (according to present law) to the 1974 rates, that is, Eq. (4-1). Consequently, Eq. (4-1) is used in future discussions that involve corporate income taxes.

Perhaps the more difficult aspect of corporate income taxes is the determination of the taxable income. For the purposes of this text, it is sufficient to say that taxable income, I_f, is

$$I_f = (G - C - D_t - I). \tag{4-2}$$

where

G = gross income (receipts); this includes sales, dividends received on stock (with certain limitations), interest royalties, etc.

C = cost of goods sold and other deductions; the cost of goods sold includes labor, materials, and other costs associated with the items sold. Other deductions might include taxes, advertising costs, executive salaries, etc.

D_t = tax depreciation allowance

I = interest paid on debt obligations (mortages, bonds, etc.)

The cost of goods sold (C) implies just that, that is, the costs associated only with the items sold and not, necessarily, the total costs in the tax period. For example, if a company manufactures 10,000 items in a particular tax period but only sells 8000 of these items, then only the costs "directly" associated with these 8000 items are deductible in computing the taxable income. The other 2000 items become a part of the company's finished-goods inventories and when they are sold, their associated costs are deductible. Actually, the cost of goods sold is tied to the problem of the valuation of inventories. A problem that is discussed quite extensively in most accounting texts and, consequently, is not discussed further in this text. Also, the 2000 items not sold in this example may result in a net increase in the working capital requirements of the company. The concept of working capital and its inclusion in economic studies is discussed in a later chapter.

Capital Gains and Losses

Under present federal tax regulations, certain types of property, called capital assets, are given special (not, necessarily, desirable) tax treatment when sold. Capital assets includes nondepreciable assets, stocks, bonds, securities, etc. Essentially, capital assets are property but does *not* include the following:

1. Inventories
2. Property held primarily for sale to customers in the ordinary course of the taxpayer's trade or business
3. Depreciable business property
4. Real property used in business (business real estate)

The tax regulations concerning some of the preceding four exclusions are discussed later.

When capital assets are sold for more or less than their original cost, a capital gain or loss occurs. If the capital asset is held for one year or less, it is called a short-term gain or loss. If the asset is held for over a year, it is called a long-term capital gain or loss. In general, long-term capital gains are given favorable tax treatment while short and long-term capital losses are given less favorable tax treatment.

Long-term gains and losses are determined for each sale of a capital asset and then added together to determine whether a net long-term gain or loss is the result. Short-term gains and losses are treated in the same manner. The net results of these two combinations are then merged to determine the tax treatment

Table 4-1
TAX TREATMENT OF CAPITAL GAINS AND LOSSES

Result of Long-Term Transactions	Result of Short-Term Transactions	Net Result	Tax Treatment: Capital Gain	Capital Loss	Ordinary Income
Long-term gain	None	Long-term gain	X		
	Short-term gain	Long-term gain	X		
		and			
		Short-term gain			X
	Short-term loss	Long-term gain	X		
		or			
		Short-term loss		X	
Long-term loss	None	Long-term loss		X	
	Short-term gain	Long-term loss		X	
		or			
		Short-term gain			X
	Short-term loss	Long-term loss		X	
		and			
		Short-term loss		X	
None	Short-term gain	Short-term gain			X
	Short-term loss	Short-term loss		X	

of the gains and losses. Obviously, there are several combinations of gains and losses that can occur. These combinations and their tax treatment are shown in Table 4-1.

For corporations, a net result denoted as a capital gain in Table 4-1 means the net result can be treated either as ordinary income and taxed at the corporation's income tax rate or the net result can be taxed at a maximum rate of 30%. The choice is up to the corporation. Obviously, corporations having a taxable income less than $25,000 will choose to treat the capital gain as ordinary income since the tax rate is 22%. Corporations having taxable incomes greater than $25,000, will choose the flat 30% rate since their incremental tax rate on ordinary income is 48%. A capital loss may be carried back to each of three years preceding the loss year. Any excess may be carried forward for five years. *When carried back or forward, the loss is treated as a short-term loss regardless of whether it was a short-term or long-term loss when it actually occurred.*

EXAMPLE 4-1

If a corporation has a taxable income of $50,000 before the following long-term and short-term capital transactions, determine the total tax. Values in parentheses indicate losses.

Transaction	Long term	Short term
1	$10,000	6000
2	5000	(7000)
3	(3000)	—
4	4000	—

The result of the long-term transactions is a net of $16,000 and the result of short-term transactions is a net loss of $1000. Therefore, the net result is a capital gain of $15,000. The tax can be computed two ways:

Solution 1:

total taxable income = 50,000 + 15,000 = $65,000
tax = (65,000)(0.22) + (65,000 − 25,000)(0.26) = $24,700

Solution 2:

tax = (50,000)(0.22) + (50,000 − 25,000)(0.26) + 15,000(0.3)
= $22,000.

Obviously, the corporation would choose Solution 2 since the taxes are the least.

As another example, consider the data given previously except the $7000 loss is now a short-term gain of $7000. In this case, the net result of the long-term transactions is a gain of $16,000 and the net result of the short-term transactions is a gain of $13,000. Therefore, the tax is

tax = (50,000 + 13,000)(0.22) + (50,000 + 13,000 − 25,000)(0.26)
+ 16,000(0.3) = $28,540

Disposal of Depreciable Property and Business Real Estate

The gains and losses from the disposal of depreciable property used in a business (company) is an important tax consideration in economic analysis, especially in the determination of net cash flows and in replacement analysis. These types of gains and losses for a corporation are treated differently than so called capital gains and losses discussed previously. The tax treatment of gains and losses from the disposal of depreciable property depends on its tax classification. There are two major tax classifications: "Sec. 1245 Property" and "Sec. 1250 Property."

Section 1245 property is depreciable property that is either (1) personal property (tangible or intangible), or (2) other tangible property (not including a building or its structural components) used as an integral part of (a) manufacturing, (b) production, (c) extraction, or (d) the furnishing of transportation, communications, electrical energy, gas, water, or sewage disposal services. This definition implies that the vast majority of equipment, facilities, and property included in the usual analysis of investment alternatives is 1245 property. The exceptions being buildings and their structural components.

Basically, the tax treatments of gains and losses from Sec. 1245 property are shown in Figure 4-1. The meaning of this figure is explained in the following example.

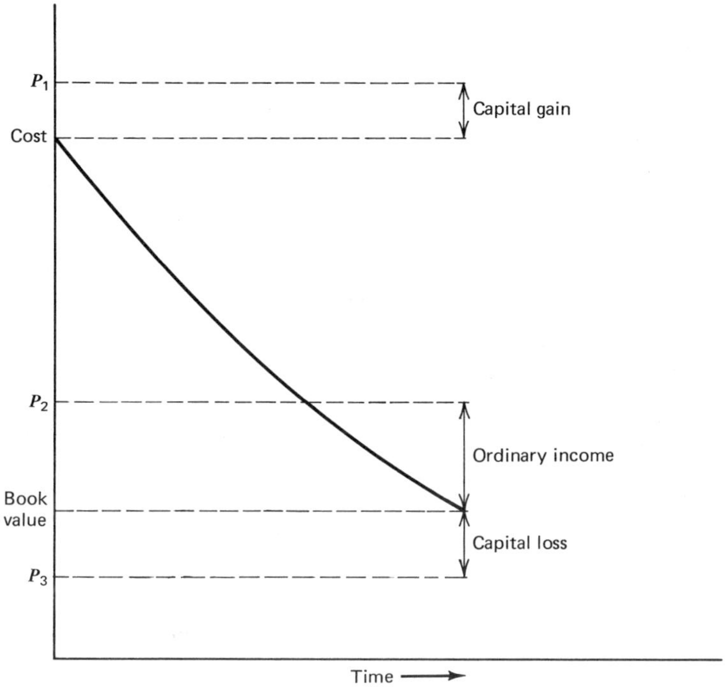

Figure 4-1.
Treatment of gains and losses from Sec. 1245 property.

EXAMPLE 4-2

Suppose a Sec. 1245 asset was purchased January 1, 1968, for \$100,000 and the total depreciation taken for tax purposes in years 1968 through 1974 was \$42,000; determine the tax effects if the selling price is \$65,000, \$110,000, and \$40,000. The present book value for this asset is

$$100{,}000 - 42{,}000 = \$58{,}000$$

Therefore, in the case of a selling price of \$65,000, the following amount is taxable as ordinary income

$$65{,}000 - 58{,}000 = \$7000$$

Actually, the tax regulations call the amount taxable as ordinary income ''excess of depreciation deducted'' and implies the following calculation: The present book value should be the selling price (\$65,000). Therefore, the total depreciation for years 1968 through 1974 should have been

$$100{,}000 - 65{,}000 = \$35{,}000$$

Consequently, the excess depreciation is

$$42{,}000 - 35{,}000 = \$7000$$

In either case, the results ($7000) are the same.

If the selling price is $110,000, then the amount

$$100{,}000 - 58{,}000 = \$42{,}000$$

is taxable as ordinary income and the amount

$$110{,}000 - 100{,}000 = \$10{,}000$$

is treated as a long-term capital gain.

If the selling price is $40,000, then the amount

$$58{,}000 - 40{,}000 = \$18{,}000$$

is treated as a long-term "capital loss." Actually, however, this loss ($18,000), under certain circumstances discussed later, can be treated as an ordinary loss.

Section 1250 property basically includes buildings and their structural components and also all other property not considered as Sec. 1245 property. Basically, the tax regulations state that any "excess depreciation" will be taxed as ordinary income and any remaining amount will be taxed as a capital gain. This regulation is explained in the following example.

EXAMPLE 4-3

On January 1, 1970, a company completed construction on a new building (1250 property) at a cost of $1,000,000 and chose sum-of-the-years depreciation with a life of 30 years and a salvage of $100,000. Determine the tax effect if the building is sold after 4 years for $850,000, $950,000, and $600,000.

First determine the depreciation amounts that have been deducted using sum-of-the-years depreciation. These amounts are shown as the second column in Table 4-2. The next step is to determine the depreciation amount based on straight-line depreciation. This value is

$$\frac{1{,}000{,}000 - 100{,}000}{30} = \$30{,}000$$

The excess depreciation in Table 4-2 is the difference, each year, between the actual depreciation deducted and the straight-line depreciation.

Table 4-2

Year	Depreciation Deduction	Straight-Line Depreciation	Excess Depreciation
1970	$ 58,065	$ 30,000	$ 28,065
1971	56,129	30,000	26,129
1972	54,194	30,000	24,194
1973	52,258	30,000	22,258
Total	220,646	120,000	100,646

The book value for the building is

$$1{,}000{,}000 - 220{,}646 = \$779{,}354$$

Therefore, the gain on the selling price of $850,000 is

$$850{,}000 - 779{,}354 = \$70{,}646$$

This total amount, $70,646, would be taxed as ordinary income since the gain is less than the total of the excess depreciation.

Now, if the selling price is $950,000, the gain is

$$950{,}000 - 779{,}354 = \$170{,}646$$

Consequently, the amount of $100,646 would be ordinary income and the amount

$$170{,}646 - 100{,}646 = \$70{,}000$$

would be a capital gain.

If the selling price is $600,000, then the amount

$$779{,}354 - 600{,}000 = \$179{,}354$$

is considered a ''capital loss.'' As with Sec. 1245 property, this loss ($179,354) can, under certain circumstances, be treated as an ordinary loss.

If straight-line depreciation is used for Sec. 1250 property, then there are no ordinary income considerations. The difference between the selling price and book value is either a capital gain or loss depending on whether or not the selling price is greater than the book value.

Losses

Section 1231 of the tax laws provides for special tax treatment of 1245 and 1250 property. This section states that after *capital gains* and losses have been determined for each disposition of depreciable property, the gains and losses should be combined. If the total gains exceed the total losses, then each gain and loss is treated as though it was derived from the sale of a long-term capital asset; that is, the capial gains and losses from the sale of 1245 and 1250 property are merged with other capital gains and losses. In effect, the net gain from the sale of 1245 or 1250 property will either be taxed as capital gain or it will, possibly, offset other capital losses.

If the total losses exceed total gains from the sale of 1245 or 1250 property, then each gain and loss is *not* treated as the sale from a capital asset. This has the effect of making the net loss a fully deductible business expense—a reduction in ordinary income.

Certain types of *nondepreciable* property are also covered by Sec. 1231. The one most often encountered in economic studies is business realty (land) held for more than one year.

Now, the preceding discussion of gains and losses from depreciable property is limited to single-asset or item depreciation. That is, each asset has its own depreciation account and depreciation on each asset is computed separately. An alternative to item depreciation is group depreciation. When group depreciation is used, assets are combined into groups and the entire group is depreciated on the basis of an average life. Perhaps, the important item, from the standpoint of economic studies, is that, in many situations, gains and losses resulting from the *normal (ordinary)* replacement of assets need not be considered when group depreciation is used.

The Class Life Asset Depreciation Range System (ADR System)

In June 1971, the ADR System was adopted by the Treasury Department and is applicable to assets placed in service after 1970. There is no requirement that a taxpayer use this system. The taxpayer may still, if desired, use depreciation regulations established prior to the adoption of the ADR System. However, because of certain desirable aspects of this system, it is fair to estimate that most companies will use the ADR System.

Basically, the ADR System establishes broad industry classes of assets and for each class specifies an asset guideline period, an asset depreciation range, and a repair allowance percentage. Some examples of these classes are shown in Table 4-3. A taxpayer may use as a depreciation period (life), for a certain class of assets, a value that is in the specified depreciation range, provided it is a whole number of years or a whole number of years plus a half year. For example, the depreciation life for Class 40.52 may be chosen from 16 through 24 years. The exception to this choice from a range is buildings. Buildings do not have a range (see Table 4-3) and, therefore, there is no choice. The depreciation period for buildings is the asset guideline period.

Essentially, the ADR System requires that all eligible assets put in service in a particular tax year be placed in a "vintage account." These accounts can be established either on a single item or multiple-asset (group) basis except that 1245 and 1250 property must be kept in separate accounts. These "vintage accounts" are then depreciated on the basis of a life chosen from the asset depreciation range of the particular class the asset (or assets) falls under. In addition, each vintage account under ADR must have a depreciation reserve to account for retirement adjustment. Certain other features of the ADR system that can have an effect on economic studies are listed below:

1. The election to employ the ADR System is an annual one. However, if ADR is used in a particular year, it must be used for all eligible property placed in service in that year.
2. The ADR System applies to Sec. 1245 and 1250 property.
3. The ADR System specifically allows the use of straight-line, sum-of-the-years-digits, and double-declining-balance (for new property) depreciation methods. It assumes that the depreciation period (life) already takes into account salvage value. Consequently, salvage values are not considered when computing the depreciation amounts using these depreciation methods. However, the total depreciation is not to exceed a "reasonable salvage value."

Table 4-3

ADR SYSTEM CLASSES

Asset Guideline Class	Description of Assets	Asset Depreciation Range (in Years) Lower Limit	Asset Guideline Period	Upper Limit	Annual Asset Guideline Repair Allowance Percentage
24.4	Manufacture of lumber, wood products, and furniture	8	10	12	6.5
26.0	Manufacture of paper and allied products:				
26.1	Manufacture of pulps from wood and other cellulose fibers and rags	13	16	19	4.5
26.2	Manufacture of paper and paperboard	9.5	12	14.5	5.5
27.0	Printing, publishing and allied industries	9	11	13	5.5
28.0	Manufacture of chemicals and allied products	9	11	13	5.5
30.0	Manufacture of rubber and plastics products:				
30.1	Manufacture of rubber products	11	14	17	5.0
30.2	Manufacture of miscellaneous finished plastics products	9	11	13	5.5
31.0	Manufacture of leather	9	11	13	5.5
32.0	Manufacture of stone, clay, glass, and concrete products:				
32.1	Manufacture of glass products	11	14	17	6.0
32.2	Manufacture of cement	16	20	24	3.0
32.3	Manufacture of other stone and clay products	12	15	18	4.5
33.0	Manufacture of primary metals:				
33.1	Ferrous metals	14.5	18	21.5	8.0
33.2	Nonferrous metals	11	14	17	4.5
34.0	Manufacture of fabricated metal products	9.5	12	14.5	6.0
35.0	Manufacture of machinery, except electrical and transportation equipment:				
35.1	Manufacture of metalworking machinery	9.5	12	14.5	5.5
35.2	Manufacture of other machines	9.5	12	14.5	5.5
	.				
	.				
	.				
	Electric generating equipment:				

(Table 4-3 *Continued*) ADR SYSTEM CLASSES

Asset Guideline Class	Description of Assets	Asset Depreciation Range (in Years) Lower Limit	Asset Guideline Period	Upper Limit	Annual Asset Guideline Repair Allowance Percentage
40.51	Hydraulic	40	50	60	1.5
40.52	Nuclear	16	20	24	3.0
40.53	Steam	22.5	28	33.5	2.5
40.54	Steam, compressed air, and other power plant equipment	22.5	28	33.5	7.5
41.0	Motor transport-passengers	6.5	8	9.5	11.5
42.0	Motor transport-freight	6.5	8	9.5	11.0
44.0	Water transportation	16	20	24	8.0
45.0	Air transport	5	6	7	14.0
46.0	Pipeline transportation	17.5	22	26.5	3.0
65.0	Building services:				
65.1	Building services—manufacturing, and machinery and equipment repair activities				
65.11	Factories		45		
65.12	Garages		45		
65.13	Machine shops		45		
65.14	Loft buildings		50		
65.2	Building services—wholesale and retail		60		
65.3	Building services—residential:				
65.31	Apartments		40		
65.32	Dwellings		45		
65.4	Building services—miscellaneous:				
65.41	Office Buildings		45		
65.42	Storage:				
65.421	Warehouses		60		
65.422	Grain elevators		60		

4. Under the ADR System, salvage values are gross salvage values; that is, salvage values unreduced by the cost of removal, dismantling, and so on. These costs are deductible as current expenses in the year paid or incurred.
5. Retirements are classified in two ways; ordinary and extraordinary retirements. All retirements of 1245 property are ordinary retirements except in the following two cases: (1) retirements of 1245 property as a direct result of fire, storm, shipwreck, or other casualty can be classified as extraordinary if the taxpayer consistently chooses

to treat such retirements as extraordinary; (2) retirements of 1245 property due to cessation, termination, curtailment, or disposition of a business, manufacturing or other income-producing process, operation, facility, or unit are classified as extraordinary provided the total original value of the retired assets is greater than 20% of the total original value in the same vintage account. All accounts (multiple-asset and single asset) having the same class and vintage are grouped together to make this 20% determination. *All retirements of 1250 property are extraordinary.*

6. All gains and losses on extraordinary retirements are recognized in the year they occur. In effect, Secs. 1245, 1250, and 1231 and their tax consequences are brought into consideration. If a single asset is retired, then the gain or loss is determined as previously discussed using the section (1245 or 1250) that the asset falls under. In the case of multiple-asset accounts, the retired asset must be removed from the vintage account and the related depreciation reserve reduced by total accumulated depreciation allowances due to this asset. The amount of accumulated depreciation is determined by using its depreciation method and ADR rate selected for the entire account. These calculations also provide the basis for determining the gain or loss on the disposal of this particular asset and the resulting tax effect.
7. Gains and losses on ordinary retirements are, in general, not recognized for multiple-asset accounts. The proceeds from the retirements are added to depreciation reserve. However, when the proceeds from the sale of an asset, added to the depreciation reserve, increases the reserve above the depreciable amount (cost minus salvage), the excess of the reserve over the unadjusted basis (original cost) must be recognized as a gain in the year in which it occurs (Sec. 1245 ordinary income). Losses are not recognized until the retirement of the last item in the multiple-asset account. For single-item accounts, gains are recognized to the extent that the depreciation reserve (as adjusted by the retirement proceeds) exceed the depreciation reserve. Since the retirement of an item in a single account is the retirement of the last item in that account, the loss can be recognized (Secs. 1245 and 1231) or taken as an additional depreciation deduction. However, if single-item accounts for Sec. 1245 property are essentially multiple-asset accounts (single-item accounts with the same class and vintage), then they are treated as if they are a multiple-asset account for purposes of determining gain and loss. The effect being, gains are recognized as they occur and losses are deferred until the last asset is retired.
8. Permission is not required, under the ADR System, to change from double-declining-balance and sum-of-the-years-digits to straight-line depreciation.
9. The ADR System provides a repair allowance in order to reduce the expense versus capitalization controversy. The taxpayer has the option of electing the percentage repair allowance or capitalizing. This option may be exercised each tax year. Basically, the percentage repair allowance provides that expenditures for repairs, maintenance, rehabilitation, or improvement (with certain restrictions) are treated as deductible repairs to the extent they do not exceed the repair allowance. The excess of such expenditures are treated as capital expenditures and placed in a vintage account and depreciated. The actual allowance for a class is obtained by multiplying the allowable percentage for the class by the unadjusted basis (original price) in that class. For example, if the total cost of the items in a vintage account is $10,000 and the class is 28.0, then the repair allowance is 5.5% (see Table 4-3). Consequently,

expenses for repairs, can be deducted to the extent they do not exceed \$550. Any expenses exceeding this amount are treated as capital expenditures.

10. Public utilities may use the ADR system provided they "normalize" tax deferrals for rate making purposes. "Normalizing" is explained later (*Chapter 7*) in this text.

Investment Tax Credit

For investments in qualified depreciable property, there is a credit against federal income tax. The amount is limited to 10% of the qualified property. Qualified property is "Sec. 38 Property," which includes Sec. 1245 property. Buildings or their components do not qualify unless they are "special buildings." In order to qualify for the full 10% deduction, the life of the depreciable property must be seven years or more. For lives less than seven years, only a certain proportion of the full allowance can be used as a tax reduction. Table 4-4 shows the allowance for various lives.

There are certain limitations imposed on the amount of investment tax credit that can be taken in any one year. The investment tax credit may not exceed the tax liability. Also, if the tax liability exceeds \$25,000, the investment tax credit may not exceed \$25,000 plus 50% of the tax liability over that amount (\$25,000). However, any part of the investment tax credit that is not applied because of these limitations can be carried back three years or forward seven years.

EXAMPLE 4-4

If a taxpayer's liability is \$50,000 and has a qualified investment of \$600,000 with a life of 10 years, then the taxpayer's investment tax credit is

$$0.10(600{,}000) = 60{,}000$$

However for the current year, the taxpayer may only claim

$$25{,}000 + (0.50)(50{,}000 - 25{,}000) = \$37{,}500$$

Table 4-4
INVESTMENT TAX CREDIT ALLOWANCES

Life	Proportion of Full Allowance
7 years or more	Full
5 and 6	$\frac{2}{3}$
3 and 4	$\frac{1}{3}$
Less than 3	None

The remaining \$22,500 may be carried back or forth as mentioned previously. Therefore, the actual taxes the taxpayer would pay for the current year are

$$50{,}000 - 37{,}500 = \$12{,}500$$

If the qualified investment had only been \$100,000, then the entire investment tax credit of

$$0.10(100{,}000) = \$10{,}000$$

is deductible from the taxpayer's tax liability of \$50,000.

In addition to the investment tax credit, there is the possibility of an investment tax *penalty* if Sec. 38 property is disposed of before the end of its useful life. Under present tax laws, the tax for the year of disposal is increased by the difference between the originally allowed investment tax credit and the investment tax credit that would have been allowed with the shorter useful life. Useful life for investment tax credit purposes is the *tax* depreciation life. As an example, an asset that is sold after 4 years of service that originally cost \$12,000 and had a 10-year useful life would result in tax increase of \$800 in the year it is sold as a result of the investment tax penalty. This value of \$800 is based in the difference between the investment tax credit originally taken, (0.10)(12,000) = \$1200, and the credit that should have been taken, $(0.10)(\frac{1}{3})(12{,}000) = \$400$, because of a useful life of only four years. The possibility of an investment tax penalty is an important consideration in replacement analysis

Depletion

The term depletion refers to the consumption of natural resources (oil, gas, minerals, etc.) in the production of products or services. To some extent, it is similar to depreciation allowances allowed manufacturers. As far as taxes are concerned, depletion amounts are a reduction in taxable income. There are two methods of computing depletion: cost depletion and percentage depletion.

Cost depletion allowances are based on the following relationships:

$$\begin{pmatrix}\text{cost depletion}\\ \text{allowance}\end{pmatrix} = \left(\frac{\text{cost of property}}{\text{number of units in property}}\right)\begin{pmatrix}\text{units sold}\\ \text{during the year}\end{pmatrix} \qquad (4\text{-}3)$$

The use of Eq. (4-3) is shown in the following example.

EXAMPLE 4-5

Suppose a taxpayer buys a mine at a cost of \$500,000 and it is estimated that there are 1,000,000 tons of ore to be extracted. If he or she extracts 25,000 tons of ore and sells 20,000 tons, then his or her cost depletion allowance is

$$\left(\frac{500{,}000}{1{,}000{,}000}\right)(20{,}000) = \$10{,}000.$$

It is important to note that only the units actually sold can be used in computing the cost depletion allowance.

Table 4-5

REPRESENTATIVE VALUES FOR PERCENTAGE DEPLETION

1. 22%—oil, gas, sulphur, uranium, graphite, cobalt, lead, mercury, nickel, platinum, tin
2. 15%—gold, silver, copper, iron ore, and oil shale
3. 14%—metal mines not included elsewhere, various clays
4. 10%—coal, lignite, sodium chloride
5. 7½%—clays used in the manufacture of sewer pipe or brick
6. 5%—gravel, peat, pumice, sand

An alternative to cost depletion is percentage depletion. Percentage depletion is based, with certain limitations, on a constant percentage of the *gross income*. An abbreviated listing of various types of property and their associated percentages are shown in Table 4-5. The limitations are (1) percentage depletion cannot exceed 50% of the taxable income of the property before depletion and (2) cost depletion must be used if it is greater than percentage depletion. These limitations suggest various combinations regarding the relative magnitudes of taxable income, cost depletion, and percentage depletion; these various combinations and their resulting deductions (assuming the taxpayer desires the highest deduction allowed) are shown in Table 4-6.

The tax regulations also state that percentage depletion will not be denied because the basis of the property has been reduced to zero. This implies that percentage depletion can still be deducted even though the total of past depletion allowances exceed the total cost of the property.

STATE INCOME TAXES

The basis on which state taxes are assessed in corporate income varies from state to state; not only do the rates vary, but also the computation of the taxable base. However, it is highly desirable to include (or, at least, approximate) state income taxes in economic studies since these taxes are, now, a significant proportion of a company's costs. The usual procedure (and the one taken in this text) is to determine a composite rate reflecting both state and federal income tax. In the

Table 4-6

DEPLETION ALLOWANCES

Taxable Income	50% of Taxable Income	Cost Depletion	Percentage Depletion	Depletion Allowance
$80,000	$40,000	$15,000	$22,000	$22,000
60,000	30,000	25,000	35,000	30,000
50,000	25,000	31,000	20,000	31,000
60,000	30,000	27,000	25,000	27,000

following pages, formulas are developed that allow for the determination of a composite tax rate using the following nomenclature.

I_b = net income before taxes
I_f = income base for federal taxes
I_s = income base for state taxes
t_1 = normal rate for federal tax = 22%
t_2 = surtax rate for federal tax = 26%
t_3 = state tax rate
T_f = federal tax
T_s = state tax
t = overall (combined) tax rate
f = incremental federal tax rate = $t_1 + t_2$
s = incremental state tax rate = t_3
D = amount deductible in computing surtax = $25,000
T = incremental overall tax rate

The basic method that is used in this section is to determine the state and federal taxes and determine the overall tax rate on the basis of

$$t = \frac{T_f + T_s}{I_b} \tag{4-4}$$

Also, under present federal tax regulations, all state taxes are deductible, therefore, the following expression is used throughout the following derivation.

$$I_f = I_b - T_s \tag{4-5}$$

The derivations consider the case where $I_f > D$. The resulting equations can then be modified to take into account the case of $I_f \leqslant D$.

Case 1: Neither state or federal tax are deductible in computing state tax. This case implies the following relations:

$$I_b = I_s \tag{4-6}$$
$$T_s = t_3 I_s = t_3 I_b \tag{4-7}$$
$$T_f = I_f\,(t_1 + t_2) - DT_2 \tag{4-8}$$

Now, substituting Eq. (4-5) into Eq. (4-8) gives

$$T_f = (I_b - T_s)(t_1 + t_2) - Dt_2 \tag{4-9}$$

and substituting Eq. (4-7) into (4-9) gives

$$T_f = (I_b - t_3 I_b)(t_1 + t_2) - Dt_2 \tag{4-10}$$

Substituting Eqs. (4-7) and (4-10) into Eq. (4-4) gives

$$t = \frac{(I_b - t_3 I_b)(t_1 + t_2) - Dt_2 + t_3 I_b}{I_b} \tag{4-11}$$

$$t = \frac{I_b(1 - t_3)(t_1 + t_2) - Dt_2 + t_3 I_b}{I_b} \tag{4-12}$$

For the case of $I_f \leqslant D$, then $D = 0$ and $t_2 = 0$, which reduces Eq. (4-12) to

$$t = \frac{I_b(1 - t_3)(t_1) + t_3 I_b}{I_b} = t_1(1 - t_3) + t_3 \tag{4-13}$$

For the incremental tax rate, $D = 0$ and Eq. (4-12) becomes

$$T = \frac{I_b(1 - s)(f) + sI_b}{I_b} = f + s - fs \tag{4-14}$$

and Eq. (4-13) is

$$T = t_1(1 - s) + s \tag{4-15}$$

Case 2: Federal tax not deductible from state tax; that is, state tax is based on federal income base. This, of course, implies that state tax is deductible in computing state tax. Therefore,

$$I_s = (I_b - T_s). \tag{4-16}$$

$$T_s = t_3 I_s = t_3(I_b - T_s) \tag{4-17}$$

$$T_s = \frac{t_3 I_3}{1 + t_3} \tag{4-18}$$

$$T_f = I_f(t_1 + t_2) - Dt_2 \tag{4-19}$$

$$T_f = (I_b - T_s)(t_1 + t_2) - Dt_2 \tag{4-20}$$

Substituting Eq. (4-18) into Eq. (4-19) gives

$$T_f = I_b\left(\frac{t_1 + t_2}{1 + t_3}\right) - Dt_2 \tag{4-20}$$

and substituting Eqs. (4-18) and (4-20) into Eq. (4-4) gives

$$t = \frac{I_b(t_1 + t_2) - Dt_2(1 + t_3) + t_3 I_b}{(1 + t_3)I_b} \tag{4-21}$$

For the case of $I_f \leqslant D$, Eq. (4-21) reduces to

$$t = \frac{I_b t_1 + t_3 I_b}{(1 + t_3)I_b} = \frac{t_1 + t_3}{1 + t_3} \tag{4-22}$$

For the incremental rates,

$$T = \frac{I_b(f) + sI_b}{(1 + s)I_b} = \frac{f + s}{1 + s} \tag{4-23}$$

for $I_f > D$. For $I_f \leqslant D$,

$$T = \frac{t_1 + s}{1 + s} \tag{4-24}$$

Case 3: Federal tax is deductible in computing state tax but state tax is not deductible in computing state tax. In this case, the following relations are implied.

$$I_s = I_b - T_f \tag{4-25}$$

$$T_s = t_3 I_s = t_3(I_b - T_f) \tag{4-26}$$

$$I_f = I_b - T_s$$

$$T_f = I_f(t_1 + t_2) - Dt_2 = (I_b - T_s)(t_1 + t_2) - Dt_2 \tag{4-27}$$

Substituting Eq. (4-26) into Eq. (4-27) gives

$$T_f = [I_b - t_3(I_b - T_f)](t_1 + t_2) - Dt_2 \tag{4-28}$$

and rearranging the result is

$$T_f = \frac{I_b\,(t_1 + t_2 - t_1t_3 - t_2t_3) - t_2D}{(1 - t_1t_3 - t_2t_3)} \tag{4-29}$$

Substituting Eq. (4-29) into Eq. (4-26) and rearranging gives

$$T_s = \frac{t_3I_b(1 - t_1 - t_2) + t_2t_3D}{(1 - t_1t_3 - t_2t_3)} \tag{4-30}$$

Therefore, the overall tax for $D > I_f$ is

$$t = \frac{I_b(t_1 + t_2 - t_1t_3 - t_2t_3) - t_2D + t_3I_b(1 - t_1 - t_2) + t_2t_3D}{I_b(1 - t_1t_3 - t_2t_3)} \tag{4-31}$$

and for $D \leqslant I_f$ is

$$t = \frac{I_b(t_1 - t_1t_3) + t_3I_b(1 - t_1)}{I_b(1 - t_1t_3)} = \frac{t_1 + t_3 - 2t_1t_3}{1 - t_1t_3} \tag{4-32}$$

The incremental tax rate for $D > I_f$ is

$$T = \frac{I_b(f - t_1s - t_2S) + sI_b(1 - f)}{I_b(1 - st_1 - st_2)} \tag{4-33}$$

and since

$$t_1 = f - t_2$$
$$T = \frac{f - s(f - t_2) - t_2s + s - sf}{1 - s(f - t_2) - st_2} = \frac{f + s - 2sf}{1 - sf} \tag{4-34}$$

and for $I_f \leqslant D$ is

$$T = \frac{t_1 + s - 2st_1}{1 - st_1}$$

Case 4: Both federal and state tax deductible in computing state tax. Therefore,

$$I_s = I_b - T_s - T_f \tag{4-35}$$
$$T_s = t_3I_s = t_3(I_b - T_s - T_f) \tag{4-36}$$
$$T_s = \frac{t_3(I_b - T_f)}{1 + t_3} \tag{4-37}$$
$$T_f = I_f(t_1 + t_2) - Dt_2 = (I_b - T_s)(t_1 + t_2) - Dt_2 \tag{4-38}$$

Substituting Eq. (4-38) into Eq. (4-37) and rearranging gives

$$T_s = \frac{t_3[I_b(1 - t_1 - t_2) - Dt_2]}{1 + t_3 - t_1t_3 - t_2t_3} \tag{4-39}$$

Substituting Eq. (4-37) into Eq. (4-38) and rearranging gives

$$T_f = \frac{I_b(t_1 + t_2) - Dt_2(1 + t_3)}{1 + t_3 - t_1t_3 - t_2t_3} \tag{4-40}$$

Using Eq. (4-4), the overall tax rate for $D > I_f$ is

$$t = \frac{I_b(t_1 + t_2) - Dt_2(1 + t_3) + t_3[I_b(1 - t_1 - t_2) - Dt_2]}{I_b\,(1 + t_3 - t_1t_3 - t_2t_3} \tag{4-41}$$

and for $D \leqslant I_f$ is

$$t = \frac{I_bt_1 + I_bt_3(1 - t_1)}{I_b(1 + t_3 - t_1t_3)}\ \frac{t_1 + t_3(1 - t_1)}{1 + t_3 - t_1t_3} \tag{4-42}$$

The incremental tax rates are

$$T = \frac{I_b(f) + sI_b(1 - f)}{I_b(1 + s - s(f - t_2) - st_2}$$
$$T = \frac{f + s - sf}{1 + s - sf} \tag{4-43}$$

and

$$T = \frac{t_1 + s(1 - t_1)}{1 + s - st_1} \tag{4-44}$$

Perhaps, from an economic study standpoint, the more important relationships are those pertaining to the incremental tax rates. Therefore, the formulas for the incremental tax rates are summarized in Table 4-7.

It must be understood that if the incremental tax rate are used in the following form to take into account federal and state taxes

$$\text{taxes} = (G - C - D - I)(T) \tag{4-44}$$

then state taxes should not be subtracted out in this expression since they are considered in the derivations of the incremental tax formulas.

Table 4-7

FORMULAS FOR INCREMENTAL COMPOSITE TAX RATES

Case	$I_f > D$	$I_f \leqslant D$
State and federal tax not deductible in computing state tax	$f + s - fs$	$t_1(1 - s) + s$
State tax based on federal income base	$\dfrac{f + s}{1 + s}$	$\dfrac{t_1 + s}{1 + s}$
Federal tax deductible but state tax not deductible	$\dfrac{f + s - 2sf}{1 - sf}$	$\dfrac{t_1 + s - 2st_1}{1 - st_1}$
Both federal and state tax deductible in computing state tax	$\dfrac{f + s - sf}{1 + s - sf}$	$\dfrac{t_1 + s - st_1}{1 + s - st_1}$

OTHER TAXES

There are other types of taxes that can become involved in economic studies. Some of these are briefly discussed below:

Property Taxes (Ad Valorem Taxes)

These taxes can be levied by states, counties, cities, and school districts. The tax is usually a percentage of the property's assessed valuation. The assessed valuation may or may not be the initial investment due to adjustments for nontaxable items. In economic studies, the usual procedure is to estimate property taxes as a constant percentage of the initial investment. The applicable percentage is, of course, dependent on the geographical location of the property.

Sales Taxes

Sales taxes are paid by a company on material and equipment purchases. The amount of the tax will vary with location. In many economic studies, sales taxes are omitted because the amount of these taxes are relatively small in comparison to other amounts in the study.

Franchise Taxes

Franchise taxes are assessed on the basis of a constant percentage of gross revenues (income). In economic studies such as cost comparisons that assume equal incomes, franchise taxes are usually omitted because they are the same for all alternatives. If gross revenues are not the same, then franchise taxes should be included, strickly speaking. However, they are often omitted because, they too, are relatively small when compared to other costs.

Payroll Taxes

The payroll component of operating and maintenance expenses is subject to taxes for Social Security, unemployment insurance, and Workmen's Compensation. These taxes can be as high (maybe higher by now) as 30% of regular labor costs. The usual procedure is to include these taxes as a part of labor and maintenance overhead. Therefore, since overhead is a part of labor and maintenance estimates, these taxes are usually implied in economic studies. In general, these taxes are deductible in computing federal taxes. However, they may or may not be deductible in computing state taxes depending upon whether or not federal taxes are deductible in computing state taxes.

ECONOMIC STUDIES AND TAXES

The majority of economic studies are concerned with the evaluation and comparison of projects (investments). In this text, this evaluation and comparison is based on the concept that each new expenditure of money must, in some manner, justify itself, implying an *incremental analysis*. Consequently, it is from this standpoint that taxes are included in future problems and examples—the incremental changes in taxes that may occur as a result of the projects under consideration.

To some extent, an incremental approach simplifies the introduction of taxes into economic studies (i.e., federal and state income taxes, investment tax credit). However, there are other tax considerations where this is not the case. For example, the true tax effect from the retirement of a particular item of depreciable property (Sec. 1245 or 1250 property) depends on such things as the gains and losses from the retirements of property other than the particular item under consideration and whether or not the retirement is classified as an ordinary or extraordinary retirement. As a result of these considerations, it is necessary, at times, to use certain conventions and assumptions when including taxes in economic studies. These assumptions and conventions are stated when specific topics are discussed in later chapters.

TAXES AND DEPRECIATION

It can be seen from Eq. (4-2) that the amount of depreciation deducted has a direct effect on the amount of taxes paid in a particular year. Consequently, accelerated depreciation models (sum-of-the-years-digits and the various forms of declining balance) offer an advantage because the depreciation amounts in the earlier years are greater than in later years and, thereby, reduce the *present worth* of taxes paid. This point is illustrated in Tables 4-8, 4-9, and 4-10. Note that the total taxes paid are the same (the slight difference is due to roundoff error) over the life of eight years.

Table 4-8

INCOME TAXES USING STRAIGHT-LINE DEPRECIATION (P = 12,000, L = 2000, n = 8)

End of Year	Depreciation	Taxable Income Before Depr.	Taxable Income After Depr.	Tax T = 52%
0				
1	$1250	$3000	$1750	$910
2	1250	3000	1750	910
3	1250	3000	1750	910
4	1250	3000	1750	910
5	1250	3000	1750	910
6	1250	3000	1750	910
7	1250	3000	1750	910
8	1250	3000	1750	910

Total taxes paid = 8(910) = $7280

Present worth of taxes = $910 \underset{(PA\ 10,8)}{5.3349} = \$4855$

Table 4-9

INCOME TAXES USING SUM-OF-THE-YEARS-DIGITS DEPRECIATION
(P = 12,000, L = 2000, n = 8)

End of Year	Depreciation	Taxable Income Before Depr.	Taxable Income After Depr.	Tax T = 52%
0				
1	$2222	$3000	$ 778	$ 405
2	1944	3000	1056	549
3	1667	3000	1333	693
4	1389	3000	1611	838
5	1111	3000	1889	982
6	833	3000	2167	1127
7	556	3000	2444	1271
8	278	3000	2722	1415

Total taxes paid = 405 + 549 + . . . 1415 = $7280
Present worth of taxes = 405(PF 10,1) + 549(PF 10,2) + . . . + 1415(PF 10,8)
= $4473

Table 4-10

INCOME TAXES USING DOUBLE-DECLINING-BALANCE DEPRECIATION

End of Year	Depreciation	Taxable Income Before Depr.	Taxable Income After Depr.	Tax T = 52%
0				
1	$3000	$3000	$ 0	$ 0
2	2250	3000	750	390
3	1687	3000	1313	683
4	1266	3000	1734	902
5	949	3000	2051	1067
6	712	3000	2288	1190
7	136[a]	3000	2864	1489
8	—	3000	3000	1560

[a]Adjusted amount.
Total taxes paid = 390 + 683 + . . . + 1560 = $7281
Present worth of taxes = 390(PF 10,2) + 683(PF 10,3) + . . . + 1560(PF 10,8)
= $4278

A company may use two depreciation methods: one for "company purposes" and one for tax purposes. Many times these two depreciations will not only differ from the standpoint of method but also the lives used in the two methods may be different.

TAXES AND INTEREST

The interest paid on externally borrowed funds is deductible as an expense in computing taxes. The net effect is to reduce the actual cost of borrowed funds. For example, if a company has a taxable income of \$200,000, its taxes are

$$t = (200{,}000)(0.22) + (200{,}000 - 25{,}000)(0.26) = \$89{,}500$$

Now, if the company borrows \$50,000 at 10% for one year, its interest is \$5000 and its taxes are reduced to

$$t = 89{,}500 - 5000(0.48) = \$87{,}100$$

Consequently, the actual cost of the borrowed money is

$$5000 - (89{,}500 - 87{,}100) = \$2600$$

meaning that the actual (after tax) interest rate on the borrowed money is

$$\frac{2600}{50{,}000} = 0.052 = 5.2\%$$

This same result can also be obtained in the following manner,

$$10(1 - 0.48) = 5.2\%$$

This simple example points out that the "cost" of borrowed funds is appreciably reduced because interest is deductible in computing taxes. However, there are other considerations besides the tax effect regarding borrowed funds. These considerations are discussed in a later chapter (ch. 6).

Problems

4-1. At the end of a tax year, the ABC Company has the following data:

gross receipts (sales)	\$100,000
cost of goods sold	25,000
tax depreciation	6,000
book depreciation	5,000
interest on debt	1,000

Determine,

(a) The tax liability.

(b) Suppose in this same year, the gross receipts increased by \$30,000, what is the tax liability?

(c) Suppose in this same year, the gross income increases by \$40,000 and the costs decrease by \$5,000, what is the tax liability?

4-2. A company has a taxable income of \$100,000 before consideration of long-term and short-term transactions. With the transactions shown in the table below, determine the following:
(a) Total tax liability for the year
(b) The tax liability if the \$4000 and \$3000 amounts under short-term are losses.
(c) The tax liability if the \$4000, \$3000, and \$8000 amounts under long-term are losses.

Transaction	Long Term	Short Term
1	\$4000	—
2	3000	—
3	8000	—
4	1000	—
5	6000	—
6	—	\$4000
7	—	3000
8	—	2000
9	—	1000

4-3. Four years ago, a company established a single depreciation account for an asset that was classified as Sec. 1245 Property. This depreciation account was based on sum of the years-digits depreciation, an initial cost of \$60,000, a salvage of \$5000, and a life of 10 years. Without consideration of the sale of this asset, the company's present taxable income is \$100,000. Determine the following:
(a) The federal tax liability of the company if the asset is sold for \$40,000 assuming there are no other gains or losses for other Sec. 1245 or 1250 property.
(b) Same as part (a) only the selling price is \$70,000.
(c) Same as part (a) only the selling price is \$20,000.

4-4. For the situation described in Problem 4-3 with the exception that the asset is classified as Sec. 1250 property, determine the following:
(a) The federal tax liability of the company if the asset is sold for \$45,000 assuming there are no other gains or losses from Sec. 1245 or 1250 property and there are no other capital gains or losses.
(b) Same as part (a) only the selling price is \$26,000.
(c) Same as part (a) only the selling price is \$30,000.
(d) Same as part (a) only the selling price is \$20,000.

4-5. A Sec. 1250 asset has been depreciated to a book value of \$20,000 using straight-line depreciation. Assuming no other Sec. 1245 or 1250 gains or losses and a taxable income (before the sale of this asset) of \$100,000, determine the following:
(a) the federal tax liability if the asset is sold for \$30,000.
(b) the federal tax liability if the asset is sold for \$15,000.

4-6. A company's taxable income is $100,000 exclusive of the data given in the table below:

SEC. 1245 TRANSACTIONS

Transactions	Purchase Price	Book Value	Selling Price
1	$30,000	$15,000	$18,000
2	20,000	14,000	22,000
3	25,000	8,000	8,000
4	50,000	32,000	20,000

SEC. 1250 TRANSACTIONS

Transactions	Purchase Price	Book Value	Selling Price	Accumulated Depreciation	Accumulated Straight-Line Dep.
1	$40,000	$32,000	$28,000	$ 8,000	$ 5,000
2	50,000	40,000	35,000	10,000	6,000
3	30,000	15,000	21,000	15,000	10,000
4[a]	10,000	5,000	7,000	5,000	5,000

[a]Assume straight-line depreciation used for this asset.

Assuming there are no other long-term or short-term capital gains, determine the company's total federal tax liability.

4-7. For a company with a taxable income of $100,000, and the Sec. 1245 and 1250 transactions shown in the table below, determine the federal tax liability for the following cases:

SEC. 1245 AND 1250 TRANSACTIONS

Transactions	Ordinary Income	Capital Gain	Loss
1	$5000	$2000	—
2	—	—	$4000
3	3000	—	—

(a) The sum of long and short-term gains (not Sec. 1231) are $2000 and $1000, respectively.
(b) Same as part (a) only the capital gain figure in the table ($2000) is $6000.

4-8. A company is considering a project that costs $10,000,000 and qualifies as Sec. 38 property. Determine the following:
(a) The investment tax credit if the estimated useful life is over seven years and the tax liability of the company is $2,000,000.
(b) Same as part (a) only the tax liability of the company is $700,000.

4-9. Four years ago an asset was purchased for \$200,000. At that time it was estimated to have a useful life of over seven years. If this asset is sold now, what is the investment tax penalty?

4-10. A company has a tract of oil wells that cost \$10,000,000. The estimated amount of oil in this tract is 2,000,000 barrels. Determine the following:
(a) The cost depletion if 100,000 barrels are sold in a particular year.*
(b) The percentage depletion if the gross income is \$2,000,000 and the company's taxable income is \$1,000,000.
(c) The same as part (b) only the company's taxable income is \$500,000.
(d) Considering parts (a) and (b), what would be the company's depletion deduction for the year.

*Note: Cost depletion is based on the number of units (barrels) sold (not extracted).

4-11. A company is considering a project that is estimated to generate the following:

yearly gross income	=	\$500,000
yearly costs	=	200,000
yearly depreciation	=	50,000

Determine the following:
(a) The incremental tax that will result from this project if the incremental federal tax rate is 48%, the state tax rate is 8%, and the state tax is based on the federal income base.
(b) Same as part (a) only both federal and state taxes are deductible in computing state tax.

4-12. Suppose that in Problem 4-6 the sum of long- and short-term transations from assets not qualifying as Sec. 1245 or 1250 property are as shown below, determine the company's tax liability.
(a) The sum of long-term transactions is a gain of \$10,000.
The sum of short-term transactions is a loss of \$5000.
(b) The sum of long-term transactions is a gain of \$5000.
The sum of short-term transactions is a loss of \$10,000.

CHAPTER 5
METHODS FOR THE EVALUATION OF PROJECTS

This chapter presents various methods used to evaluate projects (investments). The discussion considers only a *single* project having estimated gross incomes (revenues or savings). Cost comparisons, replacement analysis, and capital budgeting problems are discussed in later chapters. Although all of these topics are related, it is felt that by separating these topics the presentation is simplified and understanding enhanced.

As a result of considering only a single project, this chapter is, basically, concerned with answering the question: Is a particular project under consideration acceptable or unacceptable? In order to answer this question, three evaluation methods are presented: (1) internal rate of return, (2) net present value, and (3) payback (payout) period. All of these methods are related, either directly or indirectly, to the concept of net cash flow.

NET CASH FLOW

Simply stated, net cash flow is the algebraic sum of money flowing in and out of a company over some period of time. The usual sign convention is to make cash inflows positive and outflows negative. In the context of evaluating a single investment, the net cash flows are the *incremental* net cash flows that are expected (estimated) to occur as a result of the investment under consideration. The usual procedure (although not required) is to evaluate the net cash flows for each year of the investment's life and assume that these yearly net cash flows occur as discrete amounts at the end of their respective years. This procedure, then, allows for the use of discrete interest factors and introduces very little, if any, error in the evaluation.

Another assumption that is often made in determining net cash flows is to assume that the project under consideration is to be financed by internally generated funds. This assumption allows for certain aspects of the financing decision (debt ratio, availability of funds, interest, and principal payments) to be separated from the evaluation of the economic desirability of the project. Although this assumption cannot be justified in many circumstances, it is sufficient for the purposes of *this chapter;* that is, an explanation of the methods for evaluating single investments and the establishment of a foundation upon which to discuss more advanced topics. The relationships between the definition of cash flow and the methods of financing the project are presented in Chapter 7.

With the assumption of financing by internally generated funds, the incremental net cash flows, as a result of a particular project, can be defined by the equation

$$X_n = (G - C) - (G - C - D)T - K + L \qquad (5\text{-}1)$$

where

X_n = net cash flow in year n
G = gross income (revenue or savings) in year n estimated as a result of the project under consideration. This could be, for example, volume of yearly sales times the selling price per unit.
C = total yearly costs in year n associated with the gross income in the same year. These are "out-of-pocket" costs; that is, money actually leaving the company. For example, they would include both direct and indirect labor and material costs.
D = tax depreciation allowance in year n
T = composite state and federal tax rate
K = capital investment in year n
L = salvage value in year n

The use of Eq. (5-1) is shown in Example 5-1.

EXAMPLE 5-1

Suppose that a particular investment indicates the data given in Tables 5-1 and 5-2 and it is necessary to determine the net cash flows for each year. The first step is to determine the

Table 5-1
INVESTMENT DATA FOR EXAMPLE 5-1

End of Year	Capital Investment	Use	Life	Salvage
0	$ 200,000	Land	—	—
1	1,000,000	Building	30	$100,000
1	2,000,000	Equipment	20	200,000
2	1,000,000	Equipment	20	0
3	500,000	Equipment	20	0

Table 5-2
GROSS INCOME AND COST DATA FOR EXAMPLE 5-1

End of Year	Gross Income	Total Costs
0	—	—
1	—	$ 80,000
2	—	100,000
3	$ 400,000	500,000
4	2,000,000	500,000
.	.	.
.	.	.
.	.	.
25	2,000,000	500,000

depreciation schedules for each capital investment. Assuming straight-line depreciation (strictly a matter of convenience), gives the results shown in Table 5-3. With these results, the appropriate values are now substituted into Eq. (5-1). The calculations and results are shown in Table 5-4.

Now, in regard to the net cash flows in Table 5-4, the following comments are necessary.

Table 5-3
DEPRECIATION SCHEDULES FOR EXAMPLE 5-1

End of Year	\$1,000,000 $n = 30$ $L = 100{,}000$	\$2,000,000 $n = 20$ $L = 200{,}000$	\$1,000,000 $n = 20$ $L = 0$	\$500,000 $n = 20$ $L = 0$	Total
0	—	—	—	—	—
1	—	—	—	—	—
2	$30,000	$90,000	—	—	$120,000
3	30,000	90,000	$50,000	—	170,000
4	30,000	90,000	50,000	$25,000	195,000
.	.	.	.	.	.
.	.	.	.	.	.
.	.	.	.	.	.
21	30,000	90,000	50,000	25,000	195,000
22	30,000	—	50,000	25,000	105,000
23	30,000	—	—	25,000	55,000
24	30,000	—	—	—	30,000
.	.	.	.	.	.
.	.	.	.	.	.
.	.	.	.	.	.
31	30,000	—	—	—	30,000

Table 5-4

NET CASH FLOWS FOR EXAMPLE 5-1

End of Year	Calculation (T' = 52%)	Net Cash Flow
0	−\$200,000	−\$200,000
1	−80,000 − (−80,000)(0.52) − 3,000,000	−3,038,400
2	−100,000 − (−100,000 − 120,000)(0.52) − 1,000,000	−985,600
3	(400,000 − 500,000) − (400,000 − 500,000 − 170,000)(0.52) − 500,000	−459,600
4	(2,000,000 − 500,000) − (2,000,000 − 500,000 − 195,000)(0.52)	821,400
.	.	.
.	.	.
.	.	.
21	.	821,400
22	(2,000,000 − 500,000) − (2,000,000 − 500,000 − 105,000)(0.52)	774,600
23	(2,000,000 − 500,000) − (2,000,000 500,000 − 55,000)(0.52)	748,600
24	(2,000,000 − 500,000) − (2,000,000 − 500,000 − 30,000)(0.52)	735,600
25	(2,000,000 − 500,000) − (2,000,000 − 500,000 − 30,000)(0.52) + 200,000	935,600
26	− (−30,000)(0.52)	15,600
.	.	.
.	.	.
.	.	.
30	.	15,600
31	− (−30,000)(0.52) + 100,000 + 200,000	315,600

1. Although land is not depreciable, it is a capital asset and its purchase (cost) represents a cash flow out of the company. It has been assumed that the land can (and will) be sold for its original purchase price. Consequently, this value (\$200,000) has been added to the cash flow in year 31. If it is not believed the land will be sold, then it should not be included in year 31.
2. It is assumed, in calculating the net cash flows, that there are sufficient incomes from other investments in the company to allow for the full deductions of costs and depreciations in computing taxable income.
3. In year 25, the salvage value (\$200,000) has been added. It might be argued that it should have been added in year 21 since this is the year when the capital asset (\$2,000,000) is fully depreciated. However, since the salvage is for equipment, it is probably more realistic to assume that the earliest this salvage can be realized is when production ceases; that is, year 25 since the data implies (gross income stops) that this is when production ends. It must be recognized that the life used for determining tax depreciation schedules is established in accordance with tax regulations; these lives may or may not agree with the expected life of the project. Of course, if it is estimated that the equipment is disposed of in year 21, then the \$200,000 should be added in this year rather than year 25.
4. The statement of this example and the manner in which the depreciation schedules are determined imply multiple-asset depreciation accounts for the equipment and a single

asset depreciation account for the building. This is in accordance with the ADR System, that is, the separation of Sec. 1245 and 1250 property. If item depreciation is used, then it is necessary to establish the depreciation schedule for each item in order to determine the total yearly depreciation.

5. It might be argued that since gross incomes cease in year 25, then all cash flows should terminate at this point. Basically, there is nothing wrong with this argument provided it is *actually* believed that all the facilities are disposed of, at this time. If this is the case, then it may be necessary to modify the cash flows in year 25 due to certain tax effects.

Tax Effects in Cash Flow Calculations

In Example 5-1, the incremental state and federal income taxes are included in the net cash flows in a relatively simple manner. However, under certain circumstances, there may be other tax effects that should be included (or at least approximated) in the cash flow calculations. This is the purpose of this section: a discussion of these circumstances and the manner in which the tax effects can be included in the net cash flows.

In Example 5-1, the investment tax credit is not included. If it is desired to include the investment tax credit, the following cash flows are changed to

$$\begin{aligned} X_2 &= -100{,}000 - (-100{,}000 - 120{,}000)(0.52) - 1{,}000{,}000 \\ &\quad + (0.10)(2{,}000{,}000) = -\$785{,}600 \\ X_3 &= (400{,}000 - 500{,}000) - (400{,}000 - 500{,}000 - 170{,}000)(0.52) - 500{,}000 \\ &\quad + (0.10)(1{,}000{,}000) = \$359{,}600 \\ X_4 &= (2{,}000{,}000 - 500{,}000) - (2{,}000{,}000 - 500{,}000 - 195{,}000)(0.52) \\ &\quad + (0.10)(500{,}000) = \$871{,}400 \end{aligned}$$

It should be noted that in the previous calculations the investment tax credits are taken in the years following the capital expenditures. It can be argued that tax credits should be taken in the same year as the capital expenditure. In actuality, logical arguments can be presented for either year. This text chooses to adopt the convention that the investment tax is realized in the year following the capital expenditure. These calculations assume that the taxable income from other activities in the company is large enough to allow for the full deduction of the investment tax credit. No investment tax credit is taken for the building because it is assumed the building does not qualify as Sec. 38 property (which is the usual case). The plus sign before the investment tax terms means that this amount is a reduction in tax; consequently, there is a decrease cash outflow.

Another tax consideration is in regard to salvage values. Salvage values used to determine tax depreciation schedules are determined not only on the basis of the estimated future value of the asset (or assets) but, also, on the basis of tax regulations. Consequently, there may be differences between the salvage values used in determining the tax depreciation schedule and the actual expected salvage value. If differences are expected between these two salvage values, then the cash flows for the year in which the sale of the asset is expected to occur should be adjusted to reflect possible tax effects. As an example, the cash flow for year 25 in Table 5-4 is computed on the basis that the salvage value used for determining the tax depreciation schedule ($200,000) equals the actual selling price. Consequently, there are no additional tax effects. However, if it is estimated that the actual selling price for the equipment in year 25 is $100,000, then the net cash flow for this year should be modified to

$$X_{25} = (2{,}000{,}000 - 500{,}000) - (2{,}000{,}000 - 500{,}000 - 30{,}000)(0.52) + 100{,}000 + (200{,}000 - 100{,}000)(0.52)$$

This calculation assumes that these assets, since they are equipment, can be treated as Sec. 1245 property. Also, since there is an implicaton that production activities are being curtailed (gross income stops in year 25), the retirement of these assets (equipment) should be classified as an extraordinary retirement. Therefore, the loss (in this example) of (200,000 − 100,000) = $100,000 is considered as a reduction in taxable income and, consequently, taxes. The resulting tax effect of $52,000 is, strictly speaking, an approximation since the true effect of the loss on taxes depends on gains and losses from the disposal of other depreciable assets in the year. If other gains or losses can be determined for the year, then they should be taken into account in considering the tax effects. Quite often (if not most often), gains or losses from the disposal of other equipment cannot readily be determined. This is particularly true in the evaluation of projects because estimates are, usually, being made far into the future. Thus, it is the usual practice to approximate the tax effects of gains and losses for a project as if they are independent of other gains and losses.

As another example of the approximation of tax effects in cash flow calculations, assume that all the facilities in Example 5-1 are to be disposed of in year 25. The first step in determining the tax effect is to calculate the book values for each class of assets. In this case, this is a relatively simple matter since most of the assets are fully depreciated by year 25 and the book values are simply the salvage values. The next step is to estimate the selling prices for each class of assets and determine if there is a loss or gain. These steps are shown in Table 5-5. Now, the

Table 5-5

TAX TREATMENTS OF GAINS AND LOSSES

Asset	Book Value	Selling Price	Tax Section	Loss	Gain	Tax Treatment
(1) Equipment—$1,000,000	0	$ 30,000	1245	—	$ 30,000	Ordinary income
(2) Equipment—$2,000,000	$200,000	100,000	1245	$100,000	—	Loss
(3) Equipment—$500,000	0	20,000	1245	—	20,000	Ordinary income
(4) Building	280,000	150,000	1250	130,000	—	Loss
(5) Land	200,000	350,000	1231	—	150,000	Capital gain

gains for items 1 and 3 are designated as ordinary income since they are Sec. 1245 property (see Figure 4-1). Therefore, the cash flows from the sale of these two items are

$$30{,}000 - (30{,}000 - 0)(0.52) = 14{,}400$$
$$20{,}000 - (20{,}000 - 0)(0.52) = 9{,}600$$

The only capital gain is from the sale of the land. Consequently, the net of capital gains and losses is

$$150{,}000 - (100{,}000 + 130{,}000) = -80{,}000$$

implying a loss. Since these gains and losses are from the disposal of Sec. 1245 and 1250 property, Sec. 1231 is applicable. Therefore, the loss can be taken as a reduction in ordinary income. The cash flow from items 2, 4, and 5 is

$$100{,}000 + 150{,}000 + 350{,}000 + (80{,}000)(0.52) = \$641{,}000$$

Using these results, the cash flow for year 25 is

$$X_{25} = (2{,}000{,}000 - 500{,}000) - (2{,}000{,}000 - 500{,}000 - 30{,}000)(0.52)$$
$$+ 14{,}400 + 9600 + 641{,}600 = \$1{,}401{,}200$$

As in the previous discussion, these tax effects are approximations because they are being made without regard (knowledge) to gains and losses from the disposal of other equipment. Of course, if it is estimated that actual selling prices are equal to the tax book values, then there are no tax effects. It is emphasized that these tax considerations are true for extraordinary retirements. If the retirements are classified as ordinary retirements, there are no tax effects.

Working Capital

Many types of projects can lead to changes in working capital. The most frequent case is to have a project increase the working capital requirements as a result of additional money being tied up in such things as cash necessary to meet payrolls and other bills, inventories (raw-materials, in-process, finished goods), and receivables from customers. The exception to this case is a project that decreases the working capital requirements as a result of increasing efficiency.

Working capital increases have the effect of increasing the cash outflows in the early years of a project. These increases are usually assumed to be recovered at the end of the investment with the recovered value being the total of the working capital outflows in the earlier years.

The omission of working capital considerations in an economic evaluation can lead to incorrect conclusions because they can have a significant effect on the cash flows.

As an example of the inclusion of working capital considerations in cash flow calculations, suppose that in the project described in Example 5-1, it is expected that the project will result in additional working capital requirements of

\$200,000 and \$100,000 in years 3 and 4, respectively. The cash flows for these years would, then, be modified to

$$X_3 = -459{,}600 - 200{,}000 = -\$659{,}600$$
$$X_4 = 821{,}400 - 100{,}000 = \$721{,}400$$

In addition, the cash flow in year 25 is

$$X_{25} = 935{,}600 + 300{,}000 = \$1{,}235{,}600$$

in order to reflect the recovery of the working capital. It is assumed that the working capital is recovered in year 25 rather than year 31 because this is the year in which the gross income stops. Of course, if it is believed the recovery of working capital will be over a period of years, then the amounts expected to be recovered should be added in those years.

INTERNAL RATE OF RETURN

The internal rate of return is the interest rate that makes the sum of the net cash flows, discounted to time zero, equal zero. Or, expressed mathematically, the internal rate of return is the interest rate that makes the following equation true.

$$0 = \sum_{n=0}^{T} \frac{X_n}{(1+i)^n} \tag{5-2}$$

where

X_n = net cash flow in year n
T = number of years of cash flow for the project
n = the year in which the cash flow, X_n, occurs
i = IRR = the internal rate of return

EXAMPLE 5-2

As a simple example of the use of Eq. (5-2), suppose the net cash flows for a particular project are calculated, using Eq. (5-1), and give the results shown in Table 5-6. Therefore, using Eq. (5-2), the internal rate of return is

$$0 = -10{,}000 + \frac{3007}{(1+i)} + \frac{3007}{(1+i)^2} + \ldots + \frac{3007}{(1+i)^6}$$
$$0 = -10{,}000 + 3007(PAi,6)$$
$$PA\ i,6 = \frac{10{,}000}{3007} = 3.3256$$
$$i = 20\%$$

Fortunately, in Example 5-2 the net cash flows are a constant annual amount and, consequently, the determination of the internal rate of return is a rather simple matter. However, there are many times when the net cash flows are not constant amounts and, in these cases, the internal rate of return must be determined through a trial-and-error procedure. As an example of this proce-

Table 5-6
NET CASH FLOWS FOR EXAMPLE 5-6

End of Year	Net Cash Flow
0	−$10,000
1	3,007
2	3,007
3	3,007
4	3,007
5	3,007
6	3,007

dure, consider the cash flows in Table 5-7. The first step is to "bracket" the internal rate of return, that is, determine two interest rates that give positive and negative results for Eq. (5-2). This step is shown in Table 5-8. Once these two interest rates are determined, the usual procedure is to approximate the IRR by linear interpolation.

$$i = \text{IRR} = 12 + 3\,\frac{146.1 - 0}{146.1 - (-712)} = 12.51\%$$

obviously, the closer the two bracketing interest rates the better the approximation. This trial-and-error procedure, although not difficult, can be time consuming. Consequently, a more efficient method is to solve for the IRR with the aid of a computer program.

Decision Criterion

The acceptance of a project is based on a comparison of the IRR with a required rate of return, often called the minimum acceptable rate of return (MARR). If the IRR $\geqslant$ MARR, the project is acceptable; if not, the project is

Table 5-7
CASH FLOWS

End of Year	Net Cash Flow
0	−$10,000
1	1,000
2	2,000
3	4,000
4	4,000
5	4,000

Table 5-8

End of Year	Net Cash Flow	Discounted Cash Flows			
		PF_{12-n}		PF_{15-n}	
0	−$10,000	1.0000	−$10,000	1.0000	−$10,000
1	1,000	0.8929	892.9	0.8696	869.6
2	2,000	0.7972	1,594.4	0.7562	1,512.4
3	4,000	0.7118	2,847.2	0.6575	2,630.0
4	4,000	0.6355	2,542.0	0.5718	2,287.2
5	4,000	0.5674	2,269.6	0.4972	1,988.8
Sum	—	—	146.1	—	−712.0

rejected. The value of MARR is related to the cost of capital; a topic that is discussed in greater detail in Chapter 6. For the purposes of this chapter, it is assumed that MARR is given.

Meaning of the IRR

For some people who are unfamiliar with the concept of the IRR, the previous definition of the IRR and Eq. (5-2) do not provide a truly understandable meaning of the IRR. Perhaps a better understanding is possible by considering Table 5-9. In this table, the net cash flows from Example 5-2 are used (IRR = 20%) and it shows that the IRR is the return on the unrecovered capital which allows for the full recovery of the capital ($10,000 in this case) over the life of the project. Another way of looking at it is, the company is "loaning" $10,000 to the project and the project is returning to the company $3007 per year for a period of six years; implying a return of 20%. Table 5-9, also, points out the origin of the word "internal"; that is, the internal rate of return is a return that is generated internally by the project as a result of its net cash flows. There are no assumptions regarding the external use (or investment) of the net cash flows.

Table 5-9

MEANING OF INTERNAL RATE OF RETURN

End of Year	Net Cash Flow	Unrecovered Capital	Return on Unrecovered Capital	Capital Recovered
0	−$10,000	−$10,000	—	
1	3,007	−10,000 + 1007 = −8993	0.2(−10,000) = −$2000	3007 − 2000 = $1007
2	3,007	−8,993 + 1208 = −7785	0.2(−8,993) = −1799	3007 − 1799 = 1208
3	3,007	−7,785 + 1450 = −6335	0.2(−7,785) = −1557	3007 − 1557 = 1450
4	3,007	−6,335 + 1740 = −4595	0.2(−6,335) = −1267	3007 − 1267 = 1740
5	3,007	−4,595 + 2008 = −2507	0.2(−4,595) = −919	3007 − 919 = 2088
6	3,007	−2,507 + 2506 = 0[a]	0.2(−2,507) = −501	3007 − 501 = 2506

[a]Actual difference of −1 is due to round-off error.

There are some who are led to incorrect conclusions regarding the meaning of the IRR, by the following calculations: First, they calculate the future worth of the cash flows using the internal rate of return:

$$F = 3007(FA\ 20,6) = 3007(9.930) = \$29,860$$

Next, they perform the calculation

$$29,860 = 10,000\ FP\ i,6$$
$$FP\ i,6 = 2.986$$
$$i = 20\%$$

and come to the conclusion that, in order to obtain an IRR of 20%, the net cash flows must be ''invested'' at a return of 20%. This conclusions is, of course, not true. As an analogy, if someone loans \$10,000 to a person and this same person pays back \$3007 a year for six years, then the person who loaned the money receives a return of 20% regardless of what he or she does with the \$3007 received each year.

Now, the decision criterion does imply a certain ''investment assumption,'' that is, that money can be invested at the MARR. This, in effect, implies that a comparison is being made between investing the money required by a project into other ventures that earn the MARR and the particular project under consideration that earns its IRR. For example, if the MARR is 12% in Example 5-2 (cash flows in Table 5-6), then the project is acceptable since IRR > MARR. Actually, the implied comparison is between investing the \$10,000 at the MARR (12%) and investing the \$10,000 in the project that earns 20%. Obviously, the latter choice is the best.

For certain types of cash flows, the IRR cannot be determined, for example, when the cash flows all have the same sign. A situtation that occurs when only the costs (or revenues) are considered. For this situation, other approaches (discussed later) must be taken.

Multiple Rates of Return

Certain series of cash flows can give more than one positive interest rates that satisfy Eq. (5-2). This situation may occur when the cash flows have more than one sign change and results from a well-known rule of algebra: Descartes' rule of signs. This rule states that, for an nth-degree polynomial, the number of real positive roots is never greater than the number of sign changes in the sequence of its coefficients. The applicability of this rule to Eq. (5-2) can be seen by expanding it to the form

$$0 = X_0 + X_1p + X_2p^2 + \ . \ . \ . + X_np^n \tag{5-3}$$

where

$$p = \frac{1}{1 + i} \tag{5-4}$$

Equation (5-3) is, of course, an nth-degree polynomial. Consequently, if the number of sign changes of the cash flows ($X_0, X_1, \ldots, X_n$) is greater than one, it is possible to have more than one rate of return that satisfies Eq. (5-2). It is pointed out that more than one sign change in the cash flows is a *necessary* condition for the existence of multiple rates of return but not a *sufficient* condition.

There are numerous examples in the literature of net cash flow series with multiple sign changes. Some of these examples, their source, and net present value profiles are given in Figure 5-1. For the sake of comparison, the net present value profile for the case of ''traditional'' cash flows (negative cash flows followed by a series of positive cash flows) is shown in Figure 5-2. It can be seen in Figure 5-1 that multiple sign changes in the net cash flows can lead to several situations: multiple rates of return, no rate of return, or only one rate of return.

When multiple rates of return exist, a problem occurs from the standpoint of defining a meaningful rate of return. However, the *acceptability* of cash flows with multiple sign changes can be readily determined by calculating the net present value of the cash flows.

An alternate method that is sometimes proposed for finding an ''internal rate of return'' when multiple rates of return exist is discussed in Chapter 9. This discussion is deferred at this point because it involves concepts (cost of capital and reinvestment of funds) that are presented in subsequent chapters.

NET PRESENT VALUE

The net present value is the sum of the net cash flows discounted to time zero at the minimum attractive rate of return. Expressed mathematically, the net present value (NPV) is

$$\text{NPV} = \sum_{n=0}^{T} \frac{X_n}{(1 + k)^n} \tag{5-5}$$

Where the nomenclature is the same as in Eq. (5-2) except k is the minimum attractive rate of return (MARR).

EXAMPLE 5-3

Calculate the net present value of the cash flows given in Table 5-6 if MARR = 10%. Using Eq. (5-5), the net present value is

$$\text{NPV} = -10{,}000 + 3007(PA\ 10{,}6)$$
$$\text{NPV} = -10{,}000 + 3007(4.3553)$$
$$\text{NPV} = \$3096$$

Decision Criterion

A series of cash flows are considered acceptable if NPV ≥ 0 and unacceptable if NPV < 0. Consequently, the cash flows in Example 5-3 are acceptable. The reasoning behind this criterion is explained in the next section.

End of Year	Net Cash Flow
0	−1,600
1	10,000
2	−10,000

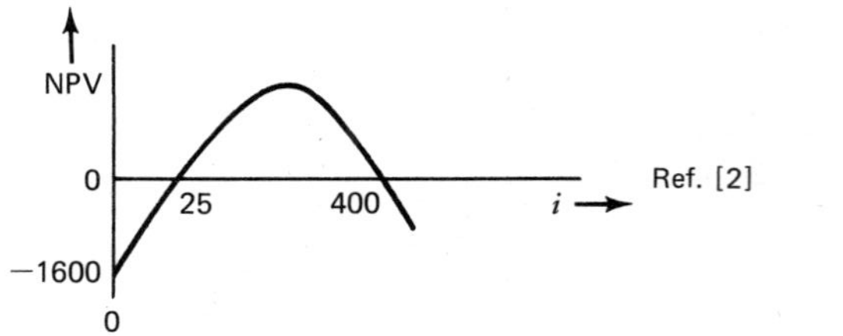

End of Year	Net Cash Flow
0	−1,000
1	6,000
2	−11,000
3	6,000

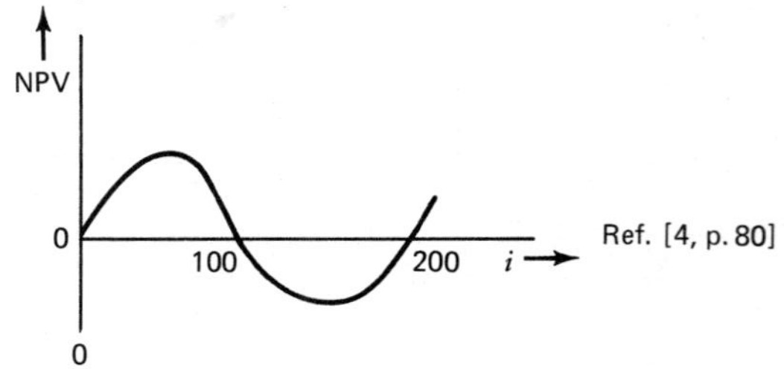

End of Year	Net Cash Flow
0	−1,000
1	1,400
2	−100

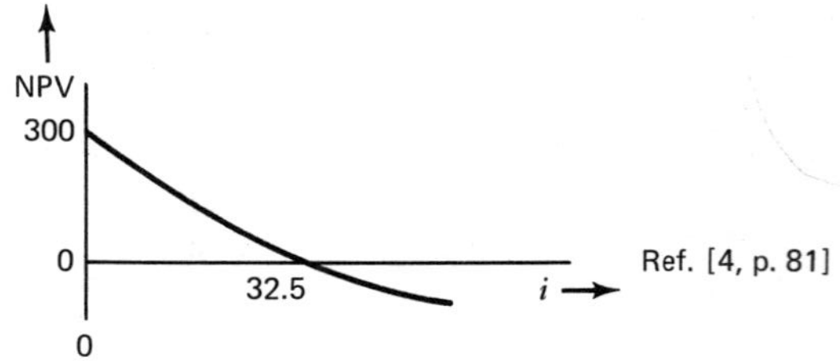

Figure 5-1.
New present value profiles for net cash flows with multiple sign changes.

End of Year	Net Cash Flow
0	—
1	$50,000
2	
.	
.	
.	
5	50,000 − 800,000
6	100,000
7	
.	
.	
.	
20	100,000

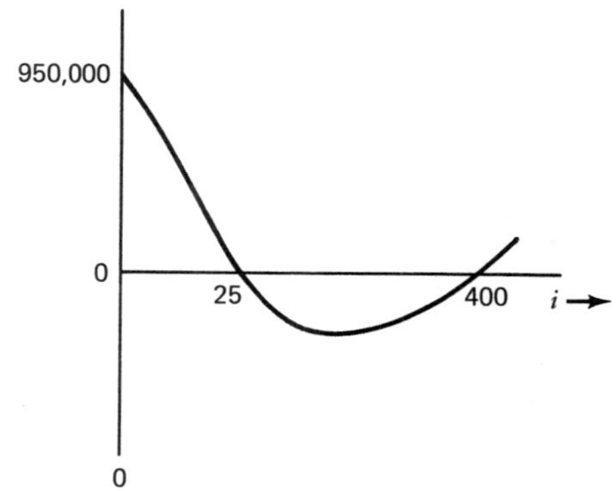

Ref. [1, p. 509]

End of Year	Net Cash Flow
0	
1	$50,000
2	50,000
.	.
.	.
5	50,000 − 700,000
6	100,000
7	100,000
.	.
.	.
.	.
20	100,000

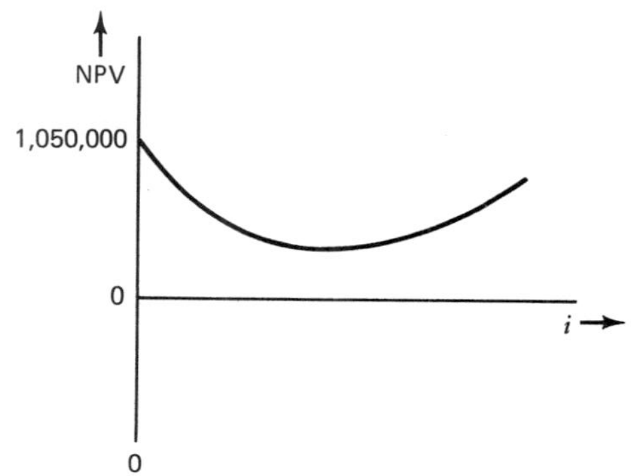

Ref. [1, p. 514
Prob. C-3]

Figure 5-1. *(Continued)*

End of Year	Net Cash Flow
0	$-100,000
1	50,000
2	.
.	.
.	.
5	50,000 – 800,000
6	100,000
.	.
.	.
.	.
20	100,000

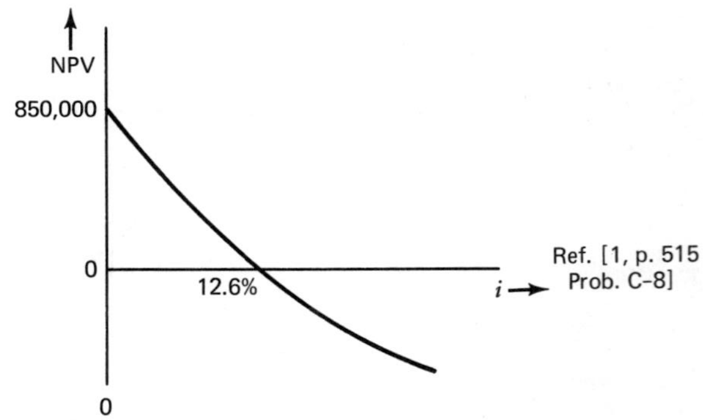

End of Year	Net Cash Flow
0	$-1000
1	4700
2	-7200
3	3600

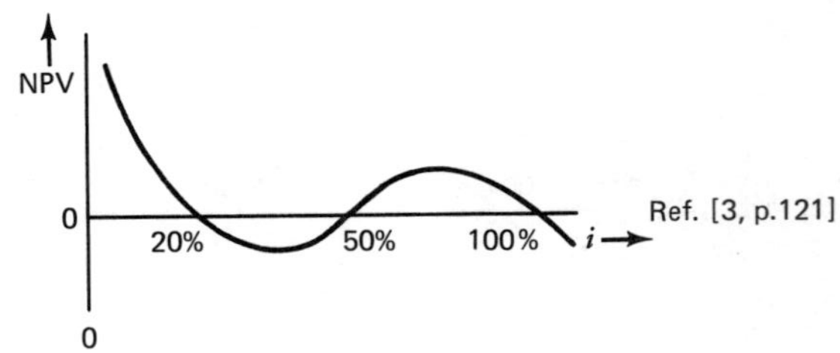

Figure 5-1. *(Continued)*

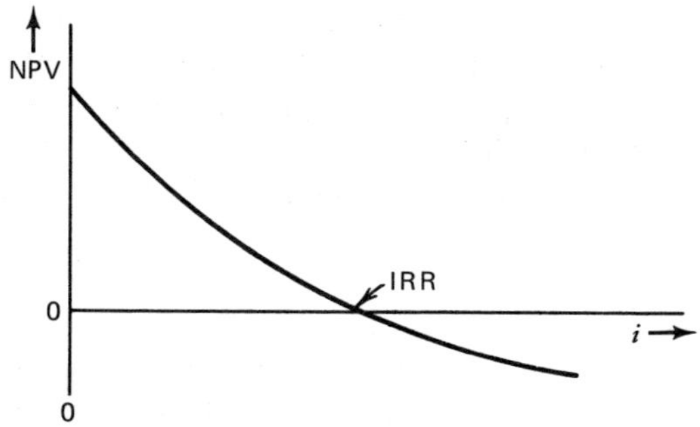

Figure 5-2.
Net present value profile for traditional net cash flows.

Meaning of NPV

A better understanding of the true meaning of net present value and the decision criterion is possible by considering Table 5-10. This table is very similar, from a computational standpoint, to Table 5-9 except MARR is used in Table 5-10 to calculate return on unrecovered capital rather than the IRR used in Table 5-9. Table 5-10 shows that all of the capital is recovered in the fifth year with a 10% (MARR) return each year on the unrecovered capital. In addition, there is a "surplus" left over after the recovery of the capital and return: \$2252 in the fifth year and \$3007 in the sixth year. Now, if this "surplus" is discounted to time zero, the result is

$$2252(PF\ 10{,}5) + 3007(PF\ 10{,}6)$$
$$2252(0.6209) + 3007(0.5645) = \$3095$$

This result (\$3095) is the same (the actual difference is due to roundoff error) as the NPV calculated using Eq. (5-5). Thus, the actual meaning and the basis of the

Table 5-10
MEANING OF NET PRESENT VALUE

End of Year	Net Cash Flow	Unrecovered Capital	Return of Unrecovered Capital	Capital Recovered
0	−\$10,000	−\$10,000	—	—
1	3,007	−10,000 + 2007 = −7993	0.1(−10,000) = −\$1000	3007 − 1000 = \$2007
2	3,007	−7993 + 2208 = −5785	0.1(−7,993) = −799	3007 − 799 = 2208
3	3,007	−5785 + 2428 = −3357	0.1(−5,785) = −579	3007 − 579 = 2428
4	3,007	−3357 + 2671 = −686	0.1(−3,357) = −336	3007 − 336 = 2671
5	3,007	−686 + 2938 = +2252	0.1(−686) = −69	3007 − 69 = 2938
6	3,007	—	—	—

decision criterion for NPV emerges—a NPV > 0 implies that all the capital is recovered over the life of the project (or a shorter period), a return (MARR) is received each year on the unrecovered capital, and a surplus or bonus (NPV) is also received. If the NPV < 0, then the preceding is not true. If the NPV = 0, this implies that MARR = IRR and, consequently, the meaning of NPV = 0 is the same as the previously discussed meaning of IRR.

RELATIONSHIP OF IRR AND NPV

For a *single project with traditional cash flows,* the internal rate of return and the net present value give the same decision regarding the acceptability of a series of cash flows. This can be seen by considering Figure 5-3. From this figure, if i_1 is MARR, then the NPV is greater than zero and IRR > MARR; consequently, the cash flows are acceptable by both criteria. On the other hand, if i_2 is MARR, then the NPV is negative and IRR < MARR; consequently, the cash flows are unacceptable by both criteria. Thus, both criterion lead to the same decision. For the case of NPV = 0, the figure shows that IRR = MARR and implies acceptability of the cash flows.

For a single project with cash flows that have more than one sign change, the net present value provides a means for determining the acceptability of the same cash flows; that is, if the NPV $\geq$ 0, the cash flows are acceptable and if the NPV < 0, the cash flows are unacceptable.

PAYBACK PERIOD

Another method that is sometimes used for the evaluation of a project is the payback period (sometimes called the payout period). Basically, the payback

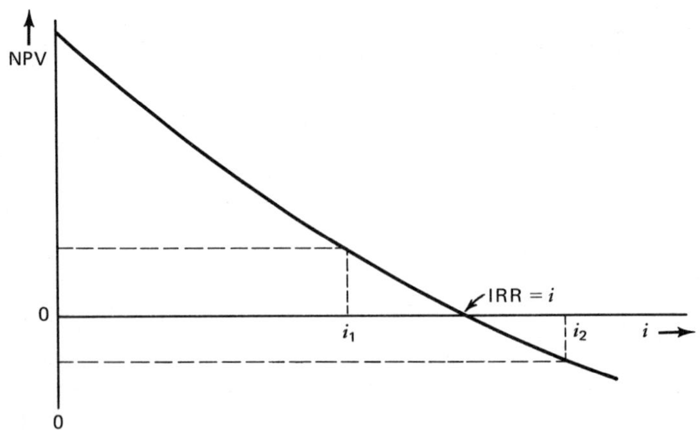

Figure 5-3.
Net present value profile for traditional cash flows.

period is the number of years required for the incoming cash flows to recover the investment. Expressed mathematically, the payback period, l, is

$$0 = \sum_{n=0}^{l} X_n \tag{5-6}$$

This definition is only one of several definitions of payback period. However, it is the one most often used and cited.

EXAMPLE 5-4

Determine the payback period for the cash flows shown in Table 5-11. Using Eq. (5-6), the sum of the cash flows for first three years is

$$-10{,}000 - 5000 + 7000 + 7000 = -1000$$

and for first four years is

$$-10{,}000 - 5000 + 7000 + 7000 + 7000 = 6000$$

Indicating that the payback period is between three and four years. Using linear interpolation, the payback period is

$$\begin{array}{cc} 3 & -1000 \\ 1 & 0 \\ 4 & 6000 \end{array}$$

$$l = 3 + \frac{1000}{7000} = 3.14 \text{ years}$$

Decision Criterion

From a decision standpoint, the calculated payback period is required to be equal to or less than some predetermined payback period for the cash flows to be considered desirable. Otherwise, the cash flows are unacceptable.

Table 5-11
CASH FLOWS FOR EXAMPLE 5-4

End of Year	Net Cash Flow
0	−$10,000
1	−5,000
2	7,000
3	7,000
4	7,000
5	7,000
6	7,000
7	7,000

As a single decision criterion, the use of payback period is questionable. Examples are possible where the IRR and the NPV indicate a desirable series of cash flows but the payback period may indicate the same cash flows are undesirable because of a long payback period. In addition, it is also possible to give examples where the payback period criterion may indicate a series of cash flows are acceptable but the NPV and IRR may indicate the same cash flows are unacceptable. This conflict results because the payback period criterion does not consider what the cash flows are after the payback period and, also, because the time value of money is not considered in the usual definition of payback period.

As a secondary criterion, the payback period may have some merit; that is, after the cash flows are acceptable, on the basis of NPV or IRR, then payback period might be considered as an additional decision criterion. Obviously, for a small company, the time required to recover an investment is an important consideration. However, to base the acceptability of a series of cash flows strictly on the basis of payback period may lead to incorrect decisions.

References

1. Grant, E. L., and W. G. Ireson, *Principles of Engineering Economy,* 4th Edition, The Ronald Press Company, New York, 1964.
2. Lorie, J. H., and L. J. Savage, "Three Problems in Rationing Capital," *Journal of Business,* October 28, 1955, pp. 236–237.
3. Thuesen, H. G., W. J. Fabrycky, and G. J. Thuesen, *Engineering Economy,* 4th Edition, Prentice-Hall, Inc., Englewood Cliffs, N.J., 1971.
4. Van Horne, J. C., *Financial Management and Policy,* 2nd Edition, Prentice-Hall, Inc., Englewood Cliffs, N.J., 1971.

Problems

5-1. A company is considering a project with an initial cost of $300,000. It is estimated that this project will generate a gross income of $200,000 for a period of 10 years and the annual costs for the same 10 years will be $70,000. The tax rate is 52% and a straight-line depreciation model will be used for tax purposes with a salvage value of zero and a life of 10 years. Determine the following:
(a) The internal rate of return.
(b) The net present value if the minimum attractive rate of return is 15%.
(c) The payback period assuming the interest rate is zero.

5-2. For a project with the data shown below, determine the following:
(a) The net cash for each year assuming the assets will be depreciated after gross income stops and the tax rate is 52%.
(b) The internal rate of return.
(c) The payback period with an interest rate of zero.

CAPITAL INVESTMENT DATA

End of Year	Capital Investment	Use	Tax-Depreciation Life	Depreciation Model	Salvage Value
0	$ 100,000	Land	—	—	—
1	2,000,000	Building	25	Straight line	$200,000
1	1,100,000	Equipment	15	Straight line	$50,000
2	3,000,000	Equipment	10	Straight line	0

NET INCREASES IN WORKING CAPITAL

End of Year	Working Capital
0	—
1	1,000,000
2	1,000,000
3	1,000,000

GROSS INCOME AND COSTS

End of Year	Gross Income, $	Costs, $
0	—	—
1	—	$ 100,000
2	—	200,000
2	300,000	500,000
4	6,000,000	1,000,000
.	.	.
.	.	.
.	.	.
16	6,000,000	1,000,000
17	2,000,000	500,000
18	2,000,000	500,000
19	1,000,000	200,000
20	1,000,000	200,000

5-3. Suppose in Problem 5-2 it is decided to include the investment tax credit. What would the cash flows become? What is the effect of including the investment tax credit on the internal rate of return?

5-4. Suppose that in Problem 5-2 it is known that the selling prices of the various assets, when sold, will be the following:

land = $250,000
building = $300,000

equipment in year one ($1,100,000) = $300,000
equipment in year two ($3,000,000) = $50,000

Assuming no other gains or losses from other assets, what effect will these selling prices have on the cash flows in the years the assets are assumed to be sold?

5-5. A project has the cash flows shown below:

End of Year	Net Cash Flow
0	−$500,000
1	75,000
.	.
.	.
.	.
15	75,000

Sketch the net present value profile for this project.

5-6. Determine if the following statements are true or false. Explain your answers.
(a) For a single project with "traditional" cash flows, the higher the IRR the better.
(b) For a single project with "traditional" cash flows, the higher the NPV the better.

5-7. Answer the following questions:
(a) For a project with "traditional" cash flows, which statement (or statements) must be satisfied for the project to be considered acceptable?
(1) NPV $<$ 0; (2) NPV $\geqslant$ 0; (3) IRR $>$ MARR; (4) IRR $<$ MARR.
(b) For a project with "traditional" cash flows that has a net present value less than zero, which of the following statements are true?
(1) MARR = IRR; (2) MARR $>$ IRR; (3) MARR $<$ IRR.

5-8. A company, in evaluating its capital investments, requires that an investment must have both a payout period of three years or less and an internal rate of return of 15% or greater. Comment on this requirement.

5-9. What is the necessary condition for the existence of multiple rates of return? How is the acceptability of a project with multiple rates of return determined?

CHAPTER 6
MINIMUM ACCEPTABLE RATE OF RETURN AND THE COST OF CAPITAL TO THE FIRM

Chapter 5 showed that the minimum acceptable rate of return (MARR) is an important factor in determining the desirability of an investment; that is, the internal rate of return is compared to MARR and the interest rate used in calculating net present value is also MARR. Consequently, establishing a value for MARR is a most important aspect in the evaluation of projects. Hence, this chapter considers a basis on which a value for MARR may be established.

Basically, the MARR is related to a well-discussed financial concept, that is, the weighted average cost of capital. In general, the cost of capital is the weighted average of the cost of capital from specific sources. Or, expressed mathematically, the cost of capital, k, is

$$k = \sum_{i=1}^{n} w_i k_i \qquad (6\text{-}1)$$

where

w_i = proportion of capital from source i
k_i = cost of capital (expressed as a percentage) from source i

Although there is widespread agreement on Eq. (6-1), there still exists some controversy about the evaluation of some of the components included in Eq. (6-1). As a result of this controversy, the determination of the cost of capital still requires some degree of judgment and estimation. This chapter presents some of the more basic considerations in determining the cost of capital in an attempt to narrow the range in which judgment and estimates are required. For those who wish to make an in-depth study of the cost of capital, references are provided at the end of this chapter.

There are two general sources of capital: (1) debt and (2) equity sources.

COST OF DEBT CAPITAL

Debt capital is money that is obtained from external sources. Some of the principal sources of debt capital are mortgages, loans, and company bonds.

Mortgages and Loans

The cost of capital for these two sources of capital is the *actual* interest rate being paid on the mortgage or loan. Many times, the stated interest rate on a mortgage or loan is not the same as the actual effective interest rate being paid due to discounting practices or the manner in which the loan is paid back. Consequently, it is most important to convert the stated interest rate on a mortgage or loan to an effective yearly compound rate, for it is in this context that it is most often used as the cost of capital in economic studies.

In general, the after-tax cost of capital from debt sources can be determined by the following relationship

$$B - \sum_{j=1}^{n} \frac{I_j(1 - T) + P_j}{(1 + k_d)^j} = 0 \qquad (6\text{-}2)$$

where

B = actual borrowed money received
I_j = interest payment in year j
P_j = principal payment in year j
n = number of years of debt payment
k_d = after tax cost of capital
T = effective tax rate

Equation (6-2) implies that the after tax cost of capital, k_d, is the interest rate that makes the present value of the interest and principal payments equal the actual money received. If the before tax cost of debt capital is desired, then $T = 0$ in Eq. (6-2).

EXAMPLE 6-1

Suppose a loan of \$10,000 is desired for five years and the stated interest rate is 8%. Also suppose that the loaning agency has a practice of discounting loans by 5%. Now, the values of five years and a 5% discount should not be considered representation of the life and discount rates found in loans used to raise debt capital. However, it is the methodology that is important here and these values simplify the calculations.

The discounting of the loan implies that if \$10,000 is borrowed, the actual money received is

$$B = 10{,}000 - 10{,}000\,(0.05) = \$9500$$

Or, in order to actually receive \$10,000, the amount of the loan must be

$$10{,}000 = P - 0.05P$$

$$P = \frac{10{,}000}{0.95} = \$10{,}526.32$$

In order to use Eq. (6-2), it is necessary to know the manner in which the loan (both interest and principal) is to be paid back. Two of the more common procedures of paying back loans are shown in Table 6-1. In this table the money received ($9500) is shown as a positive cash flow and the interest and principal payments are shown as negative cash flows. The tax adjustment is not shown in Table 6-1 but is included in subsequent calculations. The values given in Table 6-1a are based on the assumption that the principal ($10,000) is to be paid back in five equal installments of $2000 and, consequently, the interest values are the interest payments on the unpaid balance.

The values in Table 6-lb are based on the assumption that the entire principal is paid at the end of the loan.

Now, substituting the values given in Table 6-1a into Eq. (6-2) and assuming a tax rate of 52% gives

$$9500 - [800\,(1 - 0.52) + 2000]\,(PF\ k_d\ , 1) - \ldots - [160(1 - 0.52) + 2000]\,(PF\ k_d, 5) = 0 \qquad (6\text{-}3)$$

Making use of the interest factors, Eq. (6-3) can be rewritten as

$$9500 = (1 - 0.52)\,[800 - 160(AG\ k_d, 5)]\,(PA\ k_d, 5) + (2000)\,(PA\ k_d, 5) \qquad (6\text{-}4)$$

Table 6-1
POSSIBLE CASH FLOWS FOR THE DEBT PAYMENTS IN EXAMPLE 6-1 (AMOUNT BORROWED: $10000)

End of Year	Cash Flows
0	$95000
1	– 800– 2000
2	– 640– 2000
3	– 480– 2000
4	– 320– 2000
5	– 160 – 2000

(a) Equal Principal Payments

End of Year	Cash Flows
0	$9500
1	– 800
2	– 800
3	– 800
4	– 800
5	– 800 – 10,000

(b) Equal Interest Payments

and solving gives

$$k_d = 5.74\%$$

If the tax term $(1 - 0.52)$ is omitted, the result is $k_d = 10.1\%$, which is the before-tax cost of capital.

For the cash flows shown in Table 6-1b, the cost of capital is

$$9500 = (1 - 0.52)\,(800)\,(PA\ k_d, 5) + 10000\,(PF\ k_d, 5) \tag{6-5}$$
$$k_d = 5\%$$

and the before-tax cost of capital is 9.30%.

The results of the calculations shown in Eqs. (6-4) and (6-5) point out that the manner in which the loan is paid back does have an effect on the actual interest rate being paid *when discounting is involved*. However, when discounting is not involved both methods of paying back the loan shown in Table 6-1 give the same results; that is, if the \$9500 in Tables 6-1a and 6-1b is replaced by \$10000 then, 8% is obtained for both methods of debt payment. Furthermore, if the amount borrowed is \$10,526.32 in order to actually receive \$10000 after the discount, as suggested earlier in this example, then the resulting cash flows are shown in Table 6-2. If k_d is calculated using the cash flows in Table 6-2, the same results are obtained for the before and after tax values of k_d; that is 10.1% (5.74%) for Table 6-2a and 9.3% (5%) for Table 6-2b.

Bonds

The sale of bonds is another method of obtaining debt capital. There are two types of bonds that are most frequently used to obtain capital: (1) fixed maturity bonds and (2) convertible bonds.

Fixed maturity bonds mature after a specified period of time. For this type of bond, the cost of capital is the inferred (actual) interest rate being paid on the bond. It may or may not be the contractual rate because the selling price of the bond, in combination with the floatation costs, may be at a price above or below its face value.

An expression for the cost of a fixed maturity bond can be determined by considering Table 6-3 where the nomenclature refers to the following:

P = the actual money received by a company from the sale of one bond; this is selling price minus selling expenses (brokerage fees, etc.)
r = contractural rate on the bond
V = face value of the bond
$\frac{rV}{c}$ = interest paid per period
c = number of interest periods per year
n = number of years to maturity of the bond
T = incremental tax rate on corporate income

Table 6-2

POSSIBLE CASH FLOWS FOR DEBT PAYMENTS IN EXAMPLE 6-1 (AMOUNT BORROWED: $10,526.32)

End of Year	Cash Flow
0	$10,000
1	−842.11 − 2105.26
2	−673.68 − 2105.26
3	−505.26 − 2105.26
4	−336.84 − 2105.26
5	−168.42 − 2105.26

(a) Equal Principal Payment

End of Year	Cash Flow
0	$10,000
1	−842.11
2	−842.11
3	−842.11
4	−842.11
5	−842.11 − 10526.32

(b) Principal Paid at End of Loan

Table 6-3

CASH FLOWS RESULTING FROM THE SALE OF A FIXED MATURITY BOND

End of Period	Cash Flow	
0	P	
1	$-\frac{rv(1-T)}{c}$	$-\frac{(P-V)T}{cn}$
2	.	.
3	.	.
.	.	.
.	.	.
cn	$-\frac{rv(1-T)}{c}$	$-\frac{(P-V)T}{cn} - V$

$\frac{rV(1-T)}{c}$ = effective after tax interest cost

$\frac{(P-V)T}{cn}$ = amortizations of difference between selling price and face value over the life of the bond. Under present tax laws, the amount $(P - V)$ can be spread over the life of the bond and taken as an expense when $P < V$. When $P > V$ the amount of $(P - V)$ is income and is also spread over the life of the bond. This implies that the term $(P - V)$ can be either negative or positive.

Now, equating the cash flows in Table 6-3 gives the following

$$P = \frac{rv\,(1-T)}{c}\left(PA\,\frac{k_b}{c}, cn\right) + \frac{(P-V)T}{cn}\left(PA\,\frac{k_b}{c}, cn\right) + V\left(PF\,\frac{k_b}{c}, cn\right) \tag{6-6}$$

which is the expression for determining the after tax cost, k_b, of a fixed maturity bond.

EXAMPLE 6-2

Determine the cost of a \$1000 bond that matures in five years if the contractual rate is 8% paid semiannually. Assume the selling price of the bond is \$910, brokerage fees amount to \$10 per bond, and the incremental tax rate is 52%. Substituting these data into Eq. (6-6) and solving for k_b gives

$$900 = \frac{(0.08)(1000)(1-0.52)}{2}\left(PA\,\frac{k_b}{2}, 2(5)\right) + \frac{(900 - 1000)(0.52)}{2(5)}\left(PA\,\frac{k_b}{2}, 2(5)\right) + 1000\left(PF\,\frac{k_b}{2}, 2(5)\right)$$

$$k_b = 5.12\%$$

Converting this value of 5.12% to an effective annual rate gives

$$k_b = \left(1 + \frac{0.0512}{2}\right)^2 - 1 = 0.0518 = 5.18\%$$

as the after-tax cost of the bond in this example. If the before-tax cost of this bond is desired then the tax rate (T) in Eq. (6-6) is set equal to zero. If this is done for the data given in this example, the before-tax effective annual rate is 10.66%. An approximation that is often made is to compute the before tax rate and multiply it by the term $(1 - T)$. In this example, this is

$$(10.66)(1 - 0.52) = 5.12\%$$

which is in close agreement with the actual rate of 5.18%.

Bonds that include a provision for converting them to common stock are called *convertible bonds*. The conversion ratio specifies the number of common shares into which the bond is converted. Although convertible bonds have maturity dates, they are usually regarded as a means of selling common stock. Under the assumption that convertible bonds are issued to raise equity capital, the cost can be determinted by using the equation

$$P = \frac{rV(1-T)}{c}\left(PA\,\frac{k_m}{c}, cx\right) + \frac{(P-V)(T)}{cn}\left(PA\,\frac{k_m}{c}, cx\right) + FR\left(PF\,\frac{k_m}{c}, cx\right) \tag{6-7}$$

where the nomenclature is the same as before and

x = estimated time at which bond is converted to common shares

F = market value per share of common stock at the end of time x

R = conversion ratio, the number of common stock received per bond

Note that the second term in Eq. (6-7) is divided by cn. This implies that the bond discount (or premium) is not deductible (or taxable) after the bond is converted to common stock.

EXAMPLE 6-3

Suppose the data in Example 6-2 is for a convertible bond and the conversion ratio is specified as 8. Also, it is estimated that the bond will be converted in three years and, at that time, the market price of a share of common stock will be \$120.

Substituting the applicable data into Eq. (6-7) gives

$$900 = \frac{(0.08)(1000)(1-0.52)}{2}\left(PA\,\frac{k_m}{2}, 2(3)\right) + \frac{(900-1000)(0.52)}{2(5)}\left(PA\,\frac{k_m}{2}, 2(3)\right) + (8)(120)\left(PF\,\frac{k_m}{2}, 2(3)\right)$$

and solving for k_m gives

$$k_m = 5.2\%$$

and effective rate of

$$k_m = \left(1 + \frac{0.052}{2}\right)^2 - 1 = 0.053 = 5.3\%$$

Obviously, the before tax rate could be determined by setting the tax rate equal to zero as mentioned before.

COST OF EQUITY CAPITAL

The principal sources of equity funds are the sale of common stock, retained earnings, and capital surplus. The only difference in cost of funds from these sources is due to the consideration of flotation costs. If retained earnings are the source of funds, then there are no flotation costs. Consequently, the cost of retained earnings is slightly less than the cost of funds from the sale of common stock and capital surplus.

Cost of New Common Stock

One of the most often-cited models for determining the cost of a new common stock issuance is the one proposed by Gordon and Shapiro [2].

Although their model is based on some questionable assumptions (which are discussed later), it can provide a satisfactory estimate of the cost of equity funds if it is used with judgment.

The Gordon-Shapiro model is, essentially, a divident valuation model; that is, the cost of a share of common stock to a company is the present value of the expected stream of dividends paid to the stockholder. Or, expressed mathematically,

$$P_0 = D_0 + D_1(1 + k_e)^{-1} + D_2(1 + k_e)^{-2} + \dots + D_t(1 + k_e)^{-t} + \dots \infty \qquad (6\text{-}8)$$

$$P_0 = \sum_{t=0}^{\infty} \frac{D_t}{(1 + k_e)^t} \qquad (6\text{-}9)$$

where

P_0 = current market price of a share of common stock
D_t = dividend per share paid in year t
k_e = implied cost of the common stock

At this point, two assumptions are made in the Gordon-Shapiro model. One, a company is expected to retain a fraction, b, of its yearly after-tax earnings such that

$$b = \frac{Y_t - D_t}{Y_t} \qquad (6\text{-}10)$$

where Y_t is the earnings per share of common stock. Second, a company is expected to earn a return, r, on its equity book value, B_t, such that

$$r = \frac{Y_t}{B_t} \qquad (6\text{-}11)$$

For the sake of derivation purposes, the values of r and b are assumed constant; however, there is no reason why different values cannot be used in the final expression. It does not mean that r and b are permanent values.

As a result of these two assumptions, the dividends at time t are

$$D_t = (1 - b)\,Y_t \qquad (6\text{-}12)$$

Now it is required in the Gordon-Shapiro model that the retained earnings, bY_t, must earn a return, that is, $r(bY_t)$. Consequently, the earnings for time t are

$$\begin{aligned} Y_0 &= Y_0 \\ Y_1 &= Y_0 + r(bY_0) \\ Y_2 &= Y_1 + r(bY_1) \\ &\;\;\vdots \\ Y_t &= Y_{t-1} + r(bY_{t-1}) \end{aligned} \qquad (6\text{-}13)$$

If Y_1 is substituted into Y_2, the result is

$$
\begin{aligned}
Y_2 &= Y_0 + r(bY_0) + rb(Y_0 + rbY_0) \\
&= Y_0 + 2rbY_0 + r^2b^2 \\
Y_2 &= Y_0\,(1 + rb)^2
\end{aligned}
\tag{6-14}
$$

Generalizing Eq. (6-14), the earnings in time t are

$$Y_t = Y_0\,(1 + rb)^t \tag{6-15}$$

Substituting Eq. (6-15) into Eq. (6-12) gives

$$D_t = Y_0\,(1 - b)(1 + rb)^t \tag{6-16}$$

and recognizing that $Y_0\,(1-b) = D_0$, Eq. (6-16) becomes

$$D_t = D_0\,(1 + rb)^t \tag{6-17}$$

Substituting Eq. (6-17) into Eq. (6-9) gives

$$P_0 = \sum_{t=0}^{\infty} D_0(1 + rb)^t\,(1 + k_e)^{-t} \tag{6-18}$$

In order to evaluate Eq. (6-18) further, it is necessary to convert the discrete interest factors to continuous interest factors in the following manner:

$$
\begin{aligned}
(1 + rb)^t &\sim\neq e^{rbt} \\
(1 + k_e)^{-t} &\sim e^{-k_e t}
\end{aligned}
$$

Equation (6-18) can now be written as

$$P_0 = \int_0^{\infty} D_0 e^{rbt}\,e^{-k_e t}\,dt = \int_0^{\infty} D_0 e^{-t(k_e - rb)}\,dt \tag{6-19}$$

Performing the integration in Eq. (6-19) gives

$$P_0 = \frac{D_0}{k_e - rb} \tag{6-20}$$

and solving for k_e gives the result

$$k_e = \frac{D_0}{P_0} + rb \tag{6-21}$$

Equation (6-21) provides a means of estimating the cost of equity capital that is consistent with the financial objective of maximizing the value of the stockholder's equity. That is, using Eq. (6-21), sometimes called the market rate of profit, as the marginal rate of return required for investments results in the maximization of the stockholder's equity. This is also true when the source of funds are retained earnings or capital surplus.

EXAMPLE 6-4

A company is considering a new issuance of common stock to raise capital for new investments. The current market value for a share of common stock is \$120. However, to ensure the sale of this stock, it will be sold for \$115. It is estimated that brokerage fees will amount to \$3.00 per share. The company's current financial report shows the following data:

$$\begin{aligned} \text{total assets} &= \$45{,}000{,}000 \\ \text{liabilities} &= \$10{,}000{,}000 \\ \text{after-tax earnings} &= \$6{,}000{,}000 \\ \text{number of common stock} &= 500{,}000 \text{ shares} \\ \text{dividends paid} &= \$4{,}000{,}000 \end{aligned}$$

With the preceding data it is possible to estimate the cost of equity capital using Eq. (6-21). The dividends per share are

$$D_0 = \frac{4{,}000{,}000}{500{,}000} = \$8$$

and the earnings per share are

$$Y_0 = \frac{6{,}000{,}000}{500{,}000} = \$12$$

Which implies that the proportion of earnings retained is

$$b = \frac{Y_0 - D_0}{Y_0} = \frac{12 - 8}{12} = 0.333$$

The book value of a share of common stock is

$$B_0 = \frac{45{,}000{,}000 - 10{,}000{,}000}{500{,}000} = \$70$$

and the return on book value is

$$r = \frac{Y_0}{B_0} = \frac{12}{70} = 0.171$$

Substituting into Eq. (6-21) gives

$$k_e = \frac{8}{115 - 3} + (0.171)(0.333) = 0.128 = 12.8\%$$

as the incremental cost of capital.

Cost of Retained Earnings

In Example 6-4 the value substituted for P_0 is the actual amount of money received by the company from the sale of a share of stock. If the market value is substituted for P_0, then the cost of retained earnings is obtained. For example, the cost of retained earnings, using the data given in Example 6-4, is

$$k_3 = \frac{8}{120} + (0.171)(0.333) = 0.124 = 12.4\%$$

which shows that the cost of retained earnings is less than common stock because of flotation costs and, usually, underpricing of the stock.

Capital Surplus

The difference between the par value and selling price of stock is often called capital surplus. For example, if a company sells 20,000 shares of stock for \$50 a share and the par value of the stock is \$2 per share, then the company's equity portion of the balance sheet would show

stock (\$2 par value)	\$ 40,000
capital surplus	960,000
net worth	\$1,000,000

The par value of a share of stock is usually specified in the company's charter but is of little use in economic analysis. However, if the stock is sold at a value less than its par value, there are certain legal considerations—stockholders can be held liable by creditors for the difference. Consequently, par values are usually set very low. Stock can be sold without a par value. In this case, this stock is carried on company's books at some stated price. The difference between this stated price and the selling price is carried as capital surplus. In any event, funds obtained from capital surplus are equity funds and the cost of these funds is determined by using Eq. (6-21).

Another method that is often used to estimate the cost of equity capital, is to rewrite Eq. (6-21) as

$$k_e = \frac{D_0}{P_0} + g \tag{6-22}$$

Where the nomenclature is the same as before and g is the expected growth rate, that is, dividends are expected to grow at a compound rate g forever. The usual procedure is to estimate g on the basis of past dividends or earnings assuming, of course, that past dividends or earnings are felt to be meaningful for estimating the growth rate.

EXAMPLE 6-5

Suppose the company in Example 6-4 has the earnings and growth rates for the past five years as shown in Table 6-4. Where, for example, the growth rates are calculated in the following manner:

$$g_0 = \frac{6{,}000{,}000 - 5{,}700{,}000}{5{,}700{,}000} = 0.053$$

$$g_1 = \frac{5{,}700{,}000 - 5{,}500{,}000}{5{,}500{,}000} = 0.036$$

Table 6-4
EARNINGS PER YEAR FOR EXAMPLE 6-5

Year	Net Earnings	Growth Rate
0	$ 6,000,000	0.053
1	5,700,000	0.036
2	5,500,000	0.078
3	5,100,000	0.063
4	4,800,000	—
Σ	$28,900,000	0.230

Now, in calculating the cost of capital, an average growth rate might be used for estimating g, that is,

$$g = \frac{0.230}{4} = 0.058$$

Consequently, the cost of capital, using Eq. (6-22), is

$$k_e = \frac{8}{115 - 3} + 0.058 = 0.129 = 12.9\%$$

It should be noted that estimating the growth rate on the basis of br may give different results than past earnings. If $br = 5\%$, then this value might be considered representative of the growth rate. However, if $br = 30\%$, then it is extremely unlikely that a company can grow at a 30% compound rate. In all probability the growth rate will level off to a more normal rate (perhaps the growth rate of the national economy) as a company matures. What is involved, of course, is the use of judgment in the estimation of the growth rate; that is, the growth rate must be based on realistic estimates of the expected growth in earnings per share. As previously stated, Eqs. (6-21) and (6-22) involve two assumptions; namely, a constant dividend and a constant growth rate. To be sure, future dividends are uncertain and a constant growth rate is unlikely. However, it seems far better to assume a constant dividend and growth rate and estimate them on the basis of past and present data than to reject a model on the basis that the future is unpredictable.

Another Proposal

Solomon [6] suggests that the cost of a new issue of common stock is given by

$$k_e = \frac{E_a}{P} \tag{6-23}$$

where

E_a = average annual earnings per share with an adjustment for the growth of future earnings

P = net amount received by the company from the sale of a share of common stock

There are similarities between Eqs. (6-23) and (6-21) that can be seen if Eq. (6-21) is written as

$$\begin{aligned} k_e &= \frac{D_0}{P_0} + \left(\frac{Y_0}{B_0}\right)(b) \\ &= \frac{D_0}{P_0} + \frac{bY_0}{B_0} \end{aligned} \tag{6-24}$$

and Eq. (6-23) is written as

$$\begin{aligned} k_e &= \frac{D + bY}{P} \\ &= \frac{D}{P} + \frac{bY}{P} \end{aligned} \tag{6-25}$$

The basic difference between Eqs. (6-24) and (6-25) is that one has the book value as the denominator of the second term and the other has the market value. These two equations would give the same values if the book and market values are identical. However, the usual case is that the market value is greater than the book value.

Gordon and Shapiro defend the use of book value on the basis that book value and not market value is used to measure the rate of return of the corporation. For a more complete discussion of their defense of book value and their model in general, it is recommended that the Gordon-Shapiro article be studied. This book takes the point of view that the Gordon-Shapiro model does provide a satisfactory *estimate* of the cost of equity capital and, consequently, is the one that is used in this text.

COST OF PREFERRED STOCK

Preferred stock has some of the characteristics of bonds and some of the characteristics of equity capital. Like a bond, preferred stock pays a stated amount to the stockholder that is equal to its face value times a specified contractual rate. Also, preferred stockholders have a higher priority than common stockholders on earnings and in cases of liquidation. Unlike bonds, preferred stock has no maturity date and its dividends are not tax deductible. Also, the payment of a preferred stock dividend is at the discretion of the board of directors. Consequently, unlike bonds, there is no risk of legal bankruptcy if the dividends are not paid. As a result of these considerations, the cost of preferred stock is

$$k_p = \frac{rV}{P_0} \tag{6-26}$$

where

r = contractual rate
V = face value of preferred stock
P_0 = amount received from the sale of a share of preferred stock

As an example, the cost of capital for a 6%, $100 share of preferred stock selling for $98 a share with flotation costs of $4 per share is

$$k_p = \frac{(0.06)\,(100)}{98 - 4} = 0.064 = 6.4\%$$

WEIGHTED COST OF CAPITAL

Once the cost of capital is determined for each source of funds, it is possible to calculate a weighted cost of capital using Eq. (6-1) provided the weighting factors (w_i) are known. At this point, it is important to understand that it is the incremental cost of capital that is to be determined since capital is raised incrementally to invest in new projects. It is the cost of this new capital that should be used in evaluating new projects. Methods have already been presented for determining the incremented costs from different sources of funds; what needs to be done is to determine the weighting system.

Essentially, the weights must correspond to the future financing plans of the company. In actuality, new capital is raised in ''blocks'' and strict proportions from the various sources cannot be maintained. For example, a company may finance with retained earnings in one instance and entirely with debt in another instance. However, in the long run, companies are able to finance in some proportional manner. It is these future proportion of funds that should be used as the weights.

EXAMPLE 6-7

Suppose a company is going to raise $20 million for new investments and the cost of capital from each source and the planned proportions are as shown in Table 6-5. Using Eq. (6-1), the before-tax cost of capital is

$$k' = 0.3(8) + 0.4(13) + 0.1(10) + 0.2(14) = 11.4\%$$

The after-tax cost of capital can be estimated by approximating the tax consideration (T = 52%) in the cost of the debt capital by the term $(1 - T)$, that is,

$$k = 0.3(8)\,(1 - 0.52) + 0.4(13) + 0.1(10) + 0.2(14) = 10.15\%$$

Often, the future proportions of funds are estimated on the basis of the existing capital structure of the company. Implying the condition that the existing and future capital structures are expected to be the same. When this approach is taken, a problem arises; that is, should the proportions (weights) be based on market or book values? There is some disagreement on this point in the literature. This text believes that market values

Table 6-5
DATA FOR EXAMPLE 6-7

Incremental Cost of Capital		Proportion
Debt (before tax)	8%	0.30
Retained earnings	13%	0.40
Preferred stock	10%	0.10
Common stock	14%	0.20

should be used because they, rather than book values, provide a better estimate of the market's reaction to future inputs of capital.

EXAMPLE 6-8

Suppose a company has decided that its present and future financial structure are to be the same and the cost of capital from different sources are as given in Table 6-5. Also, the current market values and numbers of stocks and bonds for the company are as shown in Table 6-6. With this data the before tax cost of capital is

$$k' = 0.21(8) + 0.11(10) + 0.68(14) = 12.3\%$$

and the after tax ($T = 52\%$) cost is

$$k = 0.21(8)\,(1 - 0.52) + 0.11(10) + 0.68(14) = 11.4\%$$

Where the proportions are determined as shown in Table 6-7.

When market values are used as the basis of the weights, the following adjustment shown in Table 6-7 are necessary. Stocks and bonds are valued at their market price times the number of shares (bonds). In the case of equity, this figure includes common stock, and retained earnings (also capital surplus when applicable). The specific cost for these equity funds is the cost of common stock (14% in this example).

COST OF DEPRECIATION

Depreciation allowances are an important source of funds. However, consideration of the cost of depreciation funds has been deferred until now because its cost

Table 6-6
MARKET VALUES AND NUMBERS OF STOCKS AND BONDS

Capital	Market Value per Share (bond)	Number of Stocks or Bonds
Debt	$900	50,000
Preferred stock	80	300,000
Common stock	120	1,200,000

Table 6-7

Capital	Market Value of a Share (bond)	Number of Shares (bonds)	Amount	Proportion	Cost of Capital
Debt	\$900	50,000	$45(10)^6$	0.21	8
Preferred stock	80	300,000	$24(10)^6$	0.11	10
Equity	120	1,200,000	$144(10)^6$	0.68	14
			$213(10)^6$		

is related to the weighted cost of capital. Unfortunately, it is sometimes thought that depreciation funds do not have a cost. This, of course, is not true. Depreciation funds have an opportunity cost similar to retained earnings. Consequently, depreciation funds should not be invested into fixed assets unless they provide a return. Essentially, this return is the weighted cost of capital.

EXAMPLE 6-9

If a company plans to finance its future projects with the mix of capital shown in Table 6-8, its before-tax weighted cost of capital is 12%. The major point that needs comment in regard to Table 6-8 is the cost of depreciation. This cost is determined in the following manner.

$$\frac{0.30}{0.80}(10) + \frac{0.35}{0.80}(12) + \frac{0.15}{0.80}(16) = 12\%$$

Where, the value 0.80 is the sum of the proportions of debt, retained earnings, and common stock. It is being assumed that the alternative use to which the depreciation funds could be put to use are to repay debt (costing 10%), distribute dividends (costing 12%), and to repurchase common stock (costing 16%) implying an opportunity cost of 12%. It can be seen in Table 6-8 that the weighted cost of capital is exactly equal to the cost of depreciation funds. This is why it is often said that cost of depreciation funds does not enter into the calculation of a firm's cost of capital.

Table 6-8
COST OF CAPITAL FOR EXAMPLE 6-9

Source	Proportion of Capital	Cost Before Tax, %	Weighted Average
Debt	0.30	10	3.0
Retained earnings	0.35	12	4.2
Common stock	0.15	16	2.4
Depreciation	0.20	12.0	2.4
			12.0

ESTIMATING THE COST OF CAPITAL FOR A COMPANY

In this section an example is presented that shows how certain values required in estimating the cost of capital are obtained from a company's financial report.

EXAMPLE 6-10

A company's financial report includes the balance sheets and income statement shown in Tables 6-9 and 6-10. In addition, the following information is obtained or estimated.

1. The company's common stock has a current market value of $35 a share. However, in order to insure future sales, common stock will be offered for $33 a share with brokerage fees amounting to $2 a share. The present number of common shares is 600,000 and the present dividend is $4 a share.
2. Preferred stock has a current market value of $60 a share and there are 10,000 shares outstanding. Current market conditions indicate that in order to sell preferred stock it would require a return of 10% on a $100 share. The selling price and brokerage fee, respectively, are estimated to be $100 and $3.00 a share.
3. The company's bonds were sold sometime ago and have a current market value of $850. The number of bonds outstanding is 10,000. If new bonds are issued, they will have a face value of $1000, pay 8% semiannually, and mature in 10 years. New bonds can be sold for $950 with brokeage fees amounting to $10.00 per bonds.

With the preceding information the incremental cost of capital from various sources can be calculated in the following manner.

Table 6-9

BALANCE SHEETS FOR EXAMPLE 6-10

Assets		Capitalization	
At original cost		Common stock	$ 4,800,000
Plant A	$ 6,000,000	Retained earnings	3,000,000
Plant B	3,000,000	Capital surplus	4,200,000
Plant C	7,000,000	Total common equity	$12,000,000
Plant D	4,000,000	Preferred stock	1,000,000
	$20,000,000	Long-term debt	5,000,000
Less accumulated depreciation	3,000,000	Current liabilities	
	$17,000,000	Accounts payable	200,000
Current assets		Long-term debt due within 12 months	500,000
Cash	200,000		$18,700,000
Accounts receivable	500,000		
Inventories	1,000,000		
	$18,700,000		

Table 6-10
INCOME STATEMENT FOR EXAMPLE 6-10

Operating revenue	$12,815,000
Operating expenses and taxes	
Operation	2,500,000
Maintenance	1,000,000
Depreciation	500,000
Taxes other than federal	80,000
Federal taxes (T = 52%)	4,355,000
Total	8,435,000
Operating income	4,380,000
Interest	
Long-term debt interest	350,000
General interest	10,000
Total	360,000
Income before preferred dividends	4,020,000
Preferred dividends	900,000
Net income for common stock	3,120,000

Common Stock

The cost of new common stock is determined using Eq. (6-21). The data indicates $D_0 = \$4$ and $P_0 = 33 - 2 = \$31$. In order to determine the growth factor (rb), values for Y_0, B_0, and b must be determined. The earnings per share are

$$Y_0 = \frac{3,120,000}{600,000} = \$5.20$$

and the book value per share is

$$B_0 = \frac{12,000,000}{600,000} = \$20.00$$

The book value per share can also be determined by taking the total assets and subtracting all liabilities (current and long term) as shown below:

$$B_0 = \frac{18,700,000 - (200,000 + 500,000 + 5,000,000 + 1,000,000)}{600,000}$$
$$= \$20.00$$

With these values, the values of r and b can be determined.

$$r = \frac{Y_0}{B_0} = \frac{5.2}{20} = 0.26$$

$$b = \frac{Y_0 - D_0}{Y_0} = \frac{5.2 - 4.0}{5.2} = 0.23$$

Substituting into Eq. (6-21), the estimated cost of new common stock is

$$k_e = \frac{4}{31} + (0.26)(0.23) = 0.189 = 18.9\%$$

Retained Earnings

The cost of retained earnings is essentially the same as the new stock except P_0 is replaced by the market value. Consequently, the estimated cost of retained earnings is

$$k_e = \frac{4}{35} + (0.26)(0.23) = 0.174 = 17.4\%$$

Capital Surplus

The cost of capital surplus is the same as the cost of new common stock (18.9%) since any new capital surplus would be obtained from the sale of new common stock.

Preferred Stock

Using Eq. (6-26), the cost of preferred stock is

$$k_p = \frac{(0.1)(100)}{100 - 3} = 0.103 = 10.3\%$$

Debt

The before tax cost of new debt capital is determined using Eq. (6-6) with $T = 0$ as shown below. Evaluating this expression gives the

$$(950 - 10) = \frac{0.08(1000)}{2}\left[PA\ \frac{k'_d}{2}, 2(10)\right] + 1000\left[PF\ \frac{k'_d}{2}, 2(10)\right]$$

result $k'_d = 9\%$ compounded semiannually or an effective annual rate of 9.2%. If the after-tax ($T = 52\%$) cost of debt is desired, it can be approximaed by 9.2 (1 − 0.52) = 4.4%, or determined exactly by using Eq. (6-6) with $T = 52\%$.

Weighted Average

For comparative purposes, the before-tax weighted average cost of capital is calculated using current book values and current market values. These results are shown in Table 6-11. The book value weights used in Table 6-11a are a ratio of the respective capitalization values and the total capitalization minus current liabilities (see Table 6-9). For example, the book value weights for common stock and retained earnings, respectively, are

$$\frac{4,800,000}{18,000,000} = 0.267$$

$$\frac{3,000,000}{18,000,000} = 0.167$$

Table 6-11
WEIGHTED COST OF CAPITAL FOR EXAMPLE 6-10

Source	Weight	Cost	Weighted Cost
Common stock	0.267	18.9	5.05
Retained earnings	0.167	17.4	2.91
Capital surplus	0.233	18.9	4.40
Preferred stock	0.055	10.3	0.57
Debt	0.278	9.2	2.56
			15.49%

(a) BOOK VALUES

Source	Market Value	Weight	Cost	Weighted Cost
Equity	35 (600,000) = \$21,000,000	0.698	18.9	13.19
Preferred stock	60 (10,000) = 600,000	0.020	10.3	0.21
Debt	850 (10,000) = 8,500,000	0.282	9.2	2.59
	Total 30,100,000			15.99%

(b) MARKET VALUES

The market value weights are a ratio of the respective market values and the total market value of the company. For example the market value weight for equity is

$$\frac{21{,}000{,}000}{30{,}100{,}000} = 0.698$$

When market values are used, it is not possible to specify a weight for each equity component (common stock retained earnings, etc.). Consequently, equity components are consolidated into a single category called equity in Table 6-11b and evaluated at the current market value of a share of common stock. The "correct" answer for the weighted cost of capital depends on which proportions are expected to represent the future capital structure of the company. Of course if neither represents the future capital structure, the future capital structure must be estimated and used.

LEVERAGE

The weighted cost of capital provides a basis for a discussion of an important financial concept called leverage. Essentially, leverage is the ratio of debt to the total value of the firm at market values. A major question in regard to leverage is, "What mix of debt and equity capital minimizes the weighted cost of capital?"

The traditional view of the effect of changes in leverage upon market value is

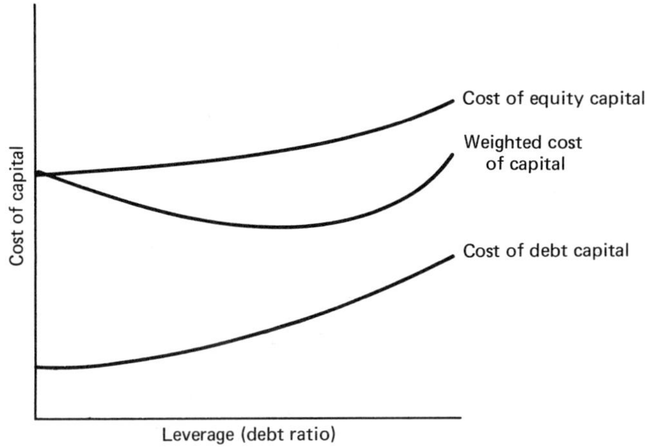

Figure 6-1.
Cost of capital versus leverage.

shown in Figure 6-1. This figure shows that as the leverage (debt ratio = w_d) increases, the cost of capital (k) decreases up to a point. Remember, the cost of capital is

$$k = w_d k_d + w_e k_e \tag{6-27}$$

After this point, the cost of capital starts to increase. These changes in the cost of capital are a direct result of the stockholders' and lenders' reactions to increases in leverage. In the beginning, the cost of capital decreases as the leverage increases because k_d is less than k_e and k_e is essentially constant. As leverage increases, past some point k_e and k_d increase because both stockholders and lenders demand a higher return as a result of the greater risk resulting in a higher weighted cost of capital. This view of the effect of leverage on the cost of capital leads to the conclusion that some optimum point does exist. That is, there is some debt ratio that provides a minimum cost of capital. Determination of this optimum point, from a practical standpoint, is a difficult task. For it would be necessary to know how both k_e and k_d change as the debt ratio changes. This would require the prediction of the stockholders' and lenders' reactions to increases in leverage, a rather difficult task.

Another view of the cost of capital is taken by Modigliani and Miller. They believe that the cost of capital is independent of the capital structure. Thus, there is no "optimal" capital structure. This point of view is not considered in this text. For those who wish to pursue this view further the references listed at the end of this chapter provide a complete discussion.

COST OF CAPITAL AND MINIMUM ATTRACTIVE RATE OF RETURN

The incremental weighted average cost of capital provides a basis on which to evaluate investments. However, it is usually not used without some adjustment.

That is, the common procedure is to adjust (increase) the cost of capital to take into account risk. This procedure of using a risk-adjusted cost of capital as the MARR is objected to by many. They take the point of view that risk should be explicitly included in the analysis of investments through the use of probabilistic models or simulation. Nevertheless, a risk-adjusted cost of capital figure is often used in practice. The amount of the risk adjustment usually depends on the type of project under consideration. Consequently, it is not unusual to find a company with several values for MARR. For example, a company might have a lower MARR for the expansion of existing facilities than the MARR for investments involving new products, the difference being a reflection of risk considerations between the two types of investments.

The cost of capital and, consequently, the MARR, may require adjustments over time as a result of changing conditions. For example, the supply and demand of funds varies with time, causing changes in the cost of debt and equity capital. For this reason and others, a company must monitor its cost of capital and make changes when necessary.

References

1. Gordon, M. J., *The Investment, Financing, and Valuation of the Corporation,* Richard D. Irwin, Homewood, Ill., 1962.
2. Gordon, M. J. and E. Shapiro, "Capital Equipment Analysis: The Required Rate of Profit," *Management Science,* Vol. 3, October 1956, pp. 102–110.
3. Haley, C. W., and L. D. Schall, *The Theory of Financial Decisions,* McGraw-Hill Book, Co., New York, 1973.
4. Lewellen, Wilbur, G., *The Cost of Capital,* Wadsworth Publishing Co., Inc., Belmont, Calif., 1969.
5. Robichek, A. A. and S. C. Myers, *Optimal Financial Decisions,* Prentice-Hall, Inc., Englewood Cliffs, N.J., 1965.
6. Solomon, Ezra, "Measuring a Company's Cost of Capital," *Journal of Business,* October 1955.
7. VanHorne, James C., *Financial Management and Policy,* 2nd ed., Prentice-Hall, Inc., Englewood Cliffs, N.J., 1971.
8. Weston, J. F., and E. F. Brighan, *Essentials of Managerial Finance,* 3rd ed., The Dryden Press, Hinsdale, Ill., 1974.

Problems

6-1. A company is negotiating a loan of $100,000 with a bank. The bank has agreed to make the loan at an interest rate of 10% and a discount rate of 2%. If the loan is made on the basis of a 10-year period, determine the following:

(a) The amount the company would actually receive as a result of the discount.

(b) The yearly payments the company would make to the bank on the basis that the principal is to be paid in 10 equal installments.
(c) The effective before tax interest rate for the loan. Assume the company receives the amount in part (a).
(d) An approximate after-tax interest rate [use part (c)] and the actual after-tax interest rate on the loan if the tax rate is 52%.

6-2. Rework Problem 6-1 on the basis that the principal will be paid back at the end of 10 years.

6-3. Rework Problem 6-1 on the basis that the company must have $100,000.

6-4. Rework Problem 6-2 on the basis that the company must have $100,000.

6-5. Based on the results of Problems 6-1 through 6-4, is it possible to make certain general comments regarding the rates of interests when money is borrowed?

6-6. A company is considering raising capital by issuing a series of fixed maturity bonds. The bonds have a face value of $1000, mature in five years, and pay 8% quarterly. It is estimated that the selling price of each bond to the general public will be $900 and brokerage fees will amount to $10 per bond. Determine the following:
(a) The company's before-tax cost of the bonds.
(b) The company's after-tax cost of the bonds if the tax rate is 52%. Compare this result with the approximate after-tax cost.

6-7. Rework Problem 6-6 assuming that the bond is a convertible bond. Also assume that the conversion ratio is 10, the bond will be converted in three years and the market value of a share of common stock at that time will be $110.

6-8. A company is considering raising some equity capital. The company's current financial reports show the following data.

total assets =	$20,000,000
liabilities =	$ 6,000,000
after-tax earnings =	$ 1,875,000
number of common stock	100,000
dividends for common stock =	$ 1,500,000
current market price of common stock =	$187 per share

Determine the following:
(a) The estimated cost of capital through the sale of common stock. It is estimated that in order to insure its sale the stock will be sold for $183 per share and brokerage fees will amount to $5.00 per share.
(b) The estimated cost of capital through the use of retained earnings.
(c) The estimated cost of capital through the use of capital surplus.
(d) From the standpoint of the company, are there any tax considerations in parts (a), (b), and (c)?

6-9. Using the data in Problem 6-8 and Solomon's approach [Eq. (6-25)], determine the cost of a new issue of common stock. Why is there a difference between this result and the result obtained in part (a) of Problem 6-8? Which one would you use? Why?

6-10. Suppose in Problem 6-8, the company's financial statements for the past six years indicated the following earnings. What is the company's average growth rate?

Year	Earnings
0	$1,875,000
1	1,780,000
2	1,675,000
3	1,625,000
4	1,540,000
5	1,510,000
6	1,437,000

Also, determine the following:

(a) The cost of capital for a new issue of common stock using the average growth rate.

(b) Compare the result of part (a) in this problem with the result of part (a) in Problem 6-8. Which one should be used?

6-11. Additional capital is to be raised by the sale of 8%, $100 preferred stock. The stock will be sold for $98 with flotation costs of $5 per share. Determine the cost of this preferred stock.

6-12. A company is going to raise capital at the costs and in the proportions shown in the following table:

Incremental Cost of Capital	Proportion
Equity 18%	0.50
Preferred stock 12%	0.10
Debt (before tax) 10%	0.40

Determine:

(a) The before-tax weighted cost of capital.

(b) The after-tax ($T = 52\%$) weighted cost of capital.

6-13. A company is going to finance future projects on the basis of the capital costs and proportions shown below. The remaining proportion (10%) is to be obtained from depreciation reserves. Determine:

(a) The cost of depreciation funds.

(b) The before-tax weighted cost of capital.

Incremental Cost of Capital	Proportion
Equity 18%	0.40
Preferred stock 12%	0.20
Debt (before tax) 10%	0.30

6-14. Given the data in the following table, estimate the after-tax weighted cost of capital if the tax rate is 52%.

Capital	Cost	Market Value	Number of Stocks or Bonds
Common stock	18%	150	2,000,000
Preferred stock	12%	90	400,000
Debt (before tax)	10%	850	100,000

6-15. A company's balance sheet shows the following capitalization.

Common stock (1,000,000 shares)	$\$ 40(10)^6$
Preferred stock (180,000 shares)	$16(10)^6$
Retained earnings	$8(10)^6$
Capital surplus (from common)	$2(10)^6$
Long-term debt (100,000 bonds)	$34(10)^6$
Total	$\$100(10)^6$

The same company's income statement and other financial reports show the following:

Net earnings after tax for common	$\$14(10)^6$
Dividends per share of common	$12
Market value per share of common	$120
Tax rate	52%

The company plans to raise additional capital for new investments under the following considerations.

(a) Common stock is expected to be sold for $115 per share in order to insure its sale and brokerage fees are expected to be $2 per share.

(b) Preferred stock is $100 per share and pays 12%. It is to be sold for $98 per share and brokerage fees are expected to be $2.00 per share. The current market value for preferred stock is $90 per share.

(c) Bonds have a face value of $1000, pay 9% semiannually, and mature in ten years. The bonds are to be sold for $950 each. Brokerage fees are expected to be $10 per bond.

Determine the following:

(1) Cost of common stock.
(2) Cost of retained earnings.
(3) Cost of capital surplus.
(4) Cost of preferred stock.
(5) Cost of bonds (before tax).
(6) The incremental before-tax weighted cost of capital. Assume that the future and existing capital structure are to be the same.
(7) Estimate the cost of depreciation funds.

CHAPTER 7
MINIMUM ANNUAL REVENUE REQUIREMENTS AND CASH FLOW DEFINITIONS

This chapter is concerned with a method, commonly used by public utilities, to evaluate capital investments—the generation of minimum annual revenue requirements. Although traditionally used by public utilities, there are no theoretical or practical reasons why the minimum annual revenue requirements can not be used by private corporations. In addition to being a useful technique to evaluate capital investments, it provides additional insight into the relationships between cost of capital, cash flow definitions, and cost comparisons. Consequently, it is for these reasons that minimum annual revenue requirements are presented at this point in this text.

THE BASICS OF MINIMUM ANNUAL REVENUE REQUIREMENTS

Basically, minimum annual revenue requirements are the minimum annual gross incomes over the life of the project (investment) that recover all expenses and provide for a stated return on the unrecovered capital. A mathematical expression for the revenue requirements in the year n can be obtained by considering the following definition of net cash flow.

$$X_n = (G - C - I) - (G - C - D_t - I)T - P \tag{7-1}$$

where

G = gross income for year n
C = expenses for year n
I = interest on borrowed money for the year n
D_t = tax depreciation for year n
T = combined state and federal tax rate
P = principal payment

The definition of cash flow given by Eq. (7-1) assumes that there are no capital requirements, salvage value considerations, or working capital changes in the particular year n. Also, the cash flows defined by Eq. (7-1) are *equity* cash flows; that is, these cash flows are the yearly returns of the equity shareholder's capital since debt payments are explicitly taken into account.

Making the substitution

$$t = (G - C - D_t - I)T \tag{7-2}$$

for the tax term in Eq. (7-1) and solving for the gross income gives

$$G = X_n + C + I + t + P \tag{7-3}$$

Since the *minimum* annual gross incomes are desired, the cash flow, X_n, must be equal to the sum of the recovered portion of the equity capital, D_e, and a return on the unrecovered equity capital, f_n, that is,

$$X_n = D_e + f_n \tag{7-4}$$

It is assumed in generating minimum annual revenue requirements that the total (equity and debt) capital recovered in a year is equal to the book depreciation, D_b, and the principal payments are a constant percentage of the book depreciation; that is,

$$P = c(D_b) \tag{7-5}$$

where c is the debt ratio.

This implies that recovery of equity capital is

$$D_e = (1 - c)D_b \tag{7-6}$$

and the annual return on equity, f_n, and interest, I, are based on the unrecovered balance, that is,

$$f_n = (1 - c)k_e(BV_{n-1}) \tag{7-7}$$
$$I_n = (c)k_d(BV_{n-1}) \tag{7-8}$$

where k_e and k_d are, respectively, the required rates of return on equity and debt capital. The term BV_{n-1} is the book value at the *beginning* of the year n.

Substituting Eqs. (7-4), (7-5), and (7-6) into Eq. (7-3) gives

$$G = D_b + f_n + C + I + t \tag{7-9}$$

and, in order to differentiate mimimum annual revenue requirements, R, from gross incomes in general, Eq. (7-9) is designated as

$$R = D_b + f_n + C + I + t \tag{7-10}$$

In order to evaluate the taxes, t, Eq. (7-10) is substituted into Eq. (7-2) for the gross income, which results in

$$t = (D_b + f + C + I + t - C - D_t - I)T \tag{7-11}$$
$$t = (D_b + f - D)T_t + tT \tag{7-12}$$
$$t(1 - T) = (D_b + f - D_t)T \tag{7-13}$$
$$t = \frac{T}{1 - T}(D_b + f - D_t) \tag{7-14}$$

The evaluation of Eqs. (7-10) and (7-14) is demonstrated in the following example.

EXAMPLE 7-1

Generate the minimum annual revenue requirements for the following data:

initial cost of project	= \$12,000
salvage value	= \$2,000
life of project	= 8 years
required equity return	= 25%
required debt return (cost of debt)	= 10%
total costs	= \$1,500 per year
debt ratio	$= \frac{1}{3}$
depreciation model for book purposes	= straight line
depreciation model for tax purposes	= sum-of-the-years digits
composite tax rate	= 52%

The minimum annual revenue and other pertinent data are shown in Table 7-1. The

Table 7-1
MINIMUM ANNUAL REVENUE REQUIREMENTS FOR EXAMPLE 7-1

End of Year	Book Depreciation D_b	Tax Depreciation D_t	Book Value BV	Equity Return f_n	Interest I	Tax t	Costs	Min. Annual Rev. Reqm't R
0	—	—	\$12,000	—	—	—	—	
1	\$1250	\$2222	10,750	\$2000	\$400	\$1114	\$1500	\$6264
2	1250	1944	9,500	1792	358	1189	1500	6089
3	1250	1667	8,250	1583	317	1263	1500	5913
4	1250	1389	7,000	1375	275	1339	1500	5739
5	1250	1111	5,750	1167	233	1415	1500	5565
6	1260	833	4,500	958	192	1490	1500	5390
7	1250	556	3,250	750	150	1564	1500	5214
8	1250	278	2,000	542	108	1640	1500	5040

procedures for obtaining the values in Table 7-1 are shown in the following sample calculations:

1. First, determine the depreciation schedules for each depreciation model; for the book depreciation

$$D_b = \frac{12{,}000 - 2000}{8} = \$1250 \text{ for each year}$$

and for the tax depreciation

$$D_{t_1} = \frac{8}{36}(12{,}000 - 2000) = \$2222,$$

$$D_{t_2} = \frac{7}{36}(12{,}000 - 2000) = \$1944.$$

2. Calculate the book value for each year

$$BV_0 = \$12{,}000$$
$$BV_1 = \$12{,}000 - 1250 = \$10{,}750$$
$$BV_2 = \$10{,}750 - 1250 = \$9500$$

3. Calculate the return on equity using Eq. (7-7).

$$f_1 = \left(1 - \frac{1}{3}\right)(0.25)\,(12{,}000) = \$2000$$

$$f_2 = \left(1 - \frac{1}{3}\right)(0.25)\,(10{,}750) = \$1792$$

4. Calculate the interest using Eq. (7-8).

$$I_1 = \left(\frac{1}{3}\right)(0.10)\,(12{,}000) = \$400$$

$$I_2 = \left(\frac{1}{3}\right)(0.10)\,(10{,}750) = \$358$$

5. Calculate the taxes using Eq. (7-14).

$$t_1 = \frac{0.52}{1 - 0.52}(1250 + 2000 - 2222) = \$1114$$

6. Finally, calculate the minimum annual revenue requirements using Eq. (7-10)

$$R_1 = 1250 + 2{,}000 + 1500 + 400 + 1114 = \$6264$$
$$R_2 = 1250 + 1{,}792 + 1500 + 358 + 1189 = \$6089$$

Decision Criterion

If the expected (estimated) gross incomes are greater than the minimum annual revenue requirements, then the project is considered acceptable. Provided, of course, the rates used for the returns on equity and debt capital are considered to represent the minimum attractive rates for capital from these two sources. In actuality, the acceptability of a project can be approached on an attribute basis; that is, are the gross incomes (revenues) expected to be higher or lower than the minimum annual revenue requirements? This approach avoids making specific estimates of the yearly gross incomes.

A Comment: At this point, some important comments are in order. First, Eq. (7-13) is not dependent on the depreciation models used for taxes and book depreciation. Any other time-dependent depreciation model can be used. Also, the yearly costs are assumed to be constant in Example 7-1. This is not a requirement in determining the minimum annual revenue requirements.

As a further explanation of some of the previous equations, consider the following: If the minimum annual revenue requirements given in Table 7-1 are used as the gross incomes in Eq. (7-2), the taxes, for the first two years, are

$$t_1 = (6264 - 1500 - 2222 - 400)\,(0.52) = \$1114$$
$$t_2 = (6089 - 1500 - 1944 - 358)\,(0.52) = \$1189$$

which are the same results obtained from Eq. (7-14). Similar results can be obtained for the remaining years in Example 7-1.

What is implied in generating the minimum annual revenue requirement is that the total investment \$12,000, using Example 7-1 for illustrative purposes, is divided between debt and equity in accordance with the debt ratio, that is,

$$\text{Equity portion of investment} = \frac{2}{3}(12{,}000) = \$8000$$

$$\text{Debt portion of investment} = \frac{1}{3}(12{,}000) = \$4000$$

Which further implies, the yearly recovery of both debt and equity capital (book depreciation) are given by [Eqs. (7-5) and (7-6)]

equity capital recovered per year = ⅔ (1,250) = \$833
debt capital recovered per year = ⅓ (1,250) = \$417

and the salvage value (\$2,000) is also proportioned between debt and equity by the debt ratio, that is,

equity capital recovered from salvage = ⅔ (2,000) = \$1,333
debt capital recovered from salvage = ⅓ (2,000) = \$667.

Consequently, the principal "payments" and recovery of equity capital assumed in generating minimum annual revenue requirements are as shown in Table 7-2.

COST OF CAPITAL AND DEFINITIONS OF CASH FLOW

The previous discussion (extensions are provided later in this chapter) of minimum annual revenue requirements provide a basis on which a discussion of the relationship between various definitions of cash flow and the cost of capital (and, consequently, MARR) is possible.

For the purposes of project analysis, there are three basic definitions of cash flow:

$$X_n = (G - C - I) - (G - C - D_t - I)T - K + L + B - P \pm W \qquad (7\text{-}15)$$
$$X_n = (G - C) - (G - C - D_t - I)T - K + L \pm W \qquad (7\text{-}16)$$
$$X_n = (G - C) - (G - C - D_t)T - K + L \pm W \qquad (7\text{-}17)$$

Table 7-2
IMPLIED PRINCIPAL PAYMENTS AND RECOVERY OF EQUITY CAPITAL FOR EXAMPLE 7-1

End of Year	Principal Payment	Recovery of Equity Capital
0	—	—
1	\$417	\$833
2	417	833
3	417	833
4	417	833
5	417	833
6	417	833
7	417	833
8	417 + 667	833 + 1333
Total	4003 (4000)[a]	7997 (8000)[a]

[a]Differences are due to roundoff error.

where the symbolism is the same as previously defined and

I = interest payment in year n
B = amount of borrowed money received in year n
W = net changes in working capital requirements in year n as a result of the project under consideration.
P = principal payment in year n

Now, all three definitions of cash flow given by Eqs. (7-15), (7-16), and (7-17) are correct under certain circumstances. The one to use depends on the definition of the cost of capital. Inspecting Eqs. (7-15), (7-16), and (7-17) reveals that the difference in these equations is due to borrowed money considerations (amount borrowed, principal payments, and interest). It is these same borrowed money considerations that determine which cost of capital definition should be associated with a particular definition of cash flow. Three definitions of cost of capital are possible: namely,

$$k = k_e \tag{7-18}$$
$$k = c(k_d) + (1 - c)k_e \tag{7-19}$$
$$k = c(1 - T)k_d + (1 - c)k_e \tag{7-20}$$

Where the symbolism is the same as given in Chapter 6 and c is the debt ratio.

Now, in order to illustrate the relationships between cost of capital and cash flow definitions, several examples are worked in the following sections.

EXAMPLE 7-2

Assume that the gross incomes for a project are estimated to be the same as the minimum annual revenue requrements given in Table 7-1, the principal payments are those given in Table 7-2, and the rest of the data given in Example 7-1 is applicable. If Eq. (7-15) is used as the definition of cash flow the results shown in Table 7-3 are obtained. The internal rate of return for these cash flows is 25%. This is not a surprising result since a 25% return on equity is provided in the generation of the minimum annual revenue requirements (see Example 7-1) which are used as the gross incomes in this example. This result also points out that the cash flows obtained from Eq. (7-15) are *equity* cash flows; that is, debt capital considerations are explicity included in Eq. (7-15).

Now, Eqs. (7-18), (7-19) and (7-20) give three possible values for the cost of capital. They are, respectively,

$$k = 25\%$$

$$k = \frac{1}{3}(10) + \frac{2}{3}(25) = 20\%$$

$$k = \frac{1}{3}(10)(1 - 0.52) + \frac{2}{3}(25) = 18.27\%$$

If these results are used as minimum acceptable rates of return, what is the correct value for determining the acceptability of the project? The answer is 25% because, as previously stated, Eq. (7-15) gives equity capital returns. Consequently, it is the required rate of return on equity that must be used in determining the project's acceptability. Using a cost of capital with an incorrect cash flow definition can lead to erroneous decisions regarding the acceptability of a project. For example, suppose the cash flows for a project are estimated on the basis of Eq. (7-15) and the resulting rate of return is 18%. Now, suppose the required returns on equity and debt are, respectively, 20% and 10%. Using a debt ratio of 30%, gives a weighted before-tax (an after tax would be even less) cost of capital of 17%. If this 17% is used to compare with the 18%, the erroneous conclusion is that the project is acceptable. The correct comparison is the required return on equity, 20%, with the 18% which indicates the project is unacceptable.

Table 7-3

NET CASH FLOWS USING EQ. (7-15)

End of Year	Net Cash Flow
0	$-12000 + 4000^{a} = -\$8000$
1	(6264 − 1500 − 400) − (6264 − 1500 − 400 − 2222)(0.52) − 417 = 2833
2	(6089 − 1500 − 358) − (6089 − 1500 − 358 − 1944)(0.52) − 417 = 2625
3	(5913 − 1500 − 317) − (5913 − 1500 − 317 − 1667)(0.52) − 417 = 2416
4	(5739 − 1500 − 275) − (5739 − 1500 − 275 − 1389)(0.52) − 417 = 2208
5	(5569 − 1500 − 233) − (5569 − 1500 − 233 − 1111)(0.52) − 417 = 2000
6	(5390 − 1500 − 192) − (5390 − 1500 − 192 − 833)(0.52) − 417 = 1791
7	(5214 − 1500 − 150) − (5214 − 1500 − 150 − 556)(0.52) − 417 = 1583
8	(5040 − 1500 − 108) − (5214 − 1500 − 108 − 278)(0.52) − 417 − 667 + 2000 = 2708

[a]Amount borrowed = ⅓(12,000) = $4000.

Table 7-4
NET CASH FLOWS USING EQ. (7-16)

End of Year	Net Cash Flow
0	$-\$12000$
1	$(6264 - 1500) - (6264 - 1500 - 2222 - 400)(0.52) = 3650$
2	$(6089 - 1500) - (6089 - 1500 - 1944 - 358)(0.52) = 3400$
3	$(5913 - 1500) - (5913 - 1500 - 1667 - 317)(0.52) = 3150$
4	$(5739 - 1500) - (5739 - 1500 - 1389 - 275)(0.52) = 2900$
5	$(5569 - 1500) - (5569 - 1500 - 1111 - 233)(0.52) = 2650$
6	$(5390 - 1500) - (5390 - 1500 - 833 - 192)(0.52) = 2400$
7	$(5214 - 1500) - (5214 - 1500 - 556 - 150)(0.52) = 2150$
8	$(5040 - 1500) - (5040 - 1500 - 278 - 108)(0.52) + 2000 = 3900$

EXAMPLE 7-3

As a further numerical example of the relationship between cash flow and cost of capital definitions, the cash flows, using Eq. (7-16) and (7-17) and the data in Example 7-1 are calculated in Table 7-4 and 7-5. If the internal rates of return are calculated for the cash flows in Tables 7-4 and 7-5, the results are, respectively, 20% and 18.27%. When these results are compared with cost of capital given by Eqs. (7-19) and (7-20), it suggests that if interest payments are explicitly included in the calculation of taxes then the cost of debt capital should not be modified by the term $(1 - T)$. On the other hand, if interest deductions are not explicitly included in the computation of taxes then the cost of debt capital should be modified by the term $(1 - T)$. Consequently, if Eq. (7-16) is used as the definition of cash flow then Eq. (7-19) should be used as the definition of cost of capital. If Eq. (7-17) is used as the definition of cash flow then Eq. (7-20) should be used as the definition of cost of capital. Basically, the internal rates of return found using Eqs. (7-16) and (7-17) as the definitions of cash flow are returns on the *total* investment.

Table 7-5
NET CASH FLOWS USING EQ. (7-17)

End of Year	Net Cash Flow
0	$-\$12000$
1	$(6264 - 1500) - (6264 - 1500 - 2222)(0.52) = 3442$
2	$(6089 - 1500) - (6089 - 1500 - 1944)(0.52) = 3214$
3	$(5913 - 1500) - (5913 - 1500 - 1667)(0.52) = 2987$
4	$(5739 - 1500) - (5739 - 1500 - 1389)(0.52) = 2757$
5	$(5569 - 1500) - (5569 - 1500 - 1111)(0.52) = 2529$
6	$(5390 - 1500) - (5390 - 1500 - 833)(0.52) = 2300$
7	$(5214 - 1500) - (5214 - 1500 - 556)(0.52) = 2072$
8	$(5040 - 1500) - (5040 - 1500 - 278)(0.52) + 2000 = 3844$

A Comment: At this point some comments are in order regarding cost of capital and cash flow definitions. In general, Eq. (7-15) should be used as the definition of cash flow since the manner in which debt obligations are paid can have an effect on the rate of return. For example, if it is decided to pay back the borrowed money in Example 7-2 in four equal installments beginning at the end of the first year, the cash flows are as shown in Table 7-6. The rate of return for these cash flows is less than the 25% found when the debt obligation was spread over the life of the project (Table 7-3).

There is a practical case for using Eq. (7-17) as the definition of cash flow: as stated in Chapter 5, it allows the separation of financing decisions from the economic desirability of a project.That is, an internal rate of return can be determined without specific knowledge of the debt payments. The rate of return determined from the cash flows defined by Eq. (7-17) is a return on the *total investment* and can be used in two ways.

First, it could be assumed that the project is to be financed entirely from equity funds. If the rate of return is greater than the required return on equity funds, this implies the project is acceptable, since, in all probability, the financing would be a combination of debt and equity capital and the actual required rate of return would be less than the required *equity* return (assuming, of course, that the cost of debt capital is less than the cost of equity capital—which is, by far, the usual case).

A second way it could be used (and often is) is to compare the rate of return to a required rate of return determined on the basis of Eq. (7-20). This requires a knowledge of the debt ratio and does imply the assumption that the debt payments (principal and interest) are spread over the life of the project. This way must be approached cautiously because it may lead to incorrect decisions if this assumption about debt payments is not correct. For example, the cash flows in Table 7-5 indicate an acceptable project when compared to the cost of capital defined by Eq. (7-20). However, the cash flows in Table 7-6 indicates the project is unacceptable since the rate of return on equity is less than 25%. This conflict is a result of the different payments of debt obligations and points out the original contention that Eq. (7-15) should be used whenever possible.

In actuality, Eq. (7-16) is of little practical or theoretical value as a definition of cash flow because it requires that interest payments be known in order to evaluate the tax term. If the interest payments are known, then the principal payments are also known or, at least, implied. Consequently, Eq. (7-15) can be used as the definition of cash flow. Equation (7-16) is considered primarily to show the relationship between the before tax, Eq. (7-19), and the after tax, Eq. (7-20), cost of capital.

MINIMUM ANNUAL REVENUE REQUIREMENT EXTENSIONS

In practice, minimum annual revenue requirements are generated using either the normalizing or flow-through methods. These two methods are presented in this section.

Normalizing Method

This method is often used by public utilities because it includes many of the accounting conventions and procedures required by regulatory agencies in establishing utility rates. This method is also required by federal tax laws in order for public utilities to qualify for the investment tax credit as mentioned in Chapter 4.

Table 7-6

NET CASH FLOW WITH FOUR PRINCIPAL PAYMENTS

End of Year	Principal Payment	Interest Payment	Net Cash Flow
0	—	—	$-12000 + 4000 = -\$8000$
1	\$1000	\$400	$(6264 - 1500 - 400) - (6264 - 1500 - 400 - 2222)(0.52) - 1000 = 2250$
2	1000	300	$(6089 - 1500 - 300) - (6089 - 1500 - 300 - 1944)(0.52) - 1000 = 2070$
3	1000	200	$(5913 - 1500 - 200) - (5913 - 1500 - 200 - 1667)(0.52) - 1000 = 1889$
4	1000	100	$(5739 - 1500 - 100) - (5739 - 1500 - 100 - 1389)(0.52) - 1000 = 1709$
5	—	—	$(5569 - 1500) - (5569 - 1500 - 1111)(0.52) = 2531$
6	—	—	$(5390 - 1500) - (5390 - 1500 - 833)(0.52) = 2300$
7	—	—	$(5214 - 1500) - (5214 - 1500 - 556)(0.52) = 2072$
8	—	—	$(5040 - 1500) - (5040 - 1500 - 278)(0.52) + 2000 = 3844$

The annual revenue requirement, R_x, includes the following components.

$$\left.\begin{aligned} R_x = {} & \text{(book depreciation)} + \text{(deferred taxes)} \\ & + \text{(investment tax credit)} \\ & - \text{(amortized investment tax credit)} \\ & + \text{(interest excluding capitalized interest)} \\ & + \text{(return on equity)} \\ & + \text{(federal and state taxes)} + \text{(costs)} \end{aligned}\right\} \quad (7\text{-}21)$$

The evaluation of the components in Eq. (7-21) is demonstrated in the following example.

EXAMPLE 7-4

Generate the minimum annual revenue requirements using the normalizing method and the following data.

first cost = \$12,000
salvage value = \$2,000
life = 8 years
required return on equity = 25%
cost of debt capital = 10%
debt ratio = ⅓
tax rate = 52%
costs per year = \$1,500
capitalized interest = 1%
investment tax credit = 4%
book depreciation = straight line
tax depreciation = sum of the years digits

Before the minimum annual revenue requirements can be determined, the following steps must be completed.

Step 1: Determine the capitalized investment (project) cost. The capitalized investment cost, K, is

$$K = P + gP \quad (7\text{-}22)$$

where

P = actual construction cost
g = capitalized interest rate

For this example,

$$K = 12{,}000 + (0.01)(12{,}000) = \$12{,}120$$

Capitalized interest (\$120) is the interest paid on money during the construction phase of the project. The usual convention in generating minimum annual revenue requirements is to take this interest as a tax deduction in the first year of the project since this is when it is actually paid, and recover the interest over the life of the project beginning in the first year. The argument (usually by the utility companies) is that, in this way, present customers are not "penalized" for the interest cost of new construction. Instead, by spreading it over the life of the project, this interest is assessed to both present and future

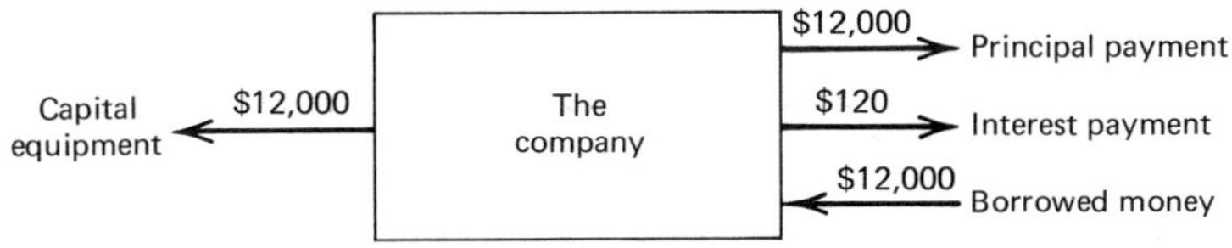

Figure 7-1.
Capitalized interest transactions.

customers. Usually, the interest rate that is used in determining the capitalized interest is set by regulatory agencies. From a cash flow standpoint, the implied transactions in regard to capitalized interest are shown in Figure 7-1.

Step 2: Calculate the depreciation schedules. There are three depreciation schedules:

1. D_b = straight-line depreciation for book purposes
2. D_s = straight-line depreciation for tax purposes
3. D_t = depreciation for tax purposes

These three depreciation models require different first costs because of the capitalized interest. These values are shown in Table 7-7. With the values in Table 7-7 and the estimated life of eight years, the depreciation schedules shown in Table 7-8 can be generated.

The depreciation schedules in Table 7-8 provide additional insight on money borrowed for construction purposes (capitalized interest)—the interest on the borrowed money is "depreciated" over the life of the project in the case of the book depreciation schedule. However, present tax laws do not recognize this procedure. Consequently, the tax depreciation schedule, D_b, is based on the initial cost that excludes the capitalized interest. The use of D_s is shown in subsequent calculations.

Step 3: Calculate the deferred taxes for each year. Mathematically, the deferred taxes in the year x are

$$d_{tx} = (D_{tx} - D_{sx})T \tag{7-23}$$

Where, T is the tax rate. As an example, the deferred taxes for years 1 and 2 are (T = 52%)

$$d_{t1} = (2222 - 1250)(0.52) = \$505$$
$$d_{t2} = (1944 - 1250)(0.52) = \$361$$

Basically, deferred taxes are the differences in taxes that result from the use of an accelerated depreciation model (sum-of-the-years digits in this example). There is a

Table 7-7
INITIAL COSTS AND SALVAGE VALUES FOR DEPRECIATION SCHEDULES

	D_b	D_s	D_t
Initial cost	$12,120	$12,000	$12,000
Salvage value	2,000	2,000	2,000

Table 7-8
DEPRECIATION SCHEDULES

End of Year	D_b	D_s	D_t
0	—	—	—
1	\$1265	\$1250	\$2222
2	1265	1250	1944
3	1265	1250	1667
4	1265	1250	1389
5	1265	1250	1111
6	1265	1250	833
7	1265	1250	556
8	1265	1250	278

reluctance to pass these "savings" on to customers due to the fact that tax rates and laws may change in the future. Consequently, deferred taxes are *normalized* by most public utilities. This means that deferred taxes are included as an expense in the early years and given as a credit in later years. Some credit is given for deferred taxes in the early years due to the manner in which the chargeable investment is calculated and used. These later two statements become evident in subsequent steps.

Step 4: Calculate the investment tax credit. Mathematically, the investment tax credit is

$$C_t = i_t P \tag{7-24}$$

and for this example is

$$C_t = (0.04)(12{,}000) = \$480$$

Step 5: Determine the amortized investment tax credit. The investment tax credit (\$480 in this example) is a reduction in taxes in the year in which construction of the project is completed. However, most public utilities treat the investment tax credit similarly to deferred taxes. That is, credit is not given for the entire amount of tax savings, as a result of the investment tax credit, in the year it occurs. Instead, the investment tax credit is spread over the life of the project. In general, the amortized investment tax credit is

$$A_t = \frac{C_t}{n} \tag{7-25}$$

and for this example is

$$A_t = \frac{480}{8} = \$60$$

Step 6: Determine (estimate) the yearly costs. These costs usually include such items as *ad valorem* taxes, insurance costs, operation costs, and maintenance costs. In this example, these costs are combined to give a total of \$1500 a year.

Step 7: Calculate the investment chargeable to the project each year. The chargeable investment, K_x, for a particular year x is

K_x = (chargeable investment in preceeding year)
− (book depreciation) − (investment tax credit)
− (deferred taxes) + (amortized investment tax credit)

Or, expressed in symbols, the chargeable investment in year x is

$$K_x = K_{x-1} - D_{bx} - C_t - d_{tx} + A_{tx} \tag{7-26}$$

For this example, some sample calculations are

$$K_0 = 12{,}120$$
$$K_1 = 12{,}120 - 1{,}265 - 480 - 505 + 60 = \$9930$$
$$K_2 = 9930 - 1265 - 0 - 361 + 60 = \$8364$$

Step 8: Calculate the debt interest. The debt interest is calculated using the relationship

$$I_x = k_d(c)\, K_{x-1} \tag{7-27}$$

where

I_x = debt interest for year x
k_d = cost of debt
c = debt ratio
K_{x-1} = chargeable investment in year $x-1$

For this example, the first two years are

$$I_1 = (0.10)\left(\frac{1}{3}\right)(12{,}120) = \$404$$
$$I_2 = (0.10)\left(\frac{1}{3}\right)(9930) = \$331$$

Step 9: Calculate the required return on equity. Mathematically, the required return is

$$f_{ex} = k_e(1 - c)K_{x-1} \tag{7-28}$$

and for this example the first two years are

$$f_{e1} = (0.25)\left(1 - \frac{1}{3}\right)(12{,}120) = \$2020$$
$$f_{e2} = (0.25)\left(1 - \frac{1}{3}\right)(9930) = \$1655$$

Step 10: Calculate the taxes. The taxes are computed using the relationship

$$t_x = \frac{T}{1-T}(f_{ex} + D_{bx} + d_{tx} + C_{tx} - A_{tx} - D_{tx} - gP) - \frac{C_{tx}}{1-T} \tag{7-29}$$

and for the first two years are

$$t_1 = \frac{0.52}{0.48}(2020 + 1265 + 505 + 480 - 60 - 2222 - 120)$$
$$= 2021 - \frac{480}{0.48} = \$1021$$

$$t_2 = \frac{0.52}{0.48}(1655 + 1265 + 361 + 0 - 60 - 1944 - 0)$$
$$= \$1383$$

The basis of Eq. (7-29) is essentially the same as Eq. (7-14). It's correctness is demonstrated later.

Step 11: Calculated the minimum annual revenue requirements. These values are calculated using the relationship

$$R_x = D_{bx} + d_{tx} + C_{tx} - A_{tx} + I_x + f_{ex} + t_x + C_x \tag{7-30}$$

For example,

$$R_1 = 1265 + 505 + 480 - 60 + 404 + 2020 + 1021 + 1500 = \$7135$$
$$R_2 = 1265 + 361 + 0 - 60 + 331 + 1655 + 1383 + 1500 = \$6433$$

This concludes the necessary steps and relationships for generating the minimum annual revenue requirements. Table 7-9 gives values for all eight years. The generation of minimum annual revenue requirements is not difficult. However, it could be considered tedious. This disadvantage can be overcome with the computer program provided in Appendix C.

Equation (7-29), the tax equation, can be verified by calculating the taxes using the relationship

$$t = (G - C - D - I)\,T - C_t \tag{7-31}$$

and letting G equal the minimum annual revenue requirement. For example, the taxes using Eq. (7-30) for the first two years

$$t_1 = (7135 - 1500 - 2222 - 404 - 120)\,(0.52) - 480 = \$1021$$
$$t_2 = (6433 - 1500 - 1944 - 331)\,(0.52) = \$1{,}383;$$

namely, the same values obtained from Eq. (7-29).

Flow-Through Method

This method for calculating minimum annual revenue requirements allocates credits and costs in the year they occur; that is, there are no deferred taxes and the investment tax credit is not amortized. Also, the interest paid during construction (previously called capitalized interest) is not capitalized but taken as an expense in the first year. This method is demonstrated in the following example.

EXAMPLE 7-5

Generate the minimum annual revenue requirements using the flow-through method and the data in Example 7-4.

As in Example 7-4, certain steps must be completed before the minimum annual revenue requirements can be determined. These steps are shown below and, for the sake of comparison, are shown in the same order as Example 7-4.

Step 1: Determine the construction interest. The amount of this interest is

$$gP = (0.01)\,(12{,}000) = \$120 \tag{7-32}$$

and is assumed to be recovered in the first year of the project.

Table 7-9

MINIMUM ANNUAL REVENUE REQUIREMENTS USING NORMALIZING METHOD

End of Year	Book Dep. D_b	Straight Line for Tax D_s	Tax Dep. D_t	Deferred Tax d_t	Invest. Tax Credit C_t	Amort. Tax Credit A_t	Chargeable Invest. K_x	Debt Interest I	Cap. Interest I_c	Return on Equity f_e	Tax t	Costs	Min. Annual Rev. Reqm't R_x
0	—	—	—	—	—	—	$12,120	—	—	—	—	—	—
1	$1265	$1250	$2222	$505	$480	$60	9,930	$404	$120	$2020	$1021	$1500	$7135
2	1265	1250	1944	361	—	60	8,364	331	—	1655	1383	1500	6433
3	1265	1250	1667	217	—	60	6,942	279	—	1394	1245	1500	5840
4	1265	1250	1389	72	—	60	5,665	231	—	1157	1132	1500	5297
5	1265	1250	1111	−72	—	60	4,532	189	—	944	1046	1500	4812
6	1265	1250	833	−217	—	60	3,544	151	—	755	986	1500	4380
7	1265	1250	556	−361	—	60	2,700	118	—	591	952	1500	4005
8	1265	1250	278	−505	—	60	2,000	90	—	450	945	1500	3685

Step 2: Calculate the depreciation schedules. In the flow-through method, the initial investment is taken as the initial cost—\$12,000 in this example. Consequently, the book depreciation (D_b) schedule is the same as the straight-line schedule for tax purposes (D_s). This schedule is given in Table 7-8. The tax depreciation schedule (D_t) is also the same as shown in Table 7-8.

Step 3: There are no deferred taxes. So the next step is to calculate the investment tax credit. This calculation is the same as in Example 7-4:

$$C_t = i_t P = 0.04\,(12{,}000) = \$480 \tag{7-33}$$

Step 4: There is no amortized investment tax credit so the next step is to estimate the yearly costs (Step 6 in Example 7-4). For this example, the yearly costs are \$1500 a year.

Step 5: Calculate the investment chargeable to the project each year. For the flow-though method, the chargeable investment, K_x, is

$$K_x = K_{(x-1)} - D_{bx} \tag{7-34}$$

For this example, some sample calculations are

$$K_0 = \$12{,}000$$
$$K_1 = 12{,}000 - 1250 = \$10{,}750$$
$$K_2 = 10{,}750 - 1250 = \$9500$$

Step 6: Calculate the debt interest. This step is essentially the same as in Example 7-4:

$$I_x = k_d(c)\,K_{x-1}$$

For the first two years, the debt interest is

$$I_1 = (0.10)\left(\frac{1}{3}\right)(12{,}000) = \$400$$
$$I_2 = (0.10)\left(\frac{1}{3}\right)(10{,}750) = \$358$$

Step 7: Calculate the required return on equity. For this example, the required returns for the first two years are

$$f_{ex} = k_e(1 - c)\,K_{x-1}$$
$$f_{e1} = (0.25)\left(1 - \frac{1}{3}\right)(12{,}000) = \$2000$$
$$f_{e2} = (0.25)\left(1 - \frac{1}{3}\right)(10{,}750) = \$1792$$

Step 8: Calculate the taxes. The taxes for the flow-through method are calculated using the relationship

$$t_x = \frac{T}{1 - T}(f_{ex} + D_{bx} - D_{tx}) - \frac{C_{tx}}{1 - T} \tag{7-35}$$

For the first two years, the taxes are

$$t_1 = \frac{0.52}{0.48}(2000 + 1250 - 2222) - \frac{480}{0.48} = \$114$$
$$t_2 = \frac{0.52}{0.48}(1792 + 1250 - 1944) = \$1189$$

Step 9: Calculate the minimum annual revenue requirements. For the flow-through method, the minimum annual revenue requirements are calculated using the relationship

$$R_x = D_{bx} + f_{ex} + I_x + gP + C_x + t_x \tag{7-36}$$

and for years one and two are

$$R_1 = 1250 + 2000 + 400 + 120 + 1500 + 114 = \$5384$$
$$R_2 = 1250 + 1792 + 358 + 0 + 1500 + 1189 = \$6089$$

All the other minimum annual revenue requirements and associated data are given in Table 7-10.

As a check on the tax equation, the taxes for the first two years are calculated using Eq. (7-31)

$$t_1 = (5384 - 1500 - 2222 - 400 - 120)\,(0.52) - 480 = \$114$$
$$t_2 = (6089 - 1500 - 1944 - 358)\,(0.52) = \$1189$$

which gives the same results as Eq. (7-35).

Levelized Minimum Annual Revenue Requirements

A word that is often used by public utilities in regard to minimum annual revenue requirements is "levelized." This word is synonymous with the words "equivalent annual amount." For example, the levelized minimum annual revenue requirement, *E,* for Example 7-4 is

$$\begin{aligned} E &= [7135\,(PF20,\,1) + 6433\,(PF20,\,2) + \ldots + 3685\,(PF20,\,8)]\,(AP20,8) \\ &= (21{,}725)\,(0.2606) = \$5661 \text{ per year} \end{aligned}$$

and for Example 7-5 is

$$\begin{aligned} E &= [5384\,(PF\,20,\,1) + 6089\,(PF\,20,\,2) + \ldots + 5040\,(PF\,20,\,8)]\,(AP\,20,8) \\ &= (20,\,574)(0.2606) = \$5622 \text{ per year} \end{aligned}$$

Showing that, the word "levelized" is just another way of saying equivalent annual amount. The discount rate used (20%) in these calculations is simply the weighted cost of capital implied in Examples 7-4 and 7-5, that is,

$$10\left(\frac{1}{3}\right) + 25\left(1\frac{1}{3}\right) = 20\%$$

These levelized minimum annual revenue requirements can also be obtained as shown in Table 7-11. To some, the calculations in Table 7-11 might be called "equivalent annual costs that include taxes." This is a correct statement and is an important point. That is, levelized minimum annual revenue requirements are synonomous with equivalent annual costs (with taxes). This point is utilized in a later chapter that considers cost comparisons. At this point, it is sufficient to say that *levelized* revenue requirements show the long-term effects of an investment while the annual revenue requirements show the short-term effects. Also, the

Table 7-10
MINIMUM ANNUAL REVENUE REQUIREMENTS USING FLOW-THROUGH METHOD

End of Year	Book Dep. D_b	Straight Line for Tax D_s	Tax Dep. D_t	Invest. Tax Credit C_t	Chargeable Invest. K_x	Debt Interest I	Cap. Interest I_c	Return on Equity f_e	Tax t	Costs	Min. Annual Rev. Reqm'ts R
0	—	—	—	—	$12,000	—	—	—	—	—	—
1	$1250	$1250	$2222	$480	10,750	$400	$120	$2000	$ 114	$1500	5384
2	1250	1250	1944	—	9,500	358	—	1792	1189	1500	6089
3	1250	1250	1667	—	8,250	317	—	1583	1263	1500	5913
4	1250	1250	1389	—	7,000	275	—	1375	1339	1500	5739
5	1250	1250	1111	—	5,750	233	—	1167	1415	1500	5565
6	1250	1250	833	—	4,500	192	—	958	1490	1500	5390
7	1250	1250	556	—	3,250	150	—	750	1564	1500	5214
8	1250	1250	278	—	2,000	108	—	542	1640	1500	5040

Table 7-11
LEVELIZED MINIMUM ANNUAL REVENUE REQUIREMENTS

	Example 7-4	Example 7-5
$(P - L)(APi, n) + L(i)$:		
$(12{,}120 - 2000)(AP\ 20, 8) + 200(0.2)$	\$3037	—
$(12{,}000 - 2000)(AP\ 20, 8) + 2000(0.2)$	—	\$3006
Costs:	1500	1500
Equivalent taxes:		
$[1021(PF\ 20, 1) + 1381(PF\ 20, 2) + \ldots + 945(PF\ 20, 8)](AP\ 20, 8)$	1124	—
$[114(PF\ 20, 1) + 1189(PF\ 20, 2) + \ldots + 1640(PF\ 20, 8)](AP\ 20, 8)$	—	1090
Capitalized interest:		
$120(PF\ 20,1)(AP\ 20, 8)$	—	26
Total	\$5661	\$5622

levelized minimum annual revenue (equivalent annual costs) provide a basis on which to compare mutually exclusive investments (projects) *under the assumption of equal gross revenues (incomes),* that is, choose the project with the smaller annual revenue requirements.

Normalizing versus Flow Through

A comparison of these two methods for generating minimum annual revenue reveals some interesting points. First, the previous section shows that the levelized minimum annual revenue requirements are less for the flow-through method. This result can be generalized. That is, for a given investment, the flow-through method will always give a smaller levelized minimum annual revenue requirement. Also, the flow-through method gives values that are a closer representation of the actual cash flows because credit and costs are taken in the year they occur. However, these points should not be interpreted to mean that the flow-through method is "best." The "best method" is dependent upon the use of the minimum annual revenue requirements. For example, most public utility companies use the normalizing method because it agrees with their accounting procedures (dictated in most instances by regulatory agencies) and it is upon values given by the accounting procedures that utility rates are argued. From the standpoint of establishing utility rates, the normalizing method is more desirable since it provides for annual revenue requirements in the early years that are greater than those provided by the flow-through method (compare Tables 7-9 and 7-10). Thus, the investment money is recovered faster. Also, the normalizing method provides another advantage from the standpoint of return on equity which is shown in the following discussion.

If the net cash flows are determined using Eq. (7-16) and the minimum annual revenues in Tables 7-9 and 7-10 as the gross incomes, the values in Table 7-12 are obtained. In obtaining the cash flows in Table 7-12 certain conventions are used in order to be consistent with the manner in which the minimum annual revenue requirements are generated. In particular, the manner in which the capitalized interest is treated. In the case of the normalizing method the capitalized interest is treated as part of the initial investment. Also, since the capitalized interest is not added into the minimum annual revenue requirements for the first year but recovered over the entire project, the capitalized interest is not considered (except in the taxes) in determining the cash flow in year 1. However, in the case of the flow-through method, the capitalized interest is not considered as part of the initial investment but as an expense in year 1 and, consequently, is included as a part of the cash flow for the same year.

Now, if the internal rate of return is determined using the two sets of cash flows given in Table 7-12, the results are 20% for both cases. That is, from the standpoint of rate of return on total investment, there is no difference between the normalizing and flow-through methods. However, this is not the case for return on equity. Using Eq. (7-15) as the definition of cash flow (returns to equity), the cash flows in Table 7-13 are obtained. If the internal rates of return are determined for both sets of cash flows in Table 7-13, the results are 26.3% (linear interpolation) for the normalizing method and exactly 25% for the flow-through method. These results can be generalized: the rates of return on the total investments will be the same for both methods of generating minimum annual revenue requirements. However, the normalizing method will give a higher rate of return on equity than the flow-through method. This difference in rates of

Table 7-12

NET CASH FLOWS FOR EXAMPLES 7-4 AND 7-5 USING EQ. (7-16)

End of year	Example 7-4	Example 7-5
0	−$12120	−$12,000
1	7135 − 1500 − 1021 = 4614	5384 − 1500 − 120 − 114 = 3650
2	6433 − 1500 − 1381 = 3550	6089 − 1500 − 1189 = 3400
3	5840 − 1500 − 1245 = 3095	5913 − 1500 − 1263 = 3150
4	5297 − 1500 − 1132 = 2665	5739 − 1500 − 1339 = 2900
5	4812 − 1500 − 1046 = 2266	5565 − 1500 − 1415 = 2650
6	4380 − 1500 − 986 = 1894	5390 − 1500 − 1490 = 2400
7	4005 − 1500 − 952 = 1553	5214 − 1500 − 1564 = 2150
8	3685 − 1500 − 945 + 2000 = 3240	5040 − 1500 − 1640 + 2000 = 3900

NOTE: The value of $2000 in year 8 is the salvage value. The capitalized interest ($120) is assumed to be paid in year 1 for the flow-through method.

Table 7-13

NET CASH FLOW FOR EXAMPLES 7-4 AND 7-5 USING EQ. (7-15)

End of Year	Example 7-4	Example 7-5
0	−12,120 + ⅓ (12,000) = −$8120	−12,000 + ⅓ (12,000) = −$8000
1	7135 − 1500 − 404 − 1021 − 417 = 3793	5384 − 1500 − 400 − 120 − 114 − 417 = 2833
2	6433 − 1500 − 331 − 1383 − 417 = 2802	6089 − 1500 − 358 − 1189 − 417 = 2625
3	5840 − 1500 − 279 − 1245 − 417 = 2399	5913 − 1500 − 317 − 1263 − 417 = 2416
4	5297 − 1500 − 231 − 1132 − 417 = 2017	5739 − 1500 − 275 − 1339 − 417 = 2208
5	4812 − 1500 − 189 − 1046 − 417 = 1660	5565 − 1500 − 233 − 1415 − 417 = 2000
6	4380 − 1500 − 151 − 986 − 417 = 1326	5390 − 1500 − 192 − 1490 − 17 = 1791
7	4005 − 1500 − 118 − 952 − 417 = 1018	5214 − 1500 − 150 − 1564 − 417 = 1583
8	3685 − 1500 − 90 − 945 − 417 − 667 + 2000 = 2066	5040 − 1500 − 108 − 1640 − 417 − 667 + 2000 = 2708

return on equity is primarily due to the treatment of deferred taxes in the normalizing method.

Summary

This chapter presented some important concepts. The first was that the definition of cash flow and cost of capital (MARR) are not independent from the standpoint of evaluating capital investments. Also, it is better by far to work with *equity* cash flows since, in evaluating capital investments, the attempt is to determine its desirability from the standpoint of *equity* stockholders. Consequently in the definition of cash flows, borrowed money (debt) should be regarded as a cash inflow thereby reducing the capital required from equity sources and debt obligations (interest and principal payments) should be explicitly included as cash outflows. In determining the net present value of these equity cash flows, the discount rate should be based only on the required return on *equity* capital (the incremental cost of new equity capital) and not the weighted (debt and equity) cost of capital. Similarly, any comparison of the rate of return on these equity cash flows should also be made with the required return on equity.

Perhaps a disadvantage of using equity cash flows is that it requires a knowledge of the amount of borrowed money and the manner in which it is paid back. In cases where there is no knowledge about the amount of debt financing, a common approach is to use the company's current debt ratio (based on market values) as the proportion of debt financing. This assumes, of course, that the future and current debt ratios are the same. If the future debt ratio is expected to be different than the current debt ratio, the future ratio should be used. As far as the principal and interest payments are concerned for a particular debt amount, it is often assumed that the principal is paid back in equal installments over the years in which gross income is received from the investment and interest is assessed on the unpaid balance.

Also presented in this chapter is the concept of minimum annual revenue requirements. An important concept from the standpoint of explaining cash flow definitions and, in its own right, for evaluating capital investments. In addition, minimum annual revenue requirements provide an excellent basis for making cost comparisons and replacement studies. These topics are discussed in Chapters 10 and 11.

In generating minimum annual revenue requirements there is often a question regarding the basis of the debt ratio. That is, should the debt ratio be based on the company's debt ratio or should it reflect the financing of a particular investment? The answer to this question is that the debt ratio should be the company's future debt ratio. As in the case of equity cash flows, it is common to use the current debt ratio in generating minimum annual revenue requirements which assumes, as previously mentioned, that the future and current debt ratios are the same.

Reference

1. Merrett, A. J., and A. Sykes, *The Finance and Analysis of Capital Projects*, 2nd ed., John Wiley & Sons, New York, 1973.

Problems

7-1. For a project with the data shown below, determine the minimum annual revenue requirements. Calculate the minimum annual revenue for the first two years by hand then use the computer program given in the appendicies to check your results and determine the revenue requirements for the remaining years.

method = normalizing
initial cost = \$100,000
salvage value (tax and book) = \$10,000
tax depreciation model = sum-of-years digits
book depreciation model = straight line
life (tax and book) = 10 years
total costs per year = \$4000
debt ratio = 40%
required return on equity = 20%
required return on debt = 10%
tax rate = 52%
capitalized interest = 0%
investment tax credit = 0%

7-2. Repeat Problem 7-1 using the flow-through method.

7-3. Repeat Problem 7-1, only in this problem the capitalized interest is 1% and the investment tax credit is 10%.

7-4. Repeat Problem 7-2, only in this problem the construction (capitalized) interest is 1% and the investment tax credit is 10%.

7-5. Repeat Problem 7-3, only in this problem base the tax depreciation on a life of six years.

7-6. Repeat Problem 7-4, only base the tax depreciation on a life of six years.

7-7. Repeat Problem 7-4, only use as the tax-depreciation model double-declining balance with the conversion to straight-line depreciation.

7-8. Repeat Problem 7-3, only use as the tax-depreciation model double-declining balance with the conversion to straight line.

7-9. Based on the results of Problems 7-1 through 7-8, what generalizations are possible regarding the effects of the following considerations on the minimum annual revenue requirements?

(a) Normalizing versus flow-through methods.
(b) Depreciation models used for tax purposes.
(c) Life used in tax-depreciation models.
(d) Investment tax credit.

7-10. Suppose that in Problems 7-1 and 7-2 both the tax depreciation and book depreciation were sum of the years-digits, what effect would this have on the minimum annual revenue requirements?

7-11. Repeat Problem 7-4 with the exception that the annual costs are not constant but are \$4000 for the first year and increase by \$500 each year for the 10-year period.

7-12. Suppose the project described in Problem 7-1 is expected to generate the gross incomes shown below, is the project acceptable?
(a) $45,000 per year for 10 years.
(b) $35,000 per year for 10 years.
(c) $25,000 per year for 10 years.

7-13. Using the minimum annual revenue requirements obtained in Problem 7-4 as gross incomes, determine the net cash flows using Eqs. (7-16), (7-17), and (7-15) and the internal rate of return for each set of cash flows. In using Egs. (7-16) and (7-17) be sure and consider the construction (capitalized) interest explicitly in year one (see Table 7-12). In using Eq. (7-15) assume the principal is paid back in 10 equal installments and a portion of the salvage value is also used as a principal payment (see Table 7-2).

7-14. A company is considering the expansion of an existing process in order to satisfy increased demand. It is estimated that the expansion will require the following capital expenditures.

End of Year	Capital Expenditure	Use	Life	Salvage
0	$100,000	Land	—	—
1	400,000	Equipment	10 years	$20,000
1	100,00	Building	15 years	10,000

In addition, it is estimated that yearly gross incomes and cost will increase by the amounts shown below:

End of Year	Gross Income	Costs
0	—	—
1	—	$100,000
2	$600,000	200,000
3	600,000	200,000
4	600,000	200,000
5	600,000	200,000
6	600,000	200,000
7	600,000	200,000
8	600,000	200,000
9	600,000	200,000
10	600,000	200,000

For the sake of calculations, assume the following:
1. The land and building are to be financed from equity sources.
2. The equipment will be financed from debt (50%) and equity (50%) sources. The debt principal will be paid in five equal installments beginning the year after the loan is made. The interest rate is 10% and assessed on the unpaid balance.

3. It is estimated that this expansion will require an increase in working capital, in year one, of $200,000 and will be obtained from equity sources.
4. The investment tax credit (10%) is applicable for the equipment.
5. The required return on equity capital is 15%.
6. Assets will be depreciation after gross income ends and there is sufficient taxable income from other operations to take advantage of the depreciation allowances.
7. The land will be sold.
8. The tax rate is 52%.
9. For tax purposes use straight line depreciation.

Determine if this expansion is economically justified.

7-15. Using the minimum annual revenue requirements as the gross incomes and the other results obtained in the solution to Problem 7-3, determine:
(a) The cash flows for each year using Eq. (7-15) and the internal rate of return.
(b) The cash flows for each year using Eq. (7 16) and the internal rate of return.

7-16. Using the minimum annual revenue requirements as the gross incomes and the other results obtained in the solution to Problem 7-4, determine:
(a) The cash flows for each year using Eq. (7-15) and the internal rate of return.
(b) The cash flows for each year using Eq. (7-16) and the internal rate of return.

7-17. Assuming Eq. (7-17) is used for the definition of cash flow in Problems 7-15 and 7-16, answer the following questions:
(a) Is the rate of return higher or lower? Why?
(b) What is an estimate of the rate of return?

CHAPTER 8
BREAKEVEN ANALYSIS

Breakeven analysis is a procedure for studying the relationships between profit, gross income (revenue), and costs. However, before these relationships can be mathematically defined, it is necessary to consider just what is meant by the words "profit" and "costs." To many people, profit is the amount left over after all costs are subtracted from the gross income. In a very general way, there is nothing wrong with this idea. However, when it is necessary to specifically define the costs and associated profit there may be differences in opinions. For example, there are decided differences between the economist's and accountant's concept of profit. The accountant (and the Internal Revenue Service) does not recognize the return on equity capital as a cost. To the accountant, this is a part of the profit. While, the economist would consider a return on equity capital as a "normal profit" or an implied charge (interest cost) on the owner's investment and, consequently, would reduce profits by this amount. Even among economists, there are differences in opinions regarding the causes and nature of profits as a result of different profit theories. In general, profit theories can be separated into five groups: risk-bearing theories, innovation theories, entrepreneurial-compensation theories, friction-and-monopolistic-imperfection theories, and Marxian theories. References 1 and 2 given at the end of this chapter provide an excellent discussion of these various profit theories.

From the standpoint of formulating breakeven models, the difference in the economist's and accountant's point of view regarding profit is, perhaps, the most significant. This difference does not pose any great problem in breakeven models as long as profit is clearly defined. In this chapter the accountants point of view is taken unless specifically stated otherwise.

LINEAR BREAKEVEN MODELS

The relationships between profit, total income, and costs in breakeven models can be either linear or nonlinear. In this section linear models are considered and in a later section nonlinear models are considered. For the linear case, the mathematical relationship is

$$P = VS - VC - F = V(S - C) - F \tag{8-1}$$

where

P = before tax (gross) profit per unit of time
V = volume of sales per unit of time
S = selling price per unit
C = variable costs per unit
F = fixed costs per unit of time
VS = gross income per unit of time
VC = variable costs per unit of time
C_t = total costs per unit of time = $VC + F$
$S - C$ = the contribution per unit. This term is the portion of the selling price that contributes to paying the fixed costs.

In practice, the usual time period is one year. That is, profit and the other time dependent variables are usually defined on an annual basis.

Equation (8-1) implies certain assumptions. First, variable costs and gross income are both linear functions of the volume of production. Second, production and sales volumes are equal. Third, all variables are known with certainty. Fourth, there is no other income. Finally, fixed costs are independent of the production rate.

Variable Costs

These are costs that are proportional to the volume of production. Some examples are direct labor, material and direct supplies, direct maintenance, and direct supervision.

Fixed Costs

These are costs that are independent of the volume of sales. That is, these costs are incurred whether the volume of production is at zero or 100% of capacity. In actuality, this may not be true since fixed costs can vary with time. However, for most studies, the assumption that the fixed costs are constant is sufficiently accurate since the time period in breakeven models is, usually, relatively short. More specifically, fixed costs are made up of such components as interest on debt, depreciation, rent, executive salaries, insurance, and *ad valorem* taxes.

It is in the fixed costs that the return on equity capital is included if the economist's point of view is taken to define profit. If the cost of equity capital, E, is included, Eq. (8-1) becomes

$$P_e = V(S - C) - (F + E) \tag{8-2}$$

Tax Considerations

Equations (8-1) and (8-2) give the before-tax profit. However, both equations can be modified to obtain the after-tax profit (P_t). In the case of Eq. (8-1) the after-tax profit is

$$\begin{aligned} P_t &= V(S - C) - (F_1 + D) - [V(S - C) - (F_1 + D_t)]\, T \\ &= [V(S - C) - F_1]\,(1 - T) - D + D_t(T) \end{aligned} \tag{8-3}$$

where the nomenclature is the same as given earlier and

F_1 = fixed expenses excluding company (book) depreciation
D = company depreciation amount
D_t = tax depreciation amount
T = tax rate

Equation (8-3) takes into account the possibility that the company's and tax depreciation amounts may be different. If they are not different (as is often assumed), Eq. (8-3) becomes

$$P_t = [V(S - C) = F]\,(1 - T) = P(1 - T) \tag{8-4}$$

Equations (8-3), (8-4), and all subsequent after-tax profit equations make the assumption that there is sufficient taxable income from other company activities to allow the full deduction of the fixed costs and tax depreciation regardless of the volume of sales for the particular product under consideration.

For Eq. (8-2), the economist's point of view, the after-tax profit, P_{te}, is

$$\begin{aligned} P_{te} &= V(S - C) - (F_1 + D + E) - [V(S - C) - (F_1 + D_t)]\,(T) \\ &= [V(S - C) = F_1]\,(1 - T) - D + D_t(T) - E \end{aligned} \tag{8-5}$$

The return on equity capital, E, does not appear in the tax term because, of course, the Internal Revenue Service does not recognize this as a cost. Comparing Eqs. (8-3) and (8-5) indicates that the relationship between the accountant's after-tax profit and the economist's after-tax profit is

$$P_{te} = P_t - E \tag{8-6}$$

Breakeven analysis derives its name from the determination of the point at which gross income exactly equals the costs, that is, the point at which the profit is zero. This point can be mathematically determined by setting the profit equal to zero in the appropriate equation and solving for the interested variable. The usual variable is volume of sales. For example, the breakeven point for Eq. (8-5) is

$$\begin{aligned} 0 &= [V(S - C) - F_1]\,(1 - T) - D + D_t(T) - E \\ V_{Be} &= \frac{F_1(1 - T) + D + E - D_t(T)}{(S - C)\,(1 - T)} \end{aligned} \tag{8-7}$$

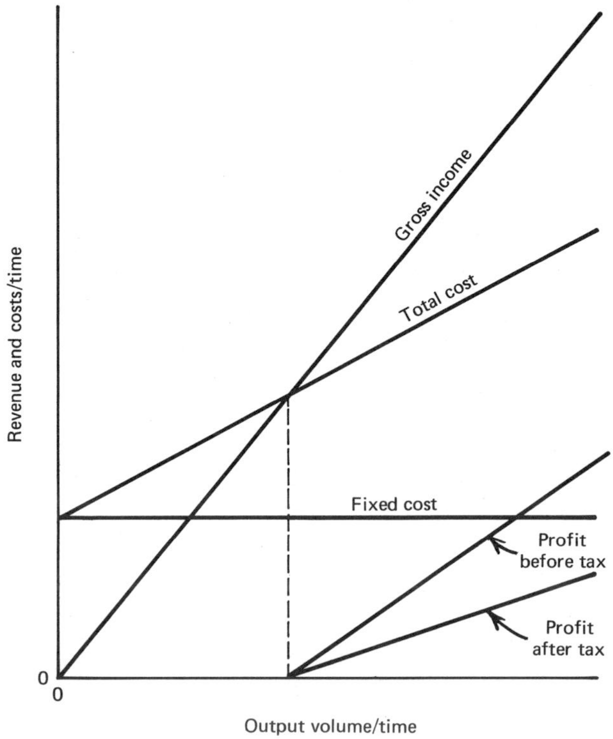

Figure 8-1.
Linear breakeven chart.

A summary of the various profit equations and associated breakeven points is given in Table 8-1.

Breakeven models are often graphically described. A generalized model is shown in Figure 8-1.

EXAMPLE 8-1

A company is making a product that it sells for \$40 per unit. Material, labor, and overhead costs are, respectively, \$7, \$8, and \$5 per unit. The depreciation for the equipment used in the production of this product amounts to \$100,000 per year. Fixed overhead expenses are \$40,000 per year and interest on borrowed money amounts to \$20,000 per year. For a production rate of 15000 units per year, determine the following:

(a) The annual before-tax profit.
(b) The annual after-tax profit if the tax rate is 52%. Assume the depreciation amounts are the same and the accountant's point of view is taken in regard to profits.
(c) The breakeven point.

Table 8-1
PROFIT AND BREAKEVEN POINT EQUATIONS

Accountant's Viewpoint	Economist's Viewpoint
$P = V(S - C) - (F_1 + D)$ Breakeven: $V_B = \frac{F_1 + D}{S - C} = \frac{F}{S - C}$	$P_e = V(S - C) - (F_1 + D + E)$ Breakeven: $V_{Be} = \frac{F_1 + D + E}{S - C} = \frac{F + E}{S - C}$
$D \neq D_t$ $P_t = V(S - C) - (F_1 + D) - [V(S - C) - (F_1 + D_t)]T$ Breakeven: $V_B = \frac{F_1(1 - T) + D - D_tT}{(S - C)(1 - T)}$	$D \neq D_t$ $P_{te} = V(S - C) - (F_1 + D + E) - [V(S - C) - (F_1 + D_t)]T$ Breakeven: $V_{Be} = \frac{F_1(1 - T) + D + E - D_tT}{(S - C)(1 - T)}$
$D = D_t$ $P_t = [V(S - C) - (F_1 + D)](1 - T) = P(1 - T)$ Breakeven: $V_B = \frac{F}{S - C}$	$D = D_t$ $P_{te} = [V(S - C) - (F_1 + D)](1 - T) - E$ Breakeven: $V_{Be} = \frac{F(1 - T) + E}{(S - C)(1 - T)}$

(d) Show the data given in this example graphically. For plotting purposes, assume the maximum plant capacity is 20,000 units per year.

Part a: The total variable cost per unit is

$$C = 7 + 8 + 5 = \$20$$

and the fixed costs per year are

$$F = 100{,}000 + 40{,}000 + 20{,}000 = \$160{,}000$$

Substituting into Eq. (8-1) gives a before-tax profit of

$$P = 15000\,(40 - 20) - 160{,}000 = \$140{,}000 \text{ per year}$$

Part b: Since $D = D_t$, the after-tax profit is

$$P_t = P(1 - T) = 140{,}000(1 - 0.52) = \$67{,}200 \text{ per year}$$

If the depreciation amounts are different, the equations given in Table 8-1 can be used. Or the equation

$$P_t = P(1 - T) + (D_t - D)T \qquad (8\text{-}8)$$

can be used. For example if $D = \$100{,}000$ and $D_t = \$120{,}000$, the after-tax profit (accountant's point of view) is

$$\begin{aligned} P_t &= 140{,}000(1 - 0.52) + (120{,}000 - 100{,}000)(0.52) \\ &= 67{,}200 + 10{,}400 = \$77{,}600 \text{ per year} \end{aligned}$$

If the economist's definition of profit is used, all that is needed to obtain the profit, P_{te}, is to subtract the equity return E, from Eq. (8-8).

Part c: The breakeven point is the number of units where the income equals the cost ($P = 0$).

$$\begin{aligned} 0 &= V(40 - 20) - 160{,}000 \\ V &= 8000 \text{ units per year} \end{aligned}$$

the breakeven point can also be expressed in terms of the gross income. In this example, the breakeven point is 8000(40) = \$320,000 per year. Note that in this example the breakeven point before and after tax are the same since the tax and book depreciation amounts are the same. This is not the case when $D \neq D_t$ as can be seen by a comparison of the equations given in Table 8-1.

Part d: The graphical representation for the data given in this problem is shown in Figure 8-2.

The abscissa in this figure is terminated at 20,000 units per year since this is the plant capacity. The implication is that, at a production rate of 15,000 units per year, the plant is being utilized at 75% (15,000/20,000) of its capacity.

In parts (c) and (d) of this example the breakeven point is based on the accountant's viewpoint of profit. If the economist's viewpoint is taken, the breakeven point is determined by including the equity return, E, as shown in the equations given in Table 8-1. A

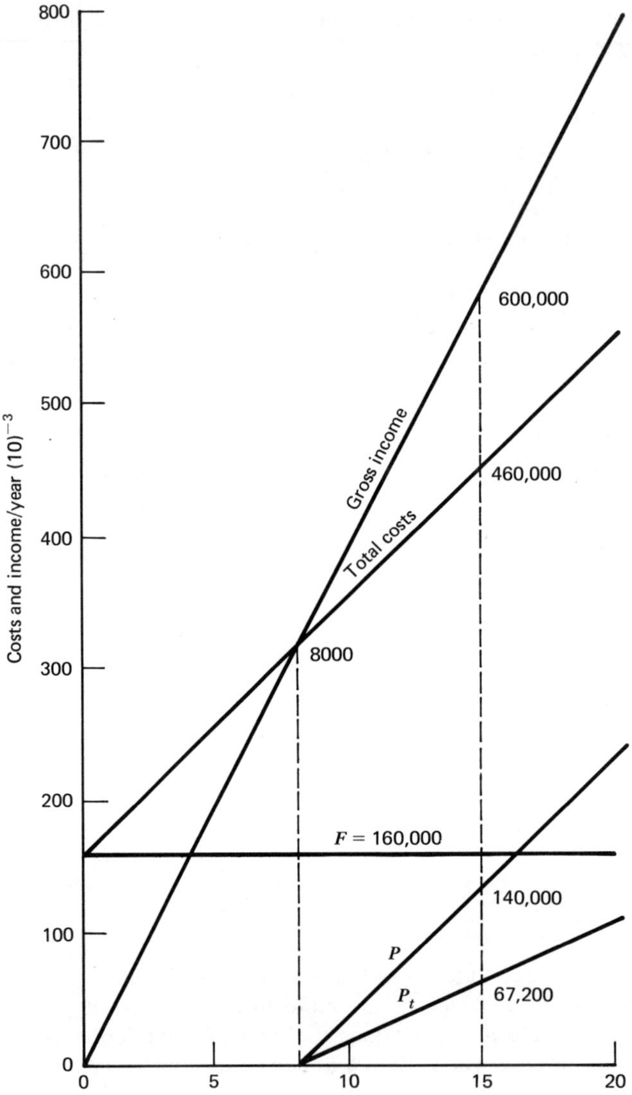

Figure 8-2.
Breakeven model for Example 8-1.

breakeven point that includes a return on equity is often referred to by the economist as the "makeout point." The implication being that, as far as the economist is concerned, profit above this point is "truly" profit.

A Comment: In the previous discussion of breakeven models, the difference between the economist's and accountant's view of profit is included. Unfortunately, in some discussions of breakeven models this difference is not explicitly stated. The following example illustrates this point.

EXAMPLE 8-2

In order to produce a certain product, a company must purchase $1,000,000 worth of equipment. This equipment is estimated to have a salvage value of $100,000 at the end of 10 years. Fixed expenses, excluding depreciation, are estimated to be $50,000 per year. The variable cost and selling price per unit are, respectively, $60 and $100. The company's minimum attractive rate of return is 12%. If the company produces 15,000 units per year, what is the profit.

The usual approach in this type of problem is to calculate the fixed costs as shown below:

$$\begin{aligned} F &= F_1 + (P - L)(APi, n) + L(i) \\ &= 50{,}000 + (1{,}000{,}000 - 100{,}000)(AP\ 12, 10) + 100{,}000\,(0.12) \\ &= \$221{,}300 \text{ per year} \end{aligned}$$

Therefore, the before-tax profit is

$$P = 15000\,(100 - 60) - 221{,}300 = \$378{,}700 \text{ per year}$$

Basically there is nothing wrong with this solution provided it is clearly understood *and stated* what is implied in this solution. A major implication in this solution is that the economist's concept of profit is being used since the term $(P - L)(APi, n) + L(i)$ is included in the fixed costs. This term is (as shown in Chapter 3) an equivalent annual amount of the capital recovered *and return* on the unrecovered capital. What is included in the return depends on the method of financing the $1,000,000 investment. If the investment is all equity, then the entire return is an equity return since the minimum attractive rate of return would be based on the required equity rate. If the investment is partially debt financed then the return is comprised of both debt interest and return on equity since the minimum attractive rate of return would be based on a composite of the required debt and equity rates of return. This latter case would also imply that debt interest should not be included in the fixed cost since it is already included in determining the minimum attractive rate of return. Finally, there is the implication that the volume of sales, variable costs, and gross income are the same for the next ten years.

Dumping

Selling a portion of a company's product at a reduced price is referred to as dumping. It implies the belief that there is a limited demand for the product at the existing selling price and that additional demand can only be generated by reducing the selling price. This, of course, is dangerous. For, if consumers become aware of the two different prices for the same product, the lower price may become the base price. However, if successful, dumping can increase the company's total profit through the utilization of unused plant capacity. Dumping is usually accomplished by selling a product in domestic markets under a different name or by selling in a foreign market. The practice of dumping is considered in the following example.

EXAMPLE 8-3

Suppose in Example 8-1 the company is currenty selling and producing 15,000 units per year and it can sell an additional 5000 units per year at a selling price of $30 per unit. This is economically attractive since the total before tax profit will increase from $140,000 per year (see Example 8-1) to

$$P = 140{,}000 + 5000(30 - 20) = \$190{,}000 \text{ per year}$$

and the after-tax total profit will increase to

$$P_t = 190{,}000(1 - 0.52) = \$91{,}200$$

This example is graphically shown in Figure 8-3.

An Extension: In this example the practice of dumping is evaluated on the basis of constant variable and fixed costs. This is not a necessary requirement. Dumping can still be economically desirable even when variable and fixed costs increase as a result of the increased output. For example, if the total variable costs increases to $25 per unit and the fixed costs increase by $20,000 per year as a result of the production of the additional 5000 units per year, the increase in before-tax profit is

$$5000(30 - 25) - 20000 = \$5000 \text{ per year}$$

and the after-tax increase in profit is

$$5000(1 - 0.52) = \$2400 \text{ per year}$$

The situation in this case is shown in Figure 8-4.

MULTIPRODUCT ANALYSIS

Up to this point the discussion of $C - V - P$ models has been limited to one product. This, usually, is an oversimplification since most manufacturers produce more than one product. Using a slightly different format, the multiproduct case can be considered both analytically and graphically. The analysis of a multiproduct case is presented in the next example.

EXAMPLE 8-4

A company is making three products. The cost and sales data for each product are given in Table 8-2. It is desired to provide a single graph that shows the economic relationship between these products. The multiproduct breakeven chart shown in Figure 8-5 satisfies this requirement.

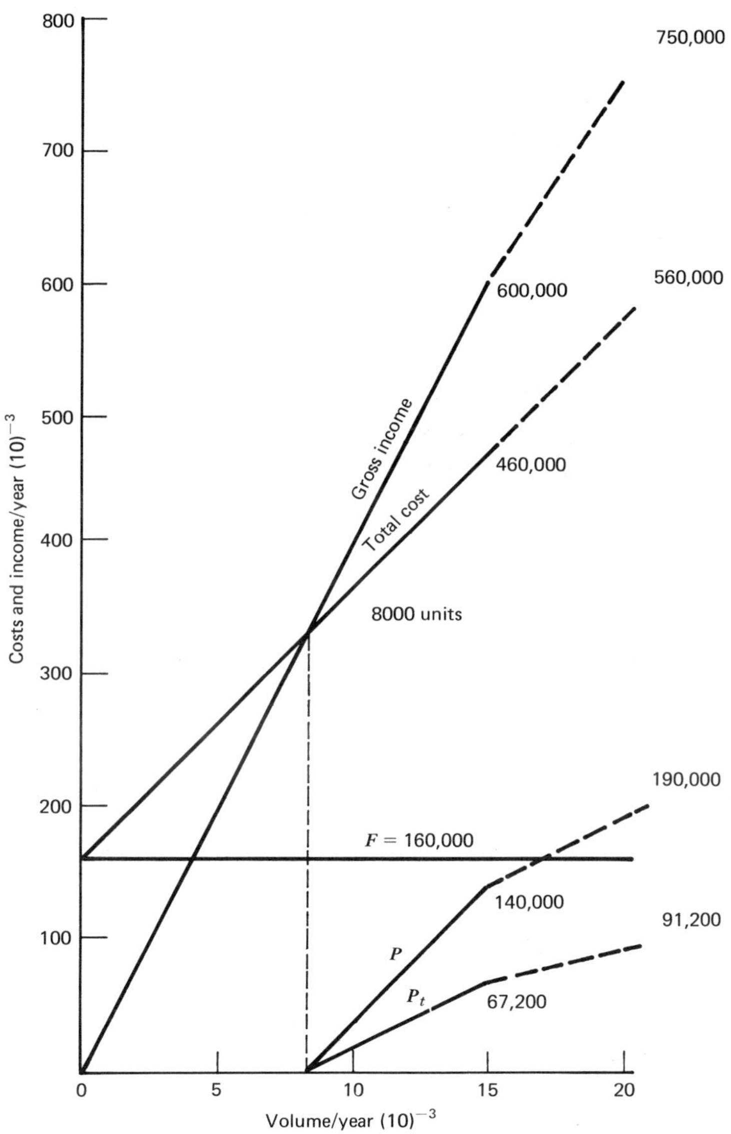

Figure 8-3.
Breakeven model with dumping (Example 8-3).

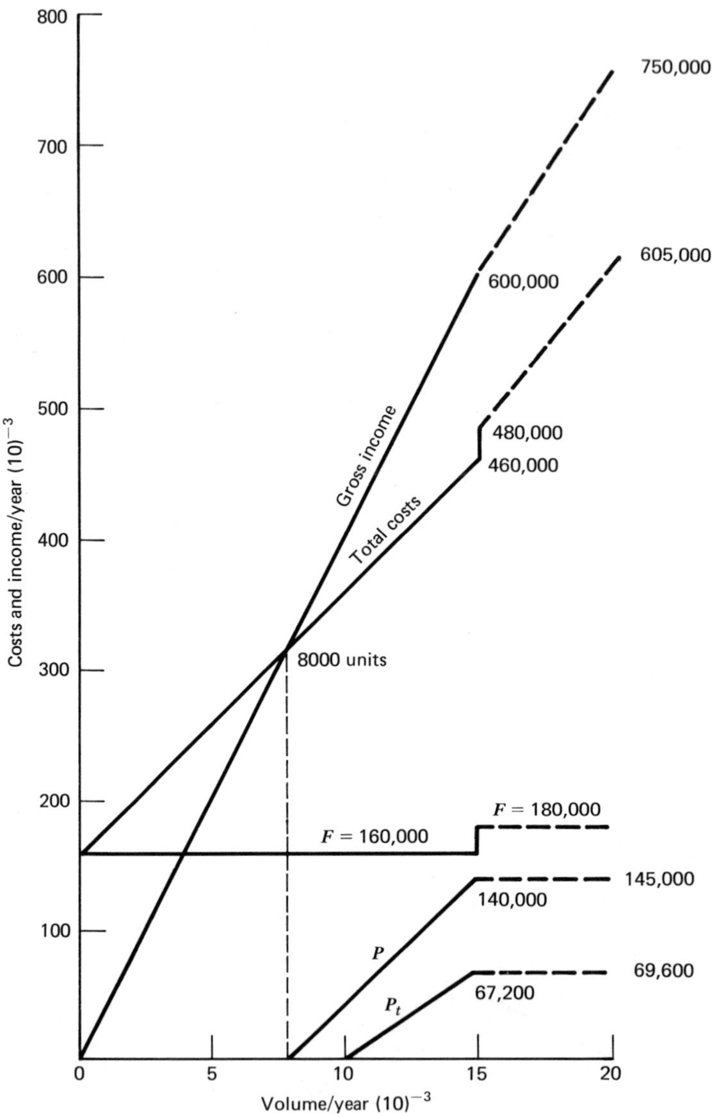

Figure 8-4.
Breakeven model with dumping and changes in costs.

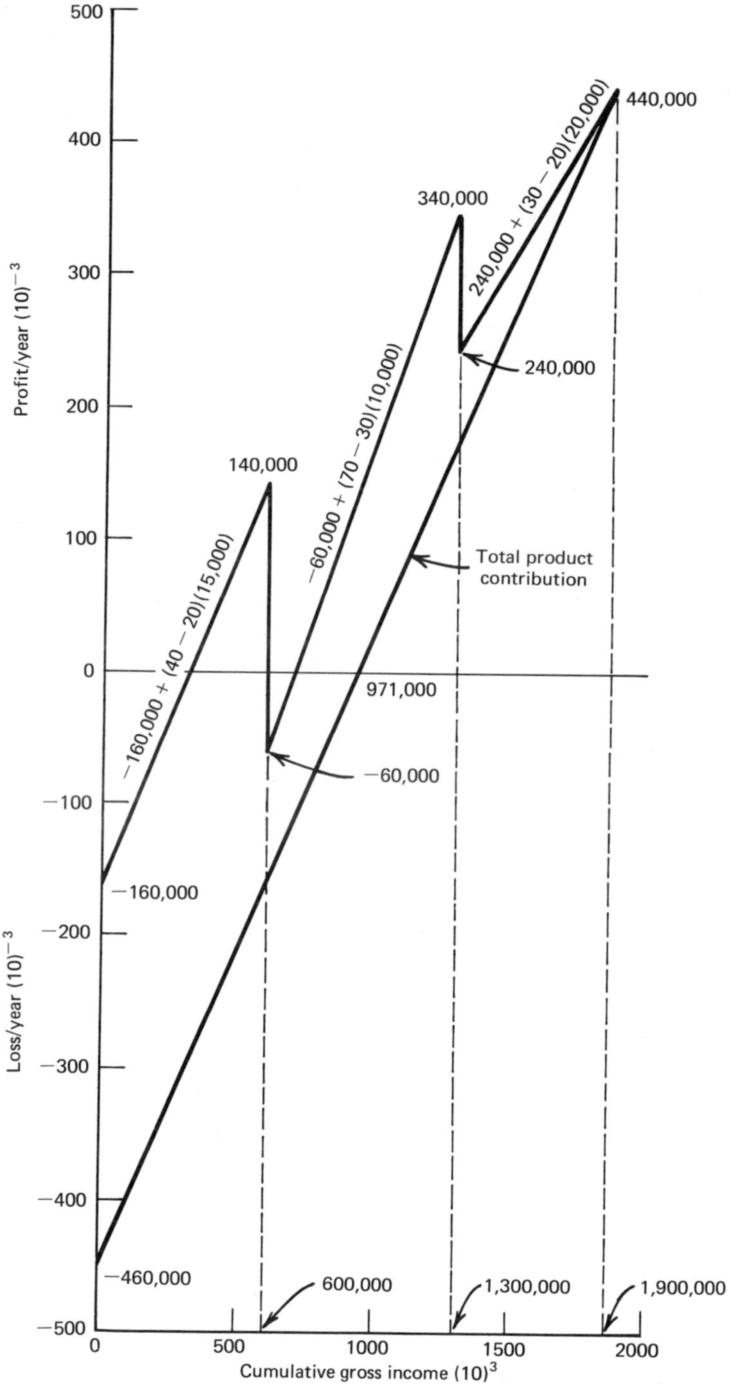

Figure 8-5.
Multiproduct breakeven chart (Example 8-4).

Table 8-2
DATA FOR EXAMPLE 8-4

	Product		
	1	2	3
Selling price per unit	$40	$70	$30
Variable cost per unit	$20	$30	$20
Fixed costs per year	$160,000	$200,000	$100,000
Volume of sales per year	15,000	10,000	20,000

This chart is constructed by plotting the cumulative total of sales income as the abscissa and loss or profit as the ordinate. The chart is started by plotting the fixed cost (a loss) for product one ($-160,000) and then adding to it

$$(40 - 20)(15000) = \$300{,}000$$

giving a total of $140,000. The remaining calculations are shown in Figure 8-5. The total product contribution line is determined by joining the end point (a profit of $440,000) with the total fixed costs ($460,000). The intersection of this line with the zero profit line gives the breakeven point in terms of sales income. In this example, this is $971,000 which means that, at this value, the costs are covered by the gross sales income, provided, the current product-sales mix does not change.

The multiproduct breakeven chart provides a means for making product comparisons. It shows the amount of fixed costs carried by each product. Also, it can be used to determine the desirability of dropping or adding products. From a comparison standpoint, a product is preferred when the slope of its contribution line is greater than the slope of the total product contribution line.

The multiproduct case can be analyzed analytically in the following manner. The first step is to determine the values shown in Table 8-3. The accuracy of the values in Table 8-3 is usually high. In general, this degree of accuracy is not required from a practical standpoint. However, for this example, it is felt to be necessary since it reduces roundoff error and, consequently, facilitates the meaning and origin of various values used in this

Table 8-3
PERCENT CONTRIBUTION PER SALES (EXAMPLE 8-4)

Product	Percent of Sales	Percent Contribution	Percent Contribution per Sales
1	31.5789	50.0000	15.7895
2	36.8421	57.1429	21.0526
3	31.5789	33.3333	10.5263
			$\Sigma = 47.3684$

discussion. For product 1 the values in Table 8-3 are determined, using the data in Table 8-2, as shown below:

$$\text{total sales revenue} = 40(15{,}000) + 70(10{,}000) + 30(20{,}000) = \$1{,}900{,}000$$

$$\text{percent of sales} = \frac{40(15{,}000)}{1{,}900{,}000}(100) = 31.5789\%$$

$$\text{percent contribution} = \frac{40 - 20}{40}(100) = 50.0000\%$$

$$\text{percent contribution of sales} = (0.315789)(0.500000)(100) = 15.7895\%$$

Corresponding values for products 2 and 3 are determined in a similar manner.

The value of 47.3684% is the total contribution per overall sales and can be used in the following way assuming that $1,900,000 is full-plant capacity.

$$\text{profit} = 1{,}900{,}000(0.473684) - 460{,}000 = \$440{,}000 \text{ per year}$$

Profits for other than full capacity can be determined using the relationship

$$P = ZXe - F \tag{8-9}$$

where

P = profit
Z = plant capacity expressed as total sales
X = percent capacity expressed as a decimal
e = total contribution per overall sales
F = total fixed cost
ZX = gross sales
$(ZX)(1 - e)$ = variable costs

In the case of 100% plant capacity, Eq. (8-9) reduces to

$$P = ZE - F \tag{8-10}$$

which is the equation used to determine the $440,000 result given earlier. Recognize that Eq. (8-9) makes the assumption that the product sales mix remains unchanged.

As an example of the use of Eq. (8-9), the profit at 70% plant capacity is

$$P = (1{,}900{,}000)(0.7)(0.473684) - 460{,}000 = \$170{,}000 \text{ per year}$$

The breakeven point, B, can be determined from the equation

$$B = \frac{F}{e} \tag{8-11}$$

and for this example

$$B = \frac{460{,}000}{0.473684} = \$971{,}111 \text{ per year}$$

This value is the same as the value determined graphically taking into consideration the problem of reading Figure 8-5 to the same accuracy as this calculation.

The breakeven point can be used to determine the volume of sales for each product in order to breakeven with the present product-sales mix. Since the product-sales mix does not change, the percent contribution per sales for each product at plant capacity, e_i, and at the breakeven point e_{ib} are equal. Consequently,

$$e_i = e_{iB}$$

$$\left(\frac{S_i - C_i}{S_i}\right)\left(\frac{S_i V_i}{Z}\right) = \left(\frac{S_i - C_i}{S_i}\right)\left(\frac{S_i V_{iB}}{B}\right) \tag{8-12}$$

$$V_{iB} = V_i\left(\frac{B}{Z}\right)$$

For this example, the breakeven volumes are

$$V_{1B} = 15{,}000\left(\frac{971{,}111}{1{,}900{,}000}\right) = 7667 \text{ units per year}$$

$$V_{2B} = 10{,}000\left(\frac{971{,}111}{1{,}900{,}000}\right) = 5111 \text{ units per year}$$

$$V_{3B} = 20{,}000\left(\frac{971{,}111}{1{,}900{,}000}\right) = 10{,}222 \text{ units per year}$$

as a check, the sum of the gross sales and total costs at the breakeven point should be zero. That is,

$$\sum_{i-1}^{3} S_i V_{iB} - \sum_{i=1}^{3} C_i V_i - F = 0 \tag{8-13}$$

Substituting into Eq. (8-13) gives

$$971{,}111 - (20)(7667) - 30(5111) - 20(10{,}222) - 460{,}000 = 0$$

which establishes the point that the breakeven point is \$971,111 per year with the current product sales mix.

It should be noted that this entire example has been worked on the basis of before tax considerations. If after-tax profits are required, the profits can be multiplied by the term $(1 - T)$, assuming that tax and book depreciation amounts are equal. If the tax and depreciation amounts are different, the after-tax profit can be obtained using Eq. (8-8).

An important point to note in this example is that the entire analysis is possible with only the total fixed costs. That is, it is not necessary to divide the fixed costs between the individual products.

NONLINEAR BREAKEVEN ANALYSIS

Up to this point the breakeven models are based on the assumption of linear relationships. This assumption in many cases provides a realistic representation. However, in some cases linear relationships are not adequate. In these cases, nonlinear relationships must be considered.

The basic approach in the analysis of nonlinear breakeven models is the same as in linear models. The major difference is that nonlinear models require the use of calculus. The analysis of a nonlinear case is shown in the next example.

EXAMPLE 8-5

A manufacturer has determined that his or her volume of weekly sales, V, is related to the selling price per unit, S, by the equation $S = (160 - V)$ and the weekly total costs, C, are given by the equation

$$C = 0.5V^2 + 100V + 200$$

The manufacturer wishes to know the following:

(a) The selling price for maximum profit and the profit at this selling price.
(b) The breakeven point.

For this case, the basic relationship is still

$$\text{Profit} = \text{income} - \text{costs} \tag{8-14}$$

and with the given data Eq. (8-14) becomes

$$\begin{aligned} P &= VS - (-0.5V^2 + 100V + 200) \\ P &= V(160 - V) + 0.5V^2 - 100V - 200 \\ P &= -0.5V^2 + 60V - 200 \end{aligned} \tag{8-15}$$

Taking the derivative of Eq. (8-15) with respect to V and setting equal to zero gives

$$\begin{aligned} \frac{dP}{dV} &= 0 = -V + 60 \\ V &= 60 \text{ units per week} \end{aligned} \tag{8-16}$$

which is the volume of sales for maximum profit. Consequently, the optimum selling price per unit is

$$S = 160 - 60 = \$100 \text{ per unit}$$

giving a weekly profit of

$$P = -0.5(60)^2 + 60(60) - 200 = \$1600$$

The breakeven points (there are more than one in this case) can be determined by setting the profit equal to zero in Eq. (8-15) and using the quadratic formula:

$$V_B = \frac{-60 \pm \sqrt{3600 - 400}}{2(-0.5)} = 3.43 \text{ and } 116.57$$

A graphical representaton of this example is given in Figure 8-6.

It is often stated in economic theory that maximum profit occurs when marginal income revenue equals marginal cost. From an instructional standpoint, it is desirable to consider this statement in the context of this example. The income, R, in this example is

$$R = V(160 - V)$$

Consequently, the marginal income, dR/dV, is

$$\frac{dR}{dV} = 160 - 2V \tag{8-17}$$

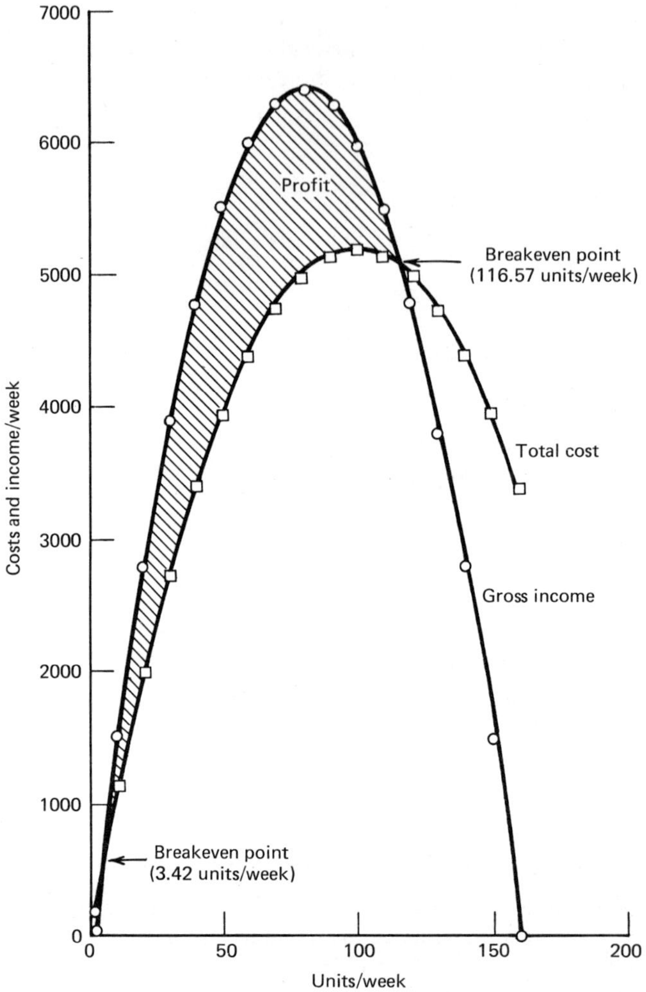

Figure 8-6.
Breakeven chart (Example 8-5).

The marginal cost, dC/dV, is

$$\frac{dC}{dV} = -V + 100 \tag{8-18}$$

Setting Eqs. (8-17) and (8-18) equal gives

$$\begin{aligned} 160 - 2V &= -V + 100 \\ V &= 60 \text{ units per week} \end{aligned} \tag{8-19}$$

which is the same result obtained previously.

This example has used a before-tax analysis. If an after-tax analysis is desired, the profit should be multiplied by the term $(1 - T)$. The breakeven points and volume for maximum profit are the same as in the before-tax analysis.

CASH FLOW AND PROFIT

At this point it is desirable to consider the relationship between cash flow and profit. Unfortunately, there are mistaken beliefs about the relationship between cash flow and profit. Generally they are not equal. One of the basic differences between profit and cash flow is due to the treatment of depreciation. Depreciation is not a cash flow. Consequently, in determining cash flow, depreciation is not deducted. However, in determining profit, depreciation is deducted since it is part of the fixed cost as pointed out earlier in this chapter.

In order to make the relationship between cash flow and profit more explicit, a comparison is made between a cash flow equation and a corresponding equation for profit. In Chapter 7, the yearly cash flow for return to equity [Eq. (7-15)] is given as

$$X_t = (G - C - I) - (G - C - I - D_t)T - K_t + L + B - P + W \qquad (8\text{-}20)$$

A corresponding formula (the accountant's point of view) for yearly profit, P_t, is

$$P_t = (G - C - I - D) - (G - C - I - D_t)T \qquad (8\text{-}21)$$

The nomenclature for Eqs. (8-20) and (8-21) is given in Chapter 7. In comparing Eqs. (8-20) and (8-21) note the following differences.

1. The book (company) depreciation, D, is subtracted in determining profit. It is not subtracted in the cash flow equation since depreciation is not a cash flow.
2. The capital investment (expenditure) is not considered in determining profit. Expenditures for capital assets are considered as an increase in the company's worth (value).
3. The recovery of the salvage value, L, does not affect profit as long as it is assumed that the actual salvage value received and the book salvage are the same.
4. The borrowed money, B, is not considered in the yearly profits. It is a liability and is offset by the increase in capital assets as a result of the capital expenditure.
5. The principal payments are not included in the yearly profit since the book depreciation, D, is based on the total capital investment and, therefore, the recovery of the borrowed money is implied in the depreciation (capital recovery) deduction.
6. Increases or decreases in working capital are not included in the yearly profits. These increases or decreases are changes in the company's assets.

References

1. Barish, Norman, N., *Economic Analysis for Engineering and Managerial Decision Making,* McGraw-Hill Book, Co., New York, N.Y. 1962 (pp 95–103).
2. Samuleson, Paul, *Economics,* 10th Ed., McGraw-Hill Book Co., New York, 1976.

Problems

8-1. A company is making a product that has the data given below:

fixed costs per year = $600,000
selling price per unit = $160
variable costs per unit = $90
equity return per year = $100,000
volume of sales per year = 20,000
tax rate = 52%

Using this data and assuming the book and tax depreciation are the same, determine the following:

(a) The before-tax profit, after-tax profit, and before-tax breakeven point using the accountant's viewpoint.
(b) Repeat part (a), using the economist's viewpoint.
(c) Plot a breakeven chart for parts (a) and (b).

8-2. Repeat parts (a) and (b) of Problem 8-1 for the case of D = $150,000 and D_t = $200,000.

8-3. A company is currently making three products A, B, and C with the data shown in the following table:

Product	Selling Price per Unit	Variable Cost per Unit	Volume of Sales per Year
A	$160	$90	20,000
B	100	70	30,000
C	50	20	50,000
Total fixed costs = $1,800,000 per year			

The company is considering eliminating Product B and replacing it with a Product D having the data shown below:

selling price per unit = $200
variable cost per unit = $100
volume of sales per year = 15,000

Should the company make this change if

(a) The fixed costs increase by $500,000 per year?
(b) The fixed costs increase by $1,000,000 per year?

8-4. A company is considering two methods, A and B, for manufacturing a particular product. Method A is a semiautomatic process that has fixed costs amounting to $260,000 a year and a variable cost of $120 per unit. Method B is a highly automated process with fixed costs of $420,000 per year and variable costs of $70 per unit. Assumng the income is the same for both methods, determine the volume of production range that is most economical for the two methods.

8-5. A company sells 15,000 units per year of a certain product. The variable cost per unit is $40 and the fixed costs are $300,000 per year. What must the selling price of the product be in order to breakeven?

8-6. Company X is presently selling 10,000 units per year of a certain product at a price of $40 per unit. The variable cost and fixed costs, respectively, for this product are

$15 per unit and $100,000 per year. Company Y has offered to purchase 5000 of these units at a price of $30 per unit and sells them under their own brand name. Company X estimates that increasing production by 5000 units will double the fixed costs. Should Company X agree to this offer?

8-7. A company is presently selling 20,000 units per year at a price of $100 per unit. The variable cost per unit is $80 and the fixed costs are $200,000 per year. The company is dissatisfied with the present yearly profit and, as a result, is considering two different alternatives. The first is to simply increase the selling price. However, this is expected to decrease sales by the amounts shown in the table:

Selling Price	Volume of Sales
$100	20,000
110	15,000
120	13,000
130	11,000
140	9,000
150	7,000

The second alternative is to increase the quality of the product and try to capture a different sector of the market. This alternative is expected to increase the variable cost to $100 per unit and the fixed costs to $300,000 per year. The selling prices and expected sales are shown below for this alternative. Which alternative should be taken?

Selling Price	Volume of Sales
130	11,000
140	10,000
150	9,000
160	8,000
170	6,000
180	5,000

8-8. For the data given below, plot a before tax multiproduct breakeven chart and determine from this plot the breakeven point in terms of gross income.

Product	Selling Price	Variable Cost	Volume/Yr	Fixed Costs/Yr
A	$160	$ 90	20,000	$ 400,000
B	100	50	30,000	600,000
C	50	20	40,000	1,000,000
D	200	100	15,000	300,000

8-9. Determine the breakeven point analytically for the data given in Problem 8-8. Also determine the following:
(a) The sales volume for each product at the breakeven point.
(b) The before-tax profit at 100% and 80% of capacity. Assume that the volumes given in Problem 8-8 are at 100% capacity.

8-10. A company is manufacturing a product that it sells for \$500 per unit. The total weekly cost of production is given by the equation

$$C_t = 0.4n^2 + 300n + 8000$$

where n is the number of units produced weekly. Determine the following:
(a) The weekly fixed costs.
(b) The variable cost if 100 units are sold in a week.
(c) The number of units for maximum profit.
(d) The maximum after tax ($T = 52\%$) profit week.
(e) The breakeven point.

8-11. A company has recorded the yearly transactions shown below for one of their products. Determine the year's after tax cash flow and profit (accountant's viewpoint) as a result of these transactions.

gross income from sales	\$1,400,000
labor costs for goods sold	300,000
overhead excluding depreciation and debt interest	100,000
expenditures for materials. Of this amount \$230,000 was used for the goods sold.	400,000
company depreciation	250,000
tax depreciation	300,000
tax rate	52%
debt interest	50,000
expenditures for capital equipment	500,000

8-12. Using the data given in Problem 7-14, generate the incremental changes in profit that would occur if this project is undertaken. Use the accountant's point of view and assume that the book and tax depreciation are the same.

CHAPTER 9
CAPITAL BUDGETING

The primary concern in the previous chapters has been the determination of the acceptability of a single project. This chapter is concerned with the capital budgeting problem (sometimes referred to as the capital rationing problem or the ranking problem). The capital budgeting problem is concerned with the choice of a group of projects from a larger group of *acceptable* projects as a result of constraints prohibiting the choice of all the projects. These constraints may result from restrictions in allowable capital expenditures, limited number of workers or materials, and mutually exclusiveness of the projects (one project precludes the choice of others). It must be understood that, in this chapter, the *deterministic* capital budgeting problem is being considered. That is, risk is not a consideration in the choice of projects.

To obtain a solution for the capital budgeting problem, a correct basis for ranking projects must be established. Many different ranking criteria and solutions to the capital budgeting problem are proposed in the literature. Unfortunately, however, some of these proposed procedures can lead to incorrect decisions. Some of these incorrect procedures are considered in a later section of this chapter.

The correct ranking criterion and solution to the capital budgeting problem is based on the financial objective of maximizing the wealth of the stockholders, which implies, from the standpoint of making capital budgeting decisions, the following:

1. Every increment of capital expenditure must justify itself.
2. An acceptable project today is better than the speculation that a better project will become available in the future. That is, "bird in hand is worth two in the bush."

Initially, it might appear that net present value or internal rate of return

Table 9-1
NET PRESENT VALUES AND INTERNAL RATES OF RETURN

End of Year	Project 1	Project 2	Project 3	Project 4
0	$−10,000	$−15,000	$−18,000	$−25,000
1	2,800	3,600	4,300	5,400
2	2,800	3,600	4,300	5,400
.	.	.	.	.
.	.	.	.	.
.	.	.	.	.
10	2,800	3,600	4,300	5,400
IRR	25.0%	20.2%	20.1%	17.4%
NPV (at 12%)	5,821	5,341	6,296	5,511

would rank a group of projects the same, since it was stated (in Chapter 5) that both net present value and internal rate of return give the same decision regarding the acceptability of a single project (with traditional cash flows). This is not the case as shown in Table 9-1. If the projects in Table 9-1 are ranked on the basis of IRR, the result is 1-2-3-4 while on the basis of NPV the result is 3-1-4-2, which clearly shows that ranking by net present value and internal rate of return are not necessarily the same. The correct ranking of these projects can be obtained and, at the same time, the basis of the solution to the capital budgeting problem can be presented by considering the special case of mutually exclusive projects.

MUTUALLY EXCLUSIVE PROJECTS

Mutually exclusive projects are a special case of the capital budgeting problem since the choice of one project precludes the choice of any other projects; that is, only one project is chosen from a group of acceptable projects. One example of mutually exclusive projects is the situation where a company is considering building a single plant but several locations are under consideration with economic advantages and disadvantages for each location and, consequently, different cash flows. Which location should be chosen? Another way of putting this question is, "Which location has the greater economic advantage?"

EXAMPLE 9-1

Suppose the cash flows given in Table 9-1 are the estimates for four different plant locations (1,2,3,4). Which location should be chosen using a minimum attractive rate of return (MARR) of 12%?

The first step in solving this problem is to determine if all the projects (plant locations) have an internal rate of return greater than MARR. Table 9-1 shows that all projects have a return greater than 12%. Therefore, all projects are initially acceptable. If a project has a return less than 12%, it should be eliminated from further consideration.

The next step is to evaluate the rate of return for each increment of investment in order to determine if each additional increment of capital justifies its expenditure. Comparing project 1 to doing nothing gives a return of 25% (Table 9-1). Or, the decision to do nothing is ruled out and the decision, at this point is to do project 1. A comparison between projects 1 and 2 shows

$$2/1 : PAi,10 = \frac{15000 - 10000}{3600 - 2800} = 6.2500$$
$$i = 9.6\%$$

Which indicates that project 2 should be eliminated since the return on the incremental capital investment of $5000 is less than the required return of 12% (MARR). The next comparison is between projects 1 and 3:

$$3/1 : PAi,10 = \frac{18000 - 10000}{4300 - 2800} = 5.333$$
$$i = 13.8\%$$

This comparison indicates that project 1 is ruled out and project 3 is the choice at this point since the return on the increment of $8000 is greater than 12%. The final comparison is between projects 3 and 4

$$4/3 : PAi,10 = \frac{25000 - 18000}{5400 - 4300} = 6.3636$$
$$i = 9.2\%$$

and rules out project 4. There, the final choice is project 3 (plant location 3).

A Comment: It is important to note that the analysis presented in Example 9-1 gives a different choice of project than if the project with the highest rate of return is chosen (project 1). It is sometimes argued (erroneously) that the project with the highest rate of return should be chosen and money could be "saved." The problem with this argument is that these "better investments" might *not* occur. Also, if a "better investment" does occur in the future, then, theoretically at least, it will be undertaken. Otherwise, the objective of maximizing the wealth of the stockholders will not be accomplished.

Of course, the choice of a project in the mutually exclusive case is dependent upon the minimum attractive rate of return as shown in Figure 9-1. This figure also shows why net present value and internal rate of return can give conflicting rankings for projects.

Also note that the method used in Example 9-1 gives the project with the highest net present value (see Table 9-1). This result can be generalized. That is, choosing the project with the highest net present value will give the same result as the incremental approach used in Example 9-1. Since the maximization of net present value is consistent with the objective of the maximizing the stockholder's wealth, either approach is, therefore, correct. This is why it is sometimes stated in the literature that the rate of return and net present value methods give the same results in capital budgeting problems. This is true, *provided* the rate of return method is correctly used; namely, an *incremental* rate of return method. Unfortunately, some have interpreted this statement incorrectly and simply choose the project with the higher rate of return. This is, of course, incorrect.

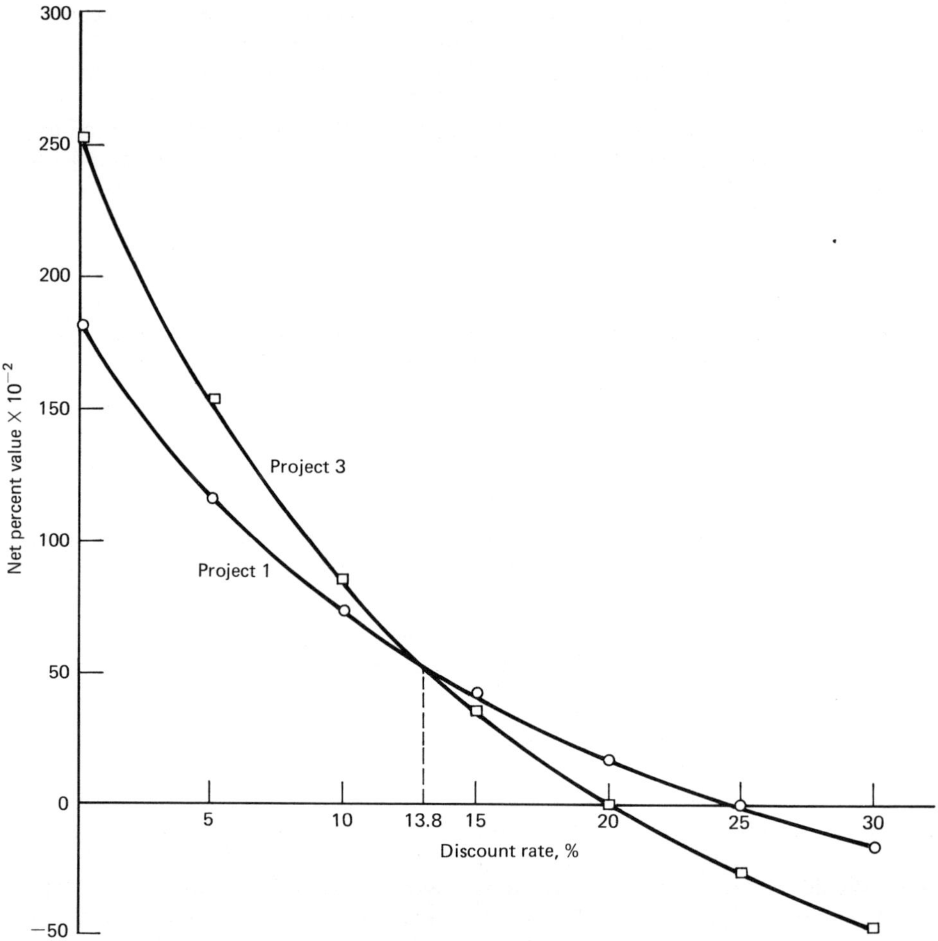

Figure 9-1.
Net present value profiles for projects 1 and 3 (Table 9-1).

It must be clearly understood that in analyzing capital budgeting problems the cash flows must be calculated using the same definition of cash flow and the corresponding minmum attractive rate of return as discussed in Chapter 7. Otherwise, completely erroneous decisions are possible.

Different Lives

Questions frequently arise about the comparison of mutually exclusive alternatives that have different lives. Consequently, it is the purpose of this section to develop a basis on which mutually exclusive alternatives with different lives may be compared. Three assumptions provide a basis for the comparison of mutually exclusive alternatives with different lives. They are:

1. Assume that money generated (cash flows) by each alternative will be invested by the company in other assets that will earn the minimum attractive rate of return for a period of time equal to the life of the longest alternative.
2. Assume that the alternatives with exactly the same characteristics will recycle for a period of time equal to the least common multiple of the alternative lives.
3. Make specific assumptions (estimates) about future investment opportunities for a period of time equal to the life of the longest alternative.

The correct assumption to use, of course, is dependent upon which one is believed to be the most accurate representative of the future. From a practical standpoint, it is better to differentiate between two cases when making a choice of an assumption; that is, (1) the comparison of mutually exclusive alternatives with primarily positive cash flows (earnings) and (2) mutually exclusive alternatives with primarily negative cash flows (cost comparisons and replacement comparisons). This statement is based on the acceptability of the assumptions for the two cases. That is, certain assumptions are more plausible than others for each case. For example, suppose the cash flows shown in Table 9-2 are for two

Table 9-2
CASH FLOWS FOR INVESTMENTS WITH DIFFERENT LIVES

End of Year	Project 1	Project 2
0	$-100,000	$-150,000
1	25,000	30,000
2	25,000	30,000
3	25,000	30,000
4	25,000	30,000
5	25,000	30,000
6	25,000	30,000
7	25,000	30,000
8	25,000	30,000
9	25,000	30,000
10	25,000	30,000
11	—	30,000
12	—	30,000
13	—	30,000
14	—	30,000
15	—	30,000
IRR	21.5%	18.6%

mutually exclusive projects that have to do with alternative ways to expand a plant's capacity. Based on the internal rate of return, project 1 is best. However, if the net present value for each alternative is calculated using 12% as the minimum attractive rate of return, the results are

$$\begin{aligned} NPV_1 &= 25{,}000(PA\,12{,}10) - 100{,}000 \\ &= 25{,}000(5.6502) - 100{,}000 = \$41{,}255 \\ NPV_2 &= 30{,}000(PA\,12{,}15) - 150{,}000 \\ &= 30{,}000(6.8109) - 150{,}000 = \$54{,}327 \end{aligned}$$

indicating that project 2 is best. It may appear that in calculating the net present value the difference in lives of the two projects has been ignored. This is not the case. What is implied in the calculations of these net present values is the use of assumption 1; that is, the cash flows will be reinvested at the minimum attractive rate of return (12%) for 15 years (life of the longest alternative). Mathematically, this would result in the amount

$$25{,}000\,(FA\,12{,}10)(FP\,12{,}5)$$

Therefore, the net present value contribution from project 1 for the 15-year period under assumption 1 is

$$NPV_1 = 25{,}000\,(FA\,12{,}10)(FP\,12{,}5)(PF\,12{,}15) - 100{,}000$$

If the interest factors in this expression are considered, an interesting fact emerges; that is,

$$\begin{aligned} (FA\,12{,}10)(FP\,12{,}5)(PF\,12{,}15) &= \left[\frac{(1 + 0.12)^{10} - 1}{0.12}\right][(1 + 0.12)^5]\left[\frac{1}{(1 + 0.12)^{15}}\right] \\ &= \frac{(1 + 0.12)^{10} - 1}{(0.12)(1 + 0.12)^{10}} = (PA\,12{,}10) \end{aligned}$$

Consequently, the net present value would be the same, using assumption 1, as the net present value previously calculated. Thus, the choice of an alternative is project 2, under assumption 1, since it maximizes net present value.

Assumption 2 implies that a comparison of the two projects in Table 9-2 is based on a period of 30 years (the least common multiple of 10 and 15 years). Furthermore, it implies that there are three cycles of project 1 and two cycles of project 2 as shown in Table 9-3. The net present value for both alternatives under assumption 2 are

$$\begin{aligned} NPV_1 &= 25{,}000\,(PA\,12{,}30) - 100{,}000\,[(PF\,12{,}10) + (PF\,12{,}20)] - 100{,}000 \\ &= 25{,}000(8.0552) - 100{,}000\,[0.3220 + 0.1037] - 100{,}000 \\ &= \$58{,}180 \\ NPV_2 &= 30{,}000(PA\,12{,}30) - 150{,}000(PF\,12{,}15) - 150{,}000 \\ &= 30{,}000(8.0552) - 150{,}000\,(0.1827) - 150{,}000 \\ &= \$64{,}251 \end{aligned}$$

Table 9-3
PROJECT CYCLES

End of Year	Project 1	Project 2
0	−\$100,000	−\$150,000
1	25,000	30,000
2	25,000	30,000
.	.	.
.	.	.
.	.	.
10	25,000 − 100,000	30,000
11	25,000	30,000
.	.	.
.	.	.
.	.	.
15	25,000	30,000 − 150,000
16	25,000	30,000
.	.	.
.	.	.
.	.	.
20	25,000 − 100,000	30,000
21	25,000	30,000
.	.	.
.	.	.
.	.	.
30	25,000	30,000

which gives the same result as assumption one—that project 2 is the choice. *However, this result cannot be generalized* because it is possible to give examples where assumptions 1 and 2 will give conflicting decisions. For example, if the net present value is calculated (MARR = 12%) for the cash flows shown in Table 9-4 under assumption one the results are

$$\begin{aligned} NPV_1 &= 10{,}000(PA\,12{,}3) - 15{,}000 \\ &= 10{,}000(2.4018) - 15{,}000 = \$9018 \\ NPV_2 &= 8000(PA\,12{,}6) - 20{,}000 \\ &= 8000(4.1114) - 20{,}000 = \$12{,}891 \end{aligned}$$

While the net present values under assumption 2 are

$$\begin{aligned} NPV_1 &= 10{,}000(PA\,12{,}6) - 15{,}000 - 15{,}000(PF\,12{,}3) \\ &= 10{,}000(4.1114) - 15{,}000 - 15{,}000(0.7118) = \$15{,}437 \\ NPV_2 &= \$12{,}891 \end{aligned}$$

and clearly shows the conflict between assumptions 1 and 2.

Table 9-4
CASH FLOWS

End of Year	Project 1	Project 2
0	−$15,000	−$20,000
1	10,000	8,000
2	10,000	8,000
3	10,000	8,000
4	—	8,000
5	—	8,000
6	—	8,000

Assumption 3 is, perhaps, the best of the three assumptions. However, although it is easy in theory, assumption 3 is very difficult in practice because it necessitates estimating investment opportunities that will occur far into the future. For example in Table 9-2, it would be necessary to estimate future investment opportunities out to the least common multiple of the lives of the two projects (30 years in this case), obviously a rather difficult task in practice and for this reason will not be discussed further.

By far, the assumption most often used (or implied) is assumption 2 because it is quite plausible when making cost and replacement comparisons (discussed in later chapters) and relatively easy. However, when it comes to projects (investments) that generate different earnings (positive cash flows), the recycling of alternatives involved in assumption 2 is not as plausible. In fact, it is most difficult to "sell" this assumption in practice. It is much easier to "sell" assumption one in the comparison of alternatives with earnings on the basis that "investing money in assets that earn the minimum attractive rate of return" can have a meaning other than investing in more capital assets. That is, the cash flows can be "invested" in retiring debt obligations and/or paying dividends to stockholders; thus, "investing" money at the minimum attractive rate of return. Consequently, this text employs assumption 1 in discussing alternatives with different earnings and assumption 2 in cost and replacement problems. This may seem inconsistent. However, in the final analysis, the correct assumption in the analysis of capital expenditures is the one that is believed to be the most accurate representation of future investment opportunities. Consequently, this text takes the point of view that this difference in assumptions for the two cases is justifiable on the basis that assumption 1 is believed to be a closer representation of the future in the comparison of alternatives with different earnings while assumption 2 provides a better representation of future events in cost and replacement comparisons at least by default. That is, there are no plausible arguments to the contrary.

THE BASIC CAPITAL BUDGETING PROBLEM

As previously stated, the solution to the mutually exclusive case provides the basis on which to formulate a solution to the capital budgeting problem. This solution is demonstrated through the use of the following example.

EXAMPLE 9-2

Suppose the projects given in Table 9-1 are not mutually exclusive and it is decided that no more than $50,000 is to be spent on new investments. Which projects should be chosen? One approach that might answer this question is to list all the possible project combinations that do not violate the budget restriction of $50,000. This is shown in Table 9-5. For the sake of completeness, all possible combinations are included in Table 9-5. However, combinations 1-3-4, 2-3-4, and 1-2-3-4 are not possible solutions because their required capital expenditures exceed $50,000.

Now, since all combinations are listed, the problem can be treated as the mutually exclusive case. That is, the choice of one combination of projects precludes the choice of any other combination of projects. Consequently, the solution is to simply choose the combination that does not violate the budget restriction and at the same time gives the highest net present value; in this example, combination 1-2-3. Although this approach gives correct answers, it does have a decided computational disadvantage, that is, the generation of all possible combinations. The number of possible combinations, excluding the one of doing nothing, is $2^n - 1$, where n is the number of projects. For example, if there are 10 projects, the number of possible combinations is 1023.

Another approach that circumvents the computational disadvantage of listing all

Table 9-5

COMBINATION OF PROJECTS IN TABLE 9-1

Project Combination	Capital Expenditure	Annual Cash Flow	Net Present Value
1	$10,000	$ 2,800	$ 5,821
2	15,000	3,600	5,341
3	18,000	4,300	6,296
4	25,000	5,400	5,511
1–2	25,000	6,400	11,162
1–3	28,000	7,100	12,117
1–4	35,000	8,200	11,332
2–3	33,000	7,900	11,637
2–4	40,000	9,000	10,852
3–4	43,000	9,700	11,807
1–2–3	43,000	10,700	17,458
1–2–4	50,000	11,800	16,673
1–3–4	53,000	12,500	17,628
2–3–4	58,000	13,300	17,148
1–2–3–4	68,000	16,100	22,969

possible combinations is to formulate the problem in terms of a mathematical programming problem, and for this example is

$$\text{maximize:} \quad z = 5821X_1 + 5341X_2 + 6296X_3 + 5511X_4 \tag{9-1}$$

subject to:

$$10{,}000X_1 + 15{,}000X_2 + 18{,}000X_3 + 25{,}000X_4 \leq 50{,}000 \tag{9-2}$$

$$X_i = 0 \text{ or } 1 \text{ for all } i \tag{9-3}$$

Since there are a small number of projects involved, Eqs. (9-1) through (9-3) can be solved by inspection: namely, $X_1 = 1$, $X_2 = 1$, $X_3 = 1$, and $X_4 = 0$. This is the same solution as obtained previously. This approach can be generalized. That is, the generalized solution to the basic capital budgeting problem is

$$\text{maximize:} \quad z = \sum_{i=1}^{n} (\text{NPV})_i X_i \tag{9-4}$$

subject to:

$$\sum_{i=1}^{n} P_i X_i \leq B \tag{9-5}$$

$$X_i = 0 \text{ or } 1 \text{ for all } i \tag{9-6}$$

where

$(\text{NPV})_i$ = net present value for project i
P_i = capital expenditure for project i
B = budget restriction
n = number of projects
X_i = decision variable for project i

The solution to Eqs. (9-4) through (9-6) is possible by the use of specialized algorithms (and a computer) in situations where there are a large number of projects. These algorithms are beyond the scope of this text and will not be discussed further. If one wishes to pursue these algorithms further, consult the many texts on mathematical programming and optimization techniques.

One final comment is needed in regard to Example 9-2. That is, it should be noted that imposing a restriction on the expenditure on capital in effect does not allow the full maximization of net present value and, consequently, does not maximize the wealth of the stockholders, for, if there is no capital expenditure restriction, all four projects in Example 9-2 should be undertaken (theoretically, at least) with a resulting net present value of \$22,969.

SOME EXTENSIONS TO FORMULATING THE CAPITAL BUDGETING PROBLEM

In this section some extensions to the formulation of the capital budgeting problem are presented. These extensions provide greater flexibility in the formulation of capital budgeting problems.

Multiperiod Budget Restrictions

It is not unusual to find that projects require capital expenditures at times other than the present ($t = 0$) and to have budget restrictions established for future years. When this is the case, the budget restriction, Eq. (9-5) is replaced by budget restrictions for each year, that is, a set, j, of budget restrictions. Expressed mathematically, this set is

$$\sum_j K_{jt} X_j \leq B_t \tag{9-7}$$

As an example, consider the data given in Table 9-6 and suppose the budget restrictions are $B_0 = \$70{,}000$, $B_1 = \$30{,}000$, and $B_2 = \$25{,}000$. The mathematical programming format is

$$\text{maximize:} \quad z = 75{,}000X_1 + 6000X_2 + 8000X_3 + 16{,}500X_4 + 18{,}500X_5 + 21{,}000X_6 + 9000X_7 + 2000X_8 \tag{9-8}$$

subject to:

$$20{,}000X_1 + 10{,}000X_2 + 5000X_3 + 15{,}000X_4 + 20{,}000X_5 + 25{,}000X_6 + 8000X_7 + 6000X_8 \leq 70{,}000 \tag{9-10}$$

$$10{,}000X_1 + 4000X_3 + 10{,}000X_4 + 15{,}000X_5 + 12{,}000X_6 + 3000X_8 \leq 30{,}000 \tag{9-11}$$

$$5000X_1 + 3000X_2 + 5000X_4 + 8000X_5 + 7000X_6 + 4000X_7 \leq 25{,}000 \tag{9-12}$$

$$X_i = 0 \text{ or } 1 \text{ for all } i$$

It is sometimes suggested that the K_{jt} and B_t values should be discounted to time zero. As long as capital funds are not allowed to shift from one year to following years (next discussion), discounting is not necessary since the discount factor, $(1 + i)^{-t}$ would be applied to both sides of the budget restrictions and, consequently, the budget restrictions are essentially unchanged.

Table 9-6

MULTIPERIOD BUDGET RESTRICTIONS

Project Number	Capital Expenditure $t = 0$	Capital Expenditure $t = 1$	Capital Expenditure $t = 2$	Net Present Value
1	\$20,000	\$10,000	\$5000	\$ 7,500
2	10,000	—	3000	6,000
3	5,000	4,000	—	8,000
4	15,000	10,000	5000	16,500
5	20,000	15,000	8000	18,500
6	25,000	12,000	7000	21,000
7	8,000	—	4000	9,000
8	6,000	3,000	—	2,000

Shifting Funds between Years

If unused funds in a particular year are allowed to be used in future time periods, the objective equation, Eq. (9-8), is the same but Eq. (9-7) must be changed to the following:

$$\sum_j \sum_{t=0}^{n} K_{jt} X_j \leq \sum_{t=0}^{n} B_t \tag{9-14}$$

Using Table 9-6 as the source of data, expanding Eq. (9-14) gives

$$\begin{aligned} &20{,}000X_1 + 10{,}000X_2 + 5000X_3 + 15{,}000X_4 \\ &+ 20{,}000X_5 + 25{,}000X_6 + 8000X_7 + 6000X_8 \leq 70{,}000 \qquad (9\text{-}15) \\ &30{,}000X_1 + 10{,}000X_2 + 9000X_3 + 25{,}000X_4 \\ &+ 35{,}000X_5 + 37{,}000X_6 + 8000X_7 + 9000X_8 \leq 100{,}000 \qquad (9\text{-}16) \\ &35{,}000X_1 + 13{,}000X_2 + 9000X_3 + 30{,}000X_4 \\ &+ 43{,}000X_5 + 44{,}000X_6 + 12{,}000X_7 + 9000X_8 \leq 125{,}000 \qquad (9\text{-}17) \end{aligned}$$

The reasoning behind Eq. (9-16) can be seen by considering that the amount of money remaining from the budget, U_0, at $t = 0$ is

$$\begin{aligned} U_0 = 70{,}000 - 20{,}000X_1 - 10{,}000X_2 - 5000X_3 - 15{,}000X_4 \\ - 20{,}000X_5 - 25{,}000X_6 - 8000X_7 - 6000X_8 \end{aligned} \tag{9-18}$$

Therefore, adding U_0 to the right-hand side of Eq. (9-11) results, after rearranging, in Eq. (9-16). The reasoning for (9-17) is basically the same: the amount remaining, U_1, from the budget after $t = 0$ and $t = 1$ is

$$\begin{aligned} U_1 = 100{,}000 - 30{,}000X_1 - 10{,}000X_2\, 9000X_3\, 25{,}000X_4 \\ - 35{,}000X_5 - 37{,}000X_6 - 8000X_7 - 9000X_8 \end{aligned} \tag{9-19}$$

Adding U_1 to the right-hand side of Eq. (9-12) gives Eq. (9-17). Now in Eqs. (9-15) through (9-17), the K_{jt} and B_t values are not discounted. If it is assumed that funds carried over can be invested at the minimum attractive rate of return, Eqs. (9-16) and (9-17) must be modified while Eq. (9-15) remains the same. Assuming that MARR = 12%, Eqs. (9-16) and (9-17) become

$$\begin{aligned} &[20{,}000 + 10{,}000\,(PF\ 12{,}1)]\ X_1 + 10{,}000X_2 + [5000 + 4000\,(PF\ 12{,}1)]\ X_3 \\ &+ [15{,}000 + 10{,}000\,(PF\ 12{,}1)]\ X_4 + [20{,}000 + 15{,}000\,(PF\ 12{,}1)]\ X_5 \\ &+ [25{,}000 + 12{,}000\,(PF\ 12{,}1)]\ X_6 + 8000X_7 \\ &+ [6000 + 3000\,(PF\ 12{,}1)]\ X_8 \leq 70{,}000 + 30{,}000\,(PF12{,}1) \qquad (9\text{-}20) \\ &[20{,}000 + 10{,}000\,(PF\ 12{,}1) + 5000\,(PF\ 12{,}2)]\ X_1 = [10{,}000 + 3000\,(PF\ 12{,}2)]\ X_2 \\ &+ [5000 + 4000\,(PF\ 12{,}1)]\ X_3 + [15{,}000 + 10{,}000\,(PF\ 12{,}1) + 5000\,(PF\ 12{,}2)]\ X_4 \\ &+ [20{,}000 + 15{,}000\,(PF\ 12{,}1) + 8000\,(PF\ 12{,}2)]\ X_5 \\ &+ [25{,}000 + 12{,}000\,(PF\ 12{,}1) + 7000\,(PF\ 12{,}2)]\ X_6 \\ &+ [8000 + 4000\,(PF\ 12{,}2)]\ X_7 \\ &+ [6000 + 3000\,(PF\ 12{,}1)]\ X_8 \leq 70{,}000 + 30{,}000\,(PF\ 12{,}1) + 25{,}000\,(PF\ 12{,}2) \\ &\qquad (9\text{-}21) \end{aligned}$$

Equations (9-20) and (9-21) are now simplified by making the appropriate numerical substitutions for the discount factors and combining values. The budget restriction for $t = 0$ is still the same as Eq. (9-10).

Mutually Exclusive Projects

If a subset, j, of projects are mutually exclusive, this situation can be handled by adding the constraint

$$\sum_j X_j \leq 1 \qquad (9\text{-}22)$$

to the formulation of the problem. Suppose, for example, that in Table 9-5, projects 2, 4, and 5 are mutually exclusive. Using Eq. (9-22), the constraint

$$X_2 + X_4 + X_5 \leq 1.0 \qquad (9\text{-}23)$$

would be added to the formulation of the problem.

There might be a tendency to simply pick the project with the highest net present value from the subset of mutually exclusive projects and include only it in the formulation of the overall capital budgeting problem. This is incorrect because it does not take into account the effect of the required capital expenditure of the chosen mutually exclusive project on the choice of other projects. That is, it is possible that a combination of one of the mutually exclusive projects with a lower net present value and lower required capital expenditure would allow a choice of another project that, in combination, with the mutually exclusive project, would give a higher total net present value.

Interdependent Projects

There are situations that can occur when the adoption of, say, project m is contingent on the adoption of another project n. However, project n can be done without m. Also, neither project m nor n are required to be adopted. This situation can be included in the formulation of the capital budgeting problem by adding the following inequality.

$$X_m \leq X_n \qquad (9\text{-}24)$$

Labor and Material Restrictions

Labor and material restrictions can be included in the formulation of the capital budgeting problem by adding constraints that are similar to the capital expenditure restrictions, that is,

$$\sum_j L_{jt} X_j \leq m_t \qquad (9\text{-}25)$$

where

L_{jt} = required number of man-hours (or amount of materials of input required for project j in year t

m_t = available man-hours (or materials) in year t

SOME INCORRECT PROPOSALS

Methods for solving the capital budgeting problem and ranking projects, other than the one (net present value) presented in this text, are often proposed in the literature. Unfortunately many of these proposals are incorrect if the point of view is taken, as in this text and most financial management texts, that maximizing stockholders' wealth is a primary long-range objective of a company. It is felt that a review of these proposals is beneficial at this point so that, when encountered, they are recognized as being incorrect from the standpoint of maximizing the stockholders' wealth.

Proposal 1: One proposal is to list the projects in descending order by rate of return and keep a cumulative total of the capital required. This procedure is shown in Table 9-7 using the data given in Table 9-1. Once the cumulative total of the required capital exceeds the budget, this determines the projects that are selected. For example in Table 9-7, if the budget is \$25,000 or greater but less than \$43,000 then projects 1 and 2 are selected. If the budget is \$43,000 or greater but less than \$68,000 then projects 1, 2, and 3 are selected. At times (but there is no way of predicting when) this proposal can give correct answers as in the case of a budget of \$50,000. This proposal and the mathematical programming format both give the same choice of projects: projects 1, 2, and 3. However, if the budget is reduced to \$30,000 then this proposal indicates projects 1 and 2 are selected but the mathematical programming format indicates 1 and 3 are selected. This latter answer is, of course, correct since it maximizes net present value. Consequently, this proposal can lead to incorrect decisions and should not be used.

Proposal 2: Because of the merits of net present value, another proposal that is sometimes advanced is to proceed as in proposal 1 only listing the projects in descending order of net present value as shown in Table 9-8. The selection of projects is then done in the same manner as described in proposal 1. On first appearances, it might seem that this proposal is a good one. Unfortunately, it can lead to incorrect decisions. For example, if a budget of \$50,000 is established, Table 9-8 suggests projects 1 and 3 are selected. This solution is, of course, incorrect because the mathematical programming formulation gave a selection of

Table 9-7
PROPOSAL 1

Project Number	Rate of Return	Cumulative Capital Required
1	25.0%	\$10,000
2	20.2	25,000
3	20.1	43,000
4	17.4	68,000

Table 9-8
PROPOSAL 2

Project Number	Net Present Value	Cumulative Capital Required
3	\$6296	\$18,000
1	5821	28,000
4	5511	53,000
2	5341	68,000

projects 1, 2, and 3 and a higher total net present value. The reason for this discrepancy is due to *discontinuities* that exist in the capital budgeting problem as a result of the requirement that a project is either selected or not selected; that is, a portion of a project is not allowed as suggested in the mathematical programming format by Eq. (9-6). This proposal does not account for the possibility of discontinuities and consequently can give incorrect decisions.

Various indices of economic worth are sometimes suggested to replace the present value in the mathematical programming format of the capital budgeting problem or for ranking projects. Some of these methods are considered below:

Proposal 3: One index that is sometimes proposed is the "benefit cost" ratio. This index is defined mathematically as

$$\frac{\sum_t X_t(1 + k)^{-t}}{\sum_t K_t(1 + k)^{-t}} \tag{9-26}$$

where

X_t = net cash flow in year t when the flow is positive
K_t = net cash flow in year t when the flow is negative
k = minimum acceptable rate of return (MARR)

The benefit cost ratio for the projects in Table 9-1 are given in Table 9-9. These ratios can lead to incorrect decisions in capital budgeting problems. For example, in the mutually exclusive case, the choice of project 1 might be suggested since it has the highest benefit-cost ratio. However, it has already been shown that the correct choice is project 3 (the one with the highest net present value). The point that is being overlooked by choosing project 1 is basically the same as previously discussed in regard to choosing a project with the highest internal rate of return; that is, an *incremental* approach must be taken to insure maximization of net present value.

The use of the benefit cost ratio in the basic capital budgeting problem can also lead to incorrect decision. If the net present values in Eq. (9-1) are replaced with the appropriate benefit cost ratios, incorrect decisions are also possible.

Table 9-9
BENEFIT COST RATIO

Project	Benefit Cost Ratio
1	$\frac{2800\,(PA\ 12{,}10)}{10{,}000} = 1.58$
2	$\frac{3600\,(PA\ 12{,}10)}{15{,}000} = 1.36$
3	$\frac{4300\,(PA\ 12{,}10)}{18{,}000} = 1.35$
4	$\frac{5400\,(PA\ 12{,}10)}{25{,}000} = 1.22$

One needs only to make these suggested substitutions in Eq. (9-1) and then solve Eqs. (9-1) through (9-3) for various budgets to see that net present value is not necessarily maximized.

Proposal 4: Another index, closely related to the benefit cost ratio, that is sometimes proposed is the premium worth percentage. Mathematically, this index is defined as

$$\frac{\text{NPV}}{K}(100) \tag{9-27}$$

and are calculated, for the projects in Table 9-1, in Table 9-10. This index has the same problems as the benefit cost ratio. That is, it can give incorrect solutions to the capital budgeting problem.

There are many more indices that are proposed (and unfortunately used) for ranking projects. For example, payback periods, proceeds per dollar of outlay, the ratio of income (profit) to investment or book value, and so on. The problems with these indices are they do not, usually, take into account the time value of money nor do they necessarily maximize net present value.

Table 9-10
PREMIUM WORTH PERCENTAGE

Project	Premium Worth Percentage
1	$\frac{2800\,(PA\ 12{,}10) - 10{,}000}{10{,}000}(100) = 58\%$
2	$\frac{3600\,(PA\ 12{,}10) - 15{,}000}{15{,}000}(100) = 36$
3	$\frac{4300\,(PA\ 12{,}10) - 18{,}000}{18{,}000}(100) = 35$
4	$\frac{5400\,(PA\ 12{,}10) - 25{,}000}{25{,}000}(100) = 22$

A DIGRESSION ON THE MULTIPLE RATE PROBLEM

At this point, it is appropriate to consider a solution that is sometimes proposed for the multiple rate problem discussed in Chapter 5. This proposed solution is based on assuming some "reinvestment rate" and converting the cash flows to a series of traditional cash flows (negative cash flows followed by positive cash flows) and then solving for a "rate of return." This text does not subscribe to this procedure for several reasons. First, the reinvestment rate used in these procedures is often implied to be an arbitrary rate or some rate that is believed to represent "reinvestment opportunities." The theoretically correct rate for reinvestment is the minimum attractive rate of return (cost of capital) since to do otherwise is not consistent with the objective of maximizing the stockholder's wealth. Consequently, and a second reason, the determination of a rate of return is not needed from the standpoint of evaluating the desirability of a series of cash flows. As mentioned in Chapter 5, the desirability of a series of cash flows can be determined by merely calculating the net present value of the cash flows using the minimum attractive rate of return as the discount rate.

As an example of the proposed solution procedure, consider the cash flows shown in Table 9-11. One method is to discount the negative cash flows in years 7 and 8 back and combine with the positive cash flows until the result is positive. In other words, using the cash flows in Table 9-11, the negative cash flows in years 7 and 8 are discounted back to year 6 and combined. If this result is positive this process stops. If the result is negative, the result is discounted and combined with the cash flow in year 5. This result is continued until the combined result is positive. Suppose, for example, that the resulting value is positive in year 5. That is, Eq. (9-28) is positive.

$$A_5' = A_5 + A_6(PFi_e,1) - A_7(PFi_e,2) - A_8(PFi_e,3) \tag{9-28}$$

where i_e is the reinvestment rate and A_5' is the adjusted cash flow in year 5. The rate of return is then determined using the cash flows shown in Table 9-12.

Table 9-11

End of Year	Cash Flows
0	$-A_0$
1	$-A_1$
2	$+A_2$
3	$+A_3$
4	$+A_4$
5	$+A_5$
6	$+A_6$
7	$-A_7$
8	$-A_8$

Table 9-12

End of Year	Cash Flows
0	$-A_0$
1	$-A_1$
2	$+A_2$
3	$+A_3$
4	$+A_4$
5	$+A_5'$

It is important to note that if the net present value of the cash flows in Table 9-12 is determined, the result is

$$\text{NPV} = -A_0 - A_1(PFk,1) + A_2(PFk,2) + A_3(PFk,3) + A_4(PFk,4) + A_5'(PFk,5) \tag{9-29}$$

where k is the minimum attractive rate of return. Now, substituting Eq. (9-28) into Eq. (9-29), the result is

$$\text{NPV} = -A_0 - A_1(PFk,1) + A_2(PFk,2) + A_3(PFk,3) + A_4(PFk,4) + [A_5 + A_6(PFi_e,1) - A_7(PFi_e,2) - A_8(PFi_e,3)](PFk,5) \tag{9-30}$$

Assuming $i_e = k$, the net present value is the same as given in Chapter 5, that is,

$$\text{NPV} = \sum_{n=0}^{T} \frac{X_n}{(1+k)^n} \tag{9-31}$$

Therefore, why use a reinvestment rate? A direct evaluation of the net present value gives the same result as the use of a reinvestment rate if this reinvestment rate is the minimum attractive rate of return (the correct choice).

Also, why begin with the negative cash flows in the later years? Why not carry the negative cash flows forward until a position combination is obtained? There are no theoretical answers to these questions. The choice of the using the cash flows in years 7 and 8 in this example as opposed to years 0 and 1 is strictly arbitrary.

There are no guarantees that the assumption of a reinvestment rate is going to provide a unique positive rate of return. For example, consider the cash flows proposed in Solomon's (see Chapter 5) often-cited example:

$$X_0 = -1600$$
$$X_1 = 10{,}000$$
$$X_2 = -10{,}000$$

If a reinvestment rate, i_e, is assumed, one method of determining a rate of return, i, is to solve the equation

$$1600 = [10{,}000 - 10{,}000(PFi_e{,}1)](PFi{,}1) \qquad (9\text{-}32)$$

Now, if i_e is assumed (or estimated) to be less than 19.05%, a unique positive interest rate cannot be determined since the value in the brackets in Eq. (9-32) is less than 1600 and implies $(PFi,1)$ is greater than 1. Which, of course, is an impossible situation from a practical standpoint. By way of showing the arbitrary nature of the multiple-return problem, suppose the negative cash flow in year zero (−1600) is carried forward at some reinvestment rate. In this case, the rate of return is given by

$$-1600(FPi_e{,}1) + 10{,}000 = 10{,}000(PFi{,}1) \qquad (9\text{-}33)$$

Equation (9-33) indicates that a solution is possible for any positive value of i_e less than 525% since a combination of the terms on the left-hand side of the equality sign in Eq. (9-33) is less than 10,000; indicating a $(PFi,1)$ value less than 1, which is acceptable.

A final comment: once the cash flows are adjusted, in some manner, to a series of traditional cash flows and a rate of return determined, what is the *financial* meaning of this rate of return? What is it compared against to determine the desirability of the original cash flows? The adjusted cash flows are pseudo cash flows and the rate of return is, consequently, a pseudo return. A comparison of this rate of return with a minimum attractive rate of return has very little meaning. From a practical standpoint, why not use the net present value? It is simple, direct, and consistent with the basic objective of maximizing the stockholder's wealth.

References

1. Bierman, H., Jr. and S. Smidt, *The Capital Budgeting Decision,* 4th Ed., Macmillan Publishing Co., New York, 1975.
2. DeGarmo, E. P. and John R. Canada, *Engineering Economy,* 5th Ed., The Macmillan Co., New York, 1973.
3. Mao, J. C. T., *Quantitative Analysis of Financial Decision,* The Macmillan Co., New York, 1969.
4. Peterson, D. E., *A Quantitative Framework For Financial Management,* Richard D. Irwin, Inc., Homewood, Ill., 1969.
5. Quirin, G. D., *The Capital Expenditure Decision,* Richard D. Irwin, Inc., Homewood, Ill., 1967.
6. Thuesen, H. G., W. J. Fabrycky, and G. J. Thuesen, *Engineering Economy,* 4th Ed., Prentice-Hall, Inc., Englewood Cliffs, N.J., 1971.

7. Van Horne, J. C., *Financial Management and Policy,* 2nd Ed., Prentice-Hall, Englewood Cliffs, N.J., 1971.

Problems

9-1. The four projects given below are mutually exclusive. Determine the following:

(a) Which project should be chosen if the required rate of return is 15%? Use an incremental approach and check your answer by calculating the net present value for each project.

(b) Which project should be chosen if the required rate of return is 25%?

	Project			
	A	B	C	D
Required investment, $	30,000	50,000	60,000	100,000
Annual cash flow, $	7,000	12,000	15,000	22,000
Life, years	10	10	10	10

9-2. Repeat Problem 9-1 assuming the life for each project is infinite.

9-3. The four projects given below are mutually exclusive. Which one should be chosen if the required rate of return is 12%?

End of	Project Cash Flows			
Year	A	B	C	D
0	−$20,000	−$40,000	−$60,000	−$90,000
1	6,000	15,000	18,000	30,000
2	6,000	20,000	18,000	40,000
3	6,000	15,000	18,000	20,000
4	6,000	10,000	18,000	20,000
5	6,000	10,000	18,000	20,000

9-4. The four projects listed below are mutually exclusive. Which one should be chosen if the required rate of return is 12%. What assumption is implied in your analysis?

	Project			
	A	B	C	D
Required investment, $	−100,000	−125,000	−150,000	−200,000
Annual cash flow,	21,000	36,000	28,000	37,000
Life, years	8	5	10	10

9-5. Suppose in Problem 9-1 the projects are not mutually exclusive but independent. Which projects should be chosen with the following budget restrictions: (a) \$200,000, (b) \$150,000, (c) \$100,000?

9-6. Write the mathematical programming formulation for determining which projects should be done using the data given below:

Projects	Capital Expenditure ($t = 0$)	Capital Expenditure ($t = 1$)	Net Present Value
1	\$10,000	\$ 8,000	\$ 6,000
2	20,000	10,000	8,000
3	30,000	20,000	15,000
4	15,000	10,000	9,000
5	25,000	20,000	10,000
6	50,000	30,000	13,000
7	10,000	5,000	7,000
8	8,000	2,000	5,000
9	7,000	4,000	3,000
10	40,000	15,000	12,000

Budget restriction ($t = 0$) = \$200,000.
Budget restriction ($t = 1$) = \$100,000.
Projects 5, 7, and 8 are mutually exclusive.
Project 2 or 4 but not both must be done.
Projects 9 and 10 are interdependent, that is, project 9 can be done without project 10 but if project 10 is done project 9 must be done.

9-7. A company has two production lines X and Y. Each production line has submitted a list of proposals as shown below:

	Production Line X	
Proposal	Initial Investment	Annual Costs
1	\$35,000	\$10,000
2	37,000	8,000
3	42,000	6,000
4	50,000	3,000
	Production Line Y	
Proposal	Initial Investment	Annual Costs
5	\$48,000	\$12,000
6	53,000	10,000
7	60,000	8,000
8	69,000	6,000

Assuming the proposals for the same production line are mutually exclusive and a proposal must be accepted for each production line, determine the following if the minimum acceptable rate of return is 15% and all proposals are expected to have a life of 10 years.

(a) What proposals should be accepted if the amount of money available for investment is not restricted?

(b) Repeat part (a) if the amount available for investment is $100,000.

9-8. Formulate part (b) of Problem 9-7 into a mathematical programming format.

9-9. Repeat Problem 9-6 with the condition that unused budget amounts can be carried forward with (a) no discounting and (b) discounting (i = 12%).

CHAPTER 10
COST COMPARISONS

This chapter is concerned with cost comparisons. Cost comparisons are another case of the capital budgeting and, consequently, could have been included in the previous chapter. However, it is believed that a separate chapter on cost comparisons is desirable because of the scope of the discussion.

Basically, cost comparisons involve a choice between mutually exclusive alternatives under the assumption that the alternatives provide equal service (or gross incomes). In cost comparisons the cash flows (except for salvage values) are negative because the elements in the cash flows (capital outlay, materials, labor, taxes, etc.) are all cash outflows. However, in this chapter the sign convention for cash flows is reversed. That is, outflows are considered positive and inflows negative. This sign convention is merely for convenience; it avoids writing the negative sign as often.

Several approaches exist in the literature for making cost comparisons. To a large extent, these different approaches result from certain simplifying assumptions made regarding tax considerations and the method of financing the alternatives. In this chapter, these "traditional approaches" are presented. Also, this chapter suggests the use of minimum annual revenue requirements as another approach in making cost comparisons; an approach that does not require the assumptions implied in the traditional approaches.

"TRADITIONAL" APPROACHES

In what this text calls "traditional" approaches, cost comparisons are based either on a before-tax or an after-tax minimum attractive rate of return and the assumption of 100% equity financing. The relationship between the before-tax, $(\text{MARR})_b$, and the after-tax, $(\text{MARR})_a$, minimum attractive rate of return is

$$(\text{MARR})_a = (\text{MARR})_b(1 - T) \tag{10-1}$$

where T is the effective tax rate. Equation (10-1) is only correct for 100% equity financing. If the asset is to be financed by some combination of debt and equity capital, Eq. (10-1) is not correct.

In addition to the MARR and financing considerations, the traditional approaches usually use either a present worth, annual cost, or capitalized cost method in making cost comparisons. These methods are presented in the following example using a before tax approach.

EXAMPLE 10-1

Two compressors that provide equal service are candidates for a particular application. The estimated cost data are shown in Table 10-1. Which compressor should be chosen?

The first step is to convert the given interest rate to a before-tax interest rate by using Eq. (10-1):

$$(\text{MARR})_b = \frac{0.12}{1 - .52} = 0.25 = 25\%$$

Present Worth Comparison: A present worth comparison is simply the cash flows discounted to the present, and for this example is

$$\begin{aligned} \text{PW}_1 &= 12{,}000 + 2000(PA\ 25{,}8) - 2000(PF\ 25{,}8) \\ &= 12{,}000 + 2000(3.3289) - 2000(0.1678) = \$18{,}322 \\ \text{PW}_2 &= 20{,}000 + 500(PA\ 25{,}8) - 2000(PF\ 25{,}8) \\ &= 20{,}000 + 500(3.3289) - 2000(0.1678) = \$21{,}329 \end{aligned}$$

Based on this comparison, the choice is compressor number one.

Annual Cost: An annual cost comparison is an equivalent annual amount of the cash flows. An equivalent annual cost comparison can be obtained by multiplying the present worths by the appropriate interest factor: namely,

$$\begin{aligned} \text{AC}_1 &= 18{,}322(AP\ 25{,}8) \\ &= 18{,}322(0.3004) = \$5504 \text{ per year} \end{aligned}$$

Table 10-1
COST DATA FOR EXAMPLE 10-1

	Compressor	
	1	2
Initial Cost	\$12,000	\$20,000
Salvage value	\$2,000	\$2,000
Annual operation and maintenance costs	\$2,000	\$500
Estimated life, years	8	8
Required return on equity after taxes	12%	12%
Tax rate	52%	52%

$$\begin{aligned} AC_2 &= 21{,}329(AP\ 25{,}8) \\ &= 21{,}329(0.3004) = \$6407 \text{ per per} \end{aligned}$$

If the present worths are not known and only an annual cost comparison is required, it can be obtained using the formula developed in a previous chapter (Chapter 3) for capital recovery with a return:

$$\begin{aligned} AC_1 &= (12{,}000 - 2000)(AP\ 25{,}8) + 2000(0.25) + 2000 \\ &= (10{,}000)(0.3004) + 500 + 2000 = \$5504 \text{ per year} \\ AC_2 &= (20{,}000 - 2000)(AP\ 25{,}8) + 2000(0.25) + 500 \\ &= (18{,}000)(0.3004) + 500 + 500 = \$6407 \text{ per year} \end{aligned}$$

This comparison gives the same choice as the present worth comparison.

Capitalized Cost Comparison: Capitalized cost is a present amount that is equivalent to a given pattern of cash flows if the given pattern is repeated forever. This definition implies the assumption that the *need* for the service provided is infinite and the costs cycle for each alternative is shown in Table 10-2. This definition also implies that the annual costs for each cycle, A, (which are the same) extend to infinity. Consequently, the capitalized cost, CC, is

$$CC = A(PAi,\ n) \tag{10-2}$$

$$= A\left[\frac{(1+i)^n - 1}{(i)(1+i)^n}\right] \tag{10-3}$$

The limit of the term in brackets as $n \to \infty$ is

$$\lim_{n>\infty}\left[\frac{(1+i)^n - 1}{(I)(1+i)^n}\right] = \frac{1}{i} \tag{10-4}$$

Therefore, the capitalized cost, in general, is

$$CC = \frac{A}{i} \tag{10-5}$$

and for this example, the capitalized costs are

$$CC_1 = \frac{5504}{0.25} = \$22{,}016$$

$$CC_2 = \frac{6407}{0.25} = \$25{,}628$$

A Comment: All three comparisons give the same choice of compressor because they are all related as shown in Example 10-1. If different methods give different decisions, there is an error in one of the methods.

Note that the sign convention previously mentioned is used in Example 10-1 That is, the salvage value is negative while all cash outflows are positive. It should be remembered that in the capital recovery formula, a negative salvage value is taken into account in its derivation.

In Example 10-1, the taxes are taken into account *implicitly* by adjusting (increasing) the required after-tax return on equity capital. This implies the assumption, commonly

Table 10-2
CAPITALIZED COST COMPARISON

End of Year	Compressor 1	Compressor 2
0	P = \$12,000	P = \$20,000
1	2000	500
2	·	·
3	·	·
4	·	·
5	·	·
6	·	·
7	·	·
8	2000; L = 2000; P = 12,000	500; L = 2000; P = 20,000
9	2000	·
10	·	·
11	·	·
12	·	·
13	·	·
14	·	·
15	·	·
16	2000; L = 2000; P = 12,000	500; L = 2000; P = 20,000
17	·	·
18	·	·
19	·	·
20	·	·
21	·	·
22	·	·
23	·	·
24	2000; L = 2000; P = 12,000	500; L = 2000; P = 20,000
·		
·		
·		
∞		

made in the traditional approach to cost comparisons that the assets are 100% equity financed. In the next example the same financing assumption is made but tax considerations are taken into account *explicitly*.

Another assumption in making the annual cost and present worth comparisons that is implied in this example is that the *need* for the service provided by the compressors is at least eight years. It is mentioned earlier that the capitalized cost comparisons assumes an infinite need. Service, depreciation, and physical life are discussed later. At present, the assumption, as in most presentations on cost comparisons, is that these three lives are equal.

Of the three methods discussed in this example, perhaps the one most often used is the annual cost comparison. There are no theoretical reasons for this preference. It appears that, for some individuals, dollars per year has a greater appeal than present-day dollars.

EXAMPLE 10-2

In this example, tax considerations are explicitly included in the cost comparisons using the data given in Table 10-1.

An annual cost comparison is given in Table 10-3. The \$260 figure in Table 10-3 needs explanation. It is based on a consideration of the respective tax deductions that would result for each alternative. For compressor 1, the tax deductions are the \$2000 per year operation cost and also the depreciation. Assuming straight-line depreciation, the depreciation deduction is \$1250 per year. For compressor 2 the tax deductions are \$500 for operation costs and \$2250 for depreciation. Therefore, the tax difference between the two compressors is

$$[(2000 + 1250) - (500 + 2250)](0.52) = \$260 \text{ per year}$$

Since the total deductions for compressor 2 is less than compressor 1, this amount, \$260, is added to compressor 2 as a yearly "cost." It is emphasized this cost comparison is strictly for decision purposes. These annual costs have no relationship to an accountant's perception of annual costs for an asset.

Note that in this example the after-tax return on equity (12%) is used as the interest rate; that is, the assumption of 100% equity financing is *still* implied.

The assumption of straight-line depreciation is not required. If some other depreciation model is believed to be more applicable, it can be included in a cost comparison by computing the equivalent annual amount of the yearly depreciation deductions for the alternatives and use the values in the same manner as the straight line values are used.

If a present worth comparison is desired, it can be obtained by using the correct interest factor. In this example, the values \$4253 and \$4623 would be multiplied by the factor (PA12,8). If a capitalized cost comparison is desired, the equivalent annual costs would be divided by the interest rate (12% in this example).

Table 10-3
EQUIVALENT ANNUAL COST COMPARISON WITH EXPLICIT TAX CONSIDERATIONS

	Compressor	
$(P-L)(APi,n) + L(i)$	1	2
(12,000–2000)(*AP* 12,8) + 2000(0.12)	\$2253	—
(20,000–2000)(*AP* 12,8) + 2000(0.12)	—	\$3863
Operation and maintenance costs	2000	500
Tax difference	—	260
Totals	4253	4623

Table 10-4
BEFORE TAX ANNUAL COSTS FOR EXAMPLE 10-1 ($12,000)

End of Year	Depreciation	Chargeable Investment	Return on Equity	Operation and Maintenance	Annual Cost	PF 25,n	Equivalent Amount at $t = 0$
0	—	$12,000	—	—	—	—	—
1	$1250	10,750	$3000	$2000	$6250	0.800	$ 5000
2	1250	9,500	2688	2000	5938	0.6400	3800
3	1250	8,250	2375	2000	5625	0.5120	2880
4	1250	7,000	2063	2000	5313	0.4096	2176
5	1250	5,750	1750	2000	5000	0.3277	1639
6	1250	4,500	1438	2000	4688	0.2622	1229
7	1250	3,250	1125	2000	4375	0.2097	917
8	1250	2,000	813	2000	4063	0.1678	682
						Total	$18,323

Equivalent annual amount = $18,323($AP$ 25,8) = $5504.

ANOTHER APPROACH

Additional insight into cost comparisons can be obtained by relating cost comparisons to the generation of minimum annual revenue requirements (discussed in Chapter 7). It should be remembered that the minimum annual revenue requirements are, in fact, annual costs with a return.

In Tables 10-4 and 10-5 the annual costs, for the data in Example 10-1, are given. The calculations for the annual cost values are based on the flow-through

Table 10-5
BEFORE TAX ANNUAL COSTS FOR EXAMPLE 10-1 ($20,000)

End of Year	Depreciation	Chargeable Investment	Return on Equity	Operation and Maintenance	Annual Cost	PF 25,n	Equivalent Amount at $t = 0$
0	—	$20,000	—	—	—	—	—
1	$2250	17,750	$5000	$500	$7750	0.8000	$ 6200
2	2250	15,500	4438	500	7188	0.6400	4600
3	2250	13,250	3875	500	6625	0.5120	3392
4	2250	11,000	3313	500	6063	0.4096	2483
5	2250	8,750	2750	500	5500	0.3277	1802
6	2250	6,500	2188	500	4938	0.2622	1295
7	2250	4,250	1625	500	4375	0.2097	917
8	2250	2,000	1063	500	3813	0.1678	640
						Total	$21,329

Equivalent annual amount = $21,329($AP$ 25,8) = $6407.

method for calculating the minimum annual revenue requirements with the conditions:

1. $D_b = D_t =$ straight line.
2. $c = 0$ (since 100% equity financing is assumed).
3. $T = 0$ (since taxes have been implicitly included by adjusting the return on equity).
4. Capitalized interest and investment tax credit are not included.

As an example of the calculations involved in Table 10-4, the chargeable investments for the first three years are

$$I_0 = \$12{,}000$$
$$I_1 = 12{,}000 - 1250 = 10{,}750$$
$$I_2 = 10{,}750 - 1250 = 9500$$

The returns on equity for the first two years are

$$fe_1 = 0.25\ (12{,}000) = 3000$$
$$fe_2 = 0.25\ (10{,}750) = 2688$$

The annual costs (minimum revenue requirement) are

$$R_1 = 1250 + 3000 + 2000 = 6250$$
$$R_2 = 1250 + 2688 + 2000 = 5938$$

The calculations for the annual cost values in Table 10-5 are obtained in a very similar manner.

Now, if an equivalent annual amount is determined for the annual costs in Tables 10-4 and 10-5, the results ($5504 and $6407) are the same as obtained in Example 10-1. The point is, again, that minimum annual revenue requirements and annual costs are the same, at least, from the standpoint of the evaluation of economic alternatives that have equal outputs.

In Tables 10-6 and 10-7 the annual costs are generated with the explicit consideration of taxes (Example 10-2). The procedure for calculating these annual costs is the same as previously mentioned: the relationships developed in Chapter 7 for the generation of the minimum annual revenue requirements using the flow-through method. The same conditions as those stated for Tables 10-4 and 10-5 are used except the tax rate is 52%. The procedure for calculating the annual cost is shown in the following sample calculations.

The chargeable investments are

$$I_0 = 12{,}000$$
$$I_1 = 12{,}000 - 1250 = 10{,}750$$
$$I_2 = 10{,}750 - 1250 = 9500$$

and the returns on equity are

$$fe_1 = (0.12)(12{,}000) = 1440$$
$$fe_2 = (0.12)(10{,}750) = 1290$$

Table 10-6
AFTER-TAX ANNUAL COSTS FOR EXAMPLE 10-2 ($12,000)

End of Year	Depreciation	Chargeable Investment	Return on Equity	Tax	Operation and Maintenance	Annual Cost	*PF* 12,*n*	Equivalent Amount at $t = 0$
0	—	$12,000	—	—	—	—	—	—
1	$1250	10,750	$1440	$1560	$2000	$6250	0.8929	$ 5581
2	1250	9,500	1290	1397	2000	5937	0.7972	4733
3	1250	8,250	1140	1235	2000	5625	0.7118	4004
4	1250	7,000	990	1072	2000	5312	0.6355	3376
5	1250	5,750	840	910	2000	5000	0.5674	2837
6	1250	4,500	690	747	2000	4687	0.5066	2374
7	1250	3,250	540	585	2000	4375	0.4524	1979
8	1250	2,000	390	422	2000	4062	0.4039	1641
							Total	26,525

Equivalent annual cost = 26,525(*AP* 12,8) = $5339.

Table 10-7

AFTER-TAX ANNUAL COSTS FOR EXAMPLE 10-2 ($20,000)

End of Year	Depreciation	Chargeable Investment	Return on Equity	Tax	Operation and Maintenance	Annual Cost	*PF* 12,*n*	Equivalent Amount at $t = 0$
0	—	$20,000	—	—	—	—	—	—
1	$2250	17,750	$2400	$2600	$500	$7750	0.8929	$ 6,920
2	2250	15,500	2130	2307	500	7187	0.7972	5,729
3	2250	13,250	1860	2015	500	6625	0.7118	4,716
4	2250	11,000	1590	1722	500	6062	0.6355	3,852
5	2250	8,750	1320	1430	500	5500	0.5674	3,121
6	2250	6,500	1050	1137	500	4937	0.5066	2,501
7	2250	4,250	780	845	500	4375	0.4524	1,979
8	2250	2,000	510	552	500	3812	0.4039	1,540
							Total	30,358

Equipment annual amount = $30,358(*AP* 12,8) = $6111.

The taxes are [see Eq. (7-34)]

$$t_1 = \frac{0.52}{1 - 0.52}(1440) = 1560$$
$$t_2 = \frac{0.52}{1 - 0.52}(1290) = 1397$$

Therefore, the annual costs are

$$R_1 = 1250 + 1440 + 1560 + 2000 = 6250$$
$$R_2' = 1250 + 1290 + 1397 + 2000 = 5937$$

A Comment: It should be noted that the yearly costs in Table 10-4 and 10-6 are the same (slight differences are due to roundoff error) and this is also true for Tables 10-5 and 10-7. This substantiates the contention made earlier that Eq. (10-1) is correct provided that 100% equity financing is assumed. The same results would not be obtained if the assets are financed by some combination of debt and equity capital. The equality of the annual costs before and after tax with 100% equity financing are not dependent upon the method of depreciation provided the same model is used in both cases and it is assumed that tax depreciation equals book depreciation—assumptions that are implied in most traditional cost comparisons.

Additional examination of the results in Tables 10-6 and 10-7 provides further insight into cost comparisons. For example, if the difference between the equivalent annual costs is determined, the result is

$$\Delta = 6111 - 5339 = \$772$$

This difference is before tax. Consequently, the after-tax difference is

$$772(1 - 0.52) = \$371$$

Now, the difference in the equivalent after tax annual costs in Example 10-2 is

$$\Delta = 4623 - 4253 = \$370$$

which is the same—taking into account roundoff error, as previously obtained by multiplying the difference of the equivalent annual costs by the term $(1 - T)$. Again, this result is only obtainable if the assumption of 100% equity financing is made. This assumption may be acceptable on occasion but, in general, is not very realistic because most assets are financed from a "pool of capital" obtained from both debt and equity sources (see Chapter 6).

The purpose of the discussion to this point has been to show the relationships between after- and before-tax cost comparisons, the interest rates used in cost comparisons, and the relationships between cost comparisons and minimum annual revenue requirements under the assumption of 100% equity financing. The next section discusses a method for making cost comparisons when the assets are financed by a combination of debt and equity capital.

ANNUAL COSTS

The most inclusive approach to making cost comparisons is to generate the annual costs (minimum annual revenue requirements) for each alternative and then compare the results. This statement is based on the fact that this approach allows for the explicit consideration of all tax effects, debt and equity financing, and different depreciation models for book and tax depreciation allowances. Furthermore, it gives the *yearly* costs as well as an equivalent annual cost, while the three methods considered in Examples 10-1 and 10-2 gave only total equivalent amounts. From a decision standpoint, the specific yearly costs provide a distinct advantage; that is, they allow consideration of the short-term effects (yearly minimum gross income requirements) in the comparison of the alternatives. Another important advantage to this approach is that the generation of the annual costs seems to have an intuitive appeal to decision makers because of the clarity of its basis. That is, it is easier to explain (and defend) this approach to individuals who have not previously been exposed to the traditional approaches. The only disadvantages to generating annual costs are (1) the depreciation models must be determined and (2) the computations are longer. Both of these disadvantages are relatively minor. In most cases, companies have established their ''favorite'' depreciation models (book and tax) while the computational disadvantage can be readily overcome with the computer programs given in the appendices.

The procedure for generating annual costs (minimum annual revenue requirements) are presented in detail in Chapter 7. Remember that two procedures are presented: flow through and normalizing. In general, the flow-through procedure is more applicable to private corporations because it comes closest to cash flow concepts while the normalizing procedure is more applicable to public utilities because it is a closer representation of their accounting procedures.

The following example is worked to show the versatility of minimum annual revenue requirements (annual costs) in making cost comparisons.

EXAMPLE 10-3

The annual costs for the two alternatives shown in Table 10-8 are given in Tables 10-9 and 10-10. The values in these tables are determined using the flow-through method that is discussed in Chapter 7. For the sake of completeness, some sample calculations are provided for alternative 1.

The book depreciation is

$$D_b = \frac{12{,}000 - 2000}{8} = \$1250 \text{ per year}$$

and the tax depreciation based on the sum of the years-digits model for the first year is

$$D_{t1} = \frac{5}{15}(12{,}000 - 2000) = \$3333$$

Table 10-8
DATA FOR EXAMPLE 10-3

	Alternative	
	1	2
First cost	\$12,000	\$20,000
Annual operation and maintenance costs	\$1,700	\$500
Tax depreciation	540	540
Life, years	5	5
Salvage value	\$2000	\$2000
Book depreciation	Straight line	Straight line
Life, years	8	8
Salvage value	\$2,000	\$2,000
Debt rates	50%	50%
Cost of debt capital	8%	8%
Required return on equity capital	12%	12%
Investment tax credit	10%	10%
Tax rate	52%	52%

The investment tax credit is

$$C_t = \left(\frac{2}{3}\right)(0.10)(12{,}000) = \$800$$

Only two thirds of the full investment tax credit is allowed since the useful life is only five years. The useful life from the standpoint of determining the investment tax credit is the depreciation life used for tax purposes (see Chapter 4).

The chargeable investments in year 0 and 1 are

$$K_0 = 12{,}000$$
$$K_1 = 12{,}000 - 1250 = \$10{,}750$$

The return on equity and debt interest for the first year are, respectively,

$$fe_1 = (0.5)(0.12)(12{,}000) = \$720$$
$$I_1 = (0.5)(0.08)(12{,}000) = \$480$$

The tax for the first year is

$$t_1 = \left(\frac{0.52}{0.48}\right)(1250 + 720 - 3333) - \frac{800}{0.48} = -\$3144$$

and the annual cost is

$$R_1 = 1250 + 720 + 480 + 1700 - 3144 = \$1006$$

From the standpoint of determining which alternative is best, both the long-term costs and short-term costs should be considered. If the annual costs in Tables 10-9 and 10-10 are converted to equivalent annual amounts the results are

$$[1006\,(PF\ 10{,}1) + 3189\,(PF\ 10{,}2) + \ldots + 4840\,(PF\ 10{,}8)]\,(AP\ 10{,}8)\,[19{,}214]\,(0.1875) = \$3745 \text{ per year}$$

$$[-789\,(PF\ 10{,}1) + 2916\,(PF\ 10{,}2) + \ldots + 5889\,(PF\ 10{,}8)]\,(AP\ 10{,}8)\,[19{,}822]\,(0.1875) = \$3954 \text{ per year}$$

Comparing these values indicated that alternative 1 should be chosen. However, if just the first two years of annual costs are compared, alternative 2 has a decided advantage as a result of the larger tax depreciation allowances and the investment tax credit. It could well be that a company's circumstances are such that these short-term effects warrant the choice of alternative 2 even though the long-term effects (equivalent annual cost) are greater than alternative 1. This is one of the primary advantages, previously mentioned, of the use of annual costs (minimum annual revenue requirements) in making comparisons. That is, an explicit consideration of the short-term effects of the alternatives are possible; an advantage not possible in the traditional approaches to cost comparisons. In addition different depreciation models can be used for book and tax depreciation amounts as shown in Tables 10-9 and 10-10.

The assumptions that are implied in Tables 10-9 and 10-10 are that (1) the actual capital recovered and the book depreciation are equal and (2) the company has sufficient taxable income in any year to make use of "excess" depreciation allowances and investment tax credit.

LIVES OF ASSETS

There are different types of asset lives involved in cost comparisons and, for that matter, in economic analysis in general. The types of asset life that can be involved in economic analysis are depreciation, physical, service, and economic life. A common assumption in a large amount of economic analysis is that all of these lives are equal. This in an extremely convenient assumption but one that is not realistic nor correct in many instances. Therefore, this section considers these different types of asset life.

Depreciation Life

Remember that there are two general types of depreciations used in economics analysis. One is for company purposes, usually called book depreciation, and the other is for tax purposes. Book depreciation is an accounting method for systematically recovering investment capital. As such, the life used in determining book depreciation amounts is based on a period of time that is considered desirable and practical for the recovery of capital. The life used in determining depreciation amounts for tax purposes are, of course, based on tax regulations and other considerations. That is, in general, the shorter the life used in a particular tax depreciation model the less the present worth of the taxes. Therefore, it is not unusual for the lives used in book depreciation and tax depreciation models to be different. If this is the case, the method used for making an economic analysis should be able to take this difference into account. Examples of this are provided in Examples 5-1 and 10-3.

Table 10-9
ANNUAL COSTS FOR ALTERNATIVE 1 (EXAMPLE 10-3)

End of Year	Book Depreciation	Tax Depreciation	Investment Tax Credit	Chargeable Investment	Return on Equity	Debt Interest	Tax	Operation and Maintenance	Annual Cost
0	—	—	—	$12,000	—	—	—	—	—
1	$1250	$3333	$800	10,750	$720	$480	$−3144	$1700	$1006
2	1250	2667	—	9,500	645	430	−836	1700	3189
3	1250	2000	—	8,250	570	380	−195	1700	3705
4	1250	1333	—	7,000	495	330	446	1700	4221
5	1250	667	—	5,750	420	280	1087	1700	4737
6	1250	—	—	4,500	345	230	1728	1700	5253
7	1250	—	—	3,250	270	180	1647	1700	5047
8	1250	—	—	2,000	195	130	1565	1700	4840

Table 10-10
ANNUAL COSTS FOR ALTERNATIVE 2 (EXAMPLE 10-3)

End of Year	Book Depreciation	Tax Depreciation	Investment Tax Credit	Chargeable Investment	Return on Equity	Debt Interest	Tax	Operation and Maintenance	Annual Cost
0	—	—	—	$20,000					
1	$2250	$6000	$1333	17,750	$1200	$800	$−5539	$500	$−789
2	2250	4800	—	15,500	1065	710	−1609	500	2916
3	2250	3600	—	13,250	930	620	−455	500	3845
4	2250	2400	—	11,000	795	530	699	500	4774
5	2250	1200	—	8,750	660	440	1852	500	5702
6	2250	—	—	6,500	525	350	3006	500	6631
7	2250	—	—	4,250	390	260	2860	500	6260
8	2250	—	—	2,000	255	170	2714	500	5889

Physical Life

In this text, the physical life is the number of years that an asset is able to perform some stated function after which a combination of the forces of deterioration and obsolescence make it impossible to operate the asset.

Service Life

In this text, the service life is the *needed* life of the asset. That is, the expected number of years that the need for some service provided by an asset is required. In this context, it is entirely possible to have a service life that is completely different than the tax depreciation life, physical life, and in some instances, the book depreciation life.

Economic Life

The economic life of an asset is the number of years that an asset should be kept for the most economical service; or, the time interval that minimizes the annual costs. The following examples are worked to show how the economic life is determined.

EXAMPLE 10-4

In this example the economic life is determined using the traditional approach. That is, the implied assumption of 100% equity financing with tax considerations not explicitly included. A later example is worked that relaxes these assumptions.

Suppose that it is required to find the economic life for an asset having the data shown in Table 10-11. Basically, what is involved in determining the economic life is the

Table 10-11
COST DATA FOR EXAMPLE 10-4

First cost		\$12,000
Required return on equity		12%
End of Year	Annual Cost	Salvage
0	—	—
1	\$1700	\$7000
2	2000	5000
3	2500	4000
4	3000	3000
5	3500	2000
6	4500	1000
7	5000	500
8	5500	500

Table 10-12
ANNUAL AND EQUIVALENT ANNUAL COSTS FOR EXAMPLE 10-4

End of Year	Salvage Value	Capital Recovery Cost	Return on Unrecovered Capital	Operation Cost	Annual Cost	Equivalent Annual Cost
0	$12,000	—	—	—	—	—
1	7000	$5000	$1440	$1700	$8140	$8140
2	5000	2000	840	2000	4840	6583
3	4000	1000	600	2500	4100	5848
4	3000	1000	480	3000	4480	5561
5	2000	1000	360	3500	4860	5451
6	1000	1000	240	4500	5740	5486
7	500	500	120	5000	5620	5500
8	500	0	60	5500	5560	5505

calculation of an annual cost for each year of the asset's life and an equivalent annual cost for a given number of years service. These calculations are shown in Table 10-12 and are explained subsequently.

It should be realized that the salvage values are the actual realizable values of the asset at various points in time. Consequently, the capital recovery costs (actual depreciation) are

$$12{,}000 - 7000 = \$5000$$
$$7000 - 5000 = \$2000$$
$$5000 - 4000 = \$1000$$

for the first three years. The interest on the unrecovered balance is determined in the following manner. For the first year, it is

$$(12{,}000)\,(0.12) = \$1440$$

and for the second year

$$(7000)\,(0.12) = \$840$$

The annual costs for the first three years are

$$5000 + 1440 + 1700 = \$8140$$
$$2000 + 840 + 2000 = \$4840$$
$$1000 + 600 + 2500 = \$4100$$

The equivalent annual cost for the first three years are

$$(8140)\,(PF\ 12{,}1)\,(AP\ 12{,}1) = \$8140,$$
$$[8140\,(PF\ 12{,}1) + 4840\,(PF\ 12{,}2)]\,(AP\ 12{,}2) = \$6583$$
$$[8140\,(PF\ 12{,}1) + 4840\,(PF\ 12{,}2) + 4100\,(PF\ 12{,}3)]\,(AP\ 12{,}3) = \$5848$$

It is interesting to note that the same values are obtained for the equivalent annual costs using the capital recovery with interest formula; that is,

$$(12{,}000 - 7000)\,(AP\ 12{,}1) + 7000\,(0.12) + 1700 = \$8140\ (12{,}000 - 5000)\,(AP\ 12{,}2)$$
$$+\ 5000\,(0.12) + [1700\,(PF\ 12{,}1) + 2000\,(PF\ 12{,}2)]\,(AP\ 12{,}2) = \$6583$$

and

$$(12{,}000 - 4000)\,(AP\ 12{,}3) + 4000\,(0.12) + [1700\,(PF\ 12{,}1)$$
$$+\ 2000\,(PF\ 12{,}2) + 2500\,(PF\ 12{,}3)]\,(AP\ 12{,}3) = \$5848$$

Now, considering the data in Table 10-12, the economic life is five years because the equivalent annual cost is the smallest *and* the annual cost for the sixth year is larger than this equivalent annual cost. That is, in general, the economic life is the life at which the equivalent annual cost is the least provided the cost of the next year, a marginal cost, is greater than this equivalent annual cost.

In theory, the asset should be replaced at the end of its economic life; a theory which is based on the assumption that the replacement will have the exact same cost data as the original asset. Also, in theory, all cost comparisons should be made on their costs over their *economic life*. There is also the implication that the need (service life) is at least equal to the economic life.

Although in theory the use of the economic life to make replacement decisions and cost comparisons is sound, rarely is it used in practice. This is a result of several considerations: (1) the assumption required about replacement with an asset having the same cost data, (2) the difficulty in estimating future cost data, (3) assets are usually replaced on the basis of factors in existence at essentially the same time as the replacement rather than at the time of the purchase of the asset, and (4) a great many companies estimate annual cost on the basis of a constant percentage of the first cost. This fourth consideration results in a constant annual cost and the economic life becomes the physical life or the service life when it is less than the physical life. For example, if Example 10-4 is repeated with the same data with the exception that the operating costs are replaced with a constant value of $1700, the results in Table 10-13 are obtained, which shows that the economic life is eight years.

EXAMPLE 10-5

In this example the economic life is determined for the data given in Table 10-14. It shows how explicit tax considerations and combinations of debt and equity financing may be included in determining the economic life. This example assumes Sec. 1245 Property and ordinary retirements under the ADR system.

The basic approach is to determine the annual cost and equivalent annual costs as before. These values are given in Table 10-15 and are obtained by using the relationships given earlier for determining the minimum annual revenue requirements using the flow-through method. For example, the capital recovery costs are

$$12{,}000 - 7000 = \$5000$$
$$7000 - 5000 = \$2000$$

and

$$5000 - 4000 = \$1000$$

Table 10-13

ANNUAL AND EQUIVALENT ANNUAL COSTS WITH A CONSTANT OPERATING COST

End of Year	Salvage Value	Capital Recovery Cost	Return on Unrecovered Capital	Operating Cost	Annual Cost	Equivalent Annual Cost
0	$12,000	—	—	—	—	—
1	7,000	$5000	$1440	$1700	$8140	$8140
2	5,000	2000	840	1700	4540	6442
3	4,000	1000	600	1700	3300	5511
4	3,000	1000	480	1700	3180	5023
5	2,000	1000	360	1700	3060	4714
6	1,000	1000	240	1700	2940	4495
7	500	500	120	1700	2820	4340
8	500	0	60	1700	1760	4075

Table 10-14

DATA FOR EXAMPLE 10-5

First cost	$12,000
Tax depreciation	SYD, $n = 5$ years, $L = \$2000$
Book depreciation	Straight line
Investment tax credit	0%
Cost of debt capital	8%
Required return on equity capital	12%
Debt ratio	50%

End of Year	Operating Cost	Salvage Value
0	—	—
1	$1700	$7000
2	2000	5000
3	2500	4000
4	3000	3000
5	3500	2000
6	4500	1000
7	5000	500
8	5500	300

Table 10-15

ANNUAL AND EQUIVALENT ANNUAL COSTS FOR EXAMPLE 10-5

End of Year	Salvage Values	Capital Recovery Cost	Tax Depreciation	Return on Equity	Debt Interest	Investment Tax Credit	Tax	Operating Cost	Annual Cost	Equivalent Annual Cost
0	$12,000	—	—	—	—	—	—	—	—	—
1	7,000	$5000	$3333	$720	$480	0	$2586	$1700	$10,486	$10,486
2	5,000	2000	2667	420	280	—	−268	2000	4,432	7,604
3	4,000	1000	2000	300	200	—	−758	2500	3,242	6,286
4	3,000	1000	1333	240	160	—	−101	3000	4,299	5,858
5	2,000	1000	667	180	120	—	556	3500	5,356	5,776
6	1,000	1000	—	120	80	—	1213	4500	6,913	5,923
7	500	500	—	60	40	—	607	5000	6,207	5,952
8	300	200	—	30	20	—	249	5500	5,999	5,959

The returns on equity for years 1 and 2, respectively, are

$$12{,}000\left(\frac{1}{2}\right)(0.12) = \$720$$
$$7000\left(\frac{1}{2}\right)(0.12) = \$420$$

The debt interest for years 1 and 2 are

$$12{,}000\left(\frac{1}{2}\right)(0.08) = \$480$$
$$7000\left(\frac{1}{2}\right)(0.08) = \$280$$

The taxes for the first two years are, respectively,

$$\frac{0.52}{1-0.52}(5000 + 720 - 3333) = \$2586$$
$$\frac{0.52}{1-0.52}(2000 + 420 - 2667) = -\$268$$

and the annual costs are

$$5000 + 720 + 480 + 2586 + 1700 = \$10{,}486$$
$$2000 + 420 + 280 - 268 + 2000 = \$4432$$

The equivalent annual costs are calculated using the weighted cost of capital (10% in this example). For the first three years, they are

$$(10{,}486)(PF\ 10{,}1)\ (AP\ 10{,}1) = \$10{,}486$$
$$[10{,}486(PF\ 10{,}1) + 4432(PF\ 10{,}2)]\ (AP\ 10{,}2) = \$7604$$
$$[10{,}486(PF\ 10{,}1) + 4432\ (PF\ 10{,}2) + 3342\ (PF\ 10{,}3)]\ (AP\ 10{,}3) = \$6286$$

Table 10-15 shows that the lowest equivalent annual cost ($5776) occurs at the end of the fifth. Consequently, the economic life in this example is five years.

In regard to tax depreciation in this example, the resulting tax depreciation amounts become fixed once the tax depreciation model is determined. That is, they cannot be changed in determining the economic life. Remember, tax depreciation amounts are determined on the basis of the tax laws. A computer program for determining the economic life is included in the appendices.

In Example 10-5 it is assumed that there is no investment tax credit and the retirements are classified as ordinary. When these assumptions are not acceptable, the annual costs must be adjusted to reflect the tax effects. If, in Example 10-5, the investment tax credit is 10% and the retirements are extraordinary, the annual costs are adjusted in the following manner. Determine the tax book values at the end of each year and the difference between the salvage value and the tax book value as shown in Eq. (10-6).

$$\text{salvage value} - \text{tax book value} \qquad (10\text{-}6)$$

If the value determined by Eq. (10-6) is positive, this indicates an increase in ordinary income (Sec. 1245 property). If the value is negative this is a reduction in ordinary income. Consequently, the change in taxes is

$$(\text{salvage value} - \text{tax book value})(T) \tag{10-7}$$

The change in annual cost, as a result of this increase or decrease in taxes, is obtained by dividing Eq. (10-7) by the term $(1 - T)$. These values are shown in Table 10-16. It should be noted that these tax effects are applicable *only* if the asset is retired at the end of a given year. With the adjustments given in Table 10-16 and consideration of the investment tax credit, the annual costs can be determined in the following manner for various periods of service. For one year of service, the annual cost is

$$10{,}486 - 1806 = \$8680$$

There are no investment tax credit considerations because the asset is only used for one year. For two years of service, the annual costs are

$$\text{year 1: } 10{,}486 - \frac{2}{3}\left(\frac{1200}{0.48}\right) = \$8819$$

$$\text{year 2: } 4432 - 1083 + \frac{2}{3}\left(\frac{1200}{0.48}\right) = \$5016$$

In these calculations the investment tax credit ($800) is taken in the first year and paid back in the second year. The tax effect of the retirement (1083) is obtained from Table 10-15a.

For three years of service, the annual costs are

$$\text{year 1: } 10{,}486 - \frac{2}{3}\left(\frac{1200}{0.48}\right) = \$8819$$

Table 10-16

RETIREMENT TAX EFFECTS

End of Year	Salvage Value	Tax Book Value	Change in Taxes	Change in Annual Cost
1	$7000	$8667	−$867	−$1806
2	5000	6000	−520	−1083
3	4000	4000	0	0
4	3000	2667	173	361
5	2000	2000	0	0
6	1000	2000	−520	−1083
7	500	2000	−780	−1625
8	300	2000	−884	−1842

year 2: \$4432

$$\text{year 3: } 3242 + 0 + \frac{1}{3}\left(\frac{1200}{0.48}\right) = \$4075$$

In year 3 there is only a \$400 investment tax penalty since a three-year life qualifies for one third of the investment tax credit and only two thirds (\$800) is originally taken.

The annual costs for the remaining service lives are given in Table 10-17. In generating the values Table 10-17 it should be remembered that there is no investment tax penalty at the end of five years of service.

After the annual costs are determined for the various service lives, the equivalent annual amounts are calculated. These values are also given in Table 10-17 and indicate an economic life of six years.

COMPARISON OF ASSETS WITH DIFFERENT LIVES

It is, of course, possible for assets to have different economic lives. Consequently, it is often necessary to compare assets with different lives. For example, consider the two assets shown in Table 10-18. The economic lives are different and one question is, "What period of time should be used in comparing these two assets?" The answer to this question depends on the answer to another question. What is the service life of the assets? That is, how long is the need for the service provided by the chosen asset required? It is on the answer to this question that a cost comparison should be based rather than the life of either asset. There are two possible answers to this latter question: (1) an infinite period of time or (2) a finite period of time.

Table 10-17

ANNUAL COSTS FOR VARIOUS SERVICE LIVES WITH EXTRAORDINARY RETIREMENTS AND INVESTMENT TAX CONSIDERATIONS

Years of Service	Annual Costs, \$								Equivalent Annual Amount
	1	2	3	4	5	6	7	8	
1	\$8680	—	—	—	—	—	—	—	\$8680
2	8819	\$5016	—	—	—	—	—	—	7008
3	8819	4432	\$4075	—	—	—	—	—	5928
4	8819	4432	3242	\$5493	—	—	—	—	5637
5	8819	4432	3242	4299	\$5356	—	—	—	5467
6	8819	4432	3242	4299	5356	\$5830	—	—	5434
7	8819	4432	3242	4299	5356	6913	\$4582	—	5470
8	8819	4432	3242	4299	5356	6913	6207	\$4157	5513

Table 10-18
ASSETS WITH DIFFERENT LIVES

	Asset	
	1	2
First cost	\$9000	\$17,000
Salvage	1000	3000
Annual operating costs	4000	2000
Economic life, years[a]	5	8
Required before tax return on equity	15%	15%

[a]Since the operating costs are constants, this implies the economic and physical lives are the same.

Infinite Service Life

If the need for the service is considered to be infinite, a before-tax cost comparison of the assets in Table 10-18, assuming 100% equity financing, is

$$\text{AC}_1 = (9000 - 1000)\,(AP\ 15{,}5) + 1000\,(0.15) + 4000 = 6536$$
$$\text{AC}_2 = (17{,}000 - 3000)\,(AP\ 15{,}8) + 3000\,(0.15) + 2000 = 5571$$

and the choice is asset 2. It may appear that, in obtaining the values of \$6536 and \$5571, the difference between the lives of the assets is neglected; this is not the case. The assumption is that the cost data for each asset cycles; that is, after five years of use, asset 1 is replaced with another asset having the exact same cost data as the first and this is repeated out to infinity. Thus, the equivalent annual cost for asset one is the same every year: \$6536. In a similar manner, the equivalent annual cost for asset two is \$5571 for every year. This situation is depicted in Table 10-19.

The assumption regarding the recycling of the cost data is made by default. That is, there is no good reason to assume otherwise. If there are reasons to assume otherwise, then it is necessary to estimate the cost data for the replacement assets. From a practical standpoint, this is a difficult task, especially in the case of infinite need, because it is necessary to estimate specific cost data far into the future. In defense of the recycling assumption, it should be realized that once a choice of an asset is made as a result of a cost comparison there is no reason, theoretically at least, that the chosen asset cannot be replaced if an economic analysis indicates a more desirable asset at some future time. If the best asset is chosen at each decision point in time, this will result in the best choice of assets over the long run. This text takes the point of view that, in the absence of data to the contrary, the recycling assumption is adequate for *cost comparisons*. There are some "difficulties" with this recycling assumption in replacement studies (discussed in the next chapter).

In this discussion of cost comparisons, under the requirement of infinite

Table 10-19
COST COMPARISON WITH INFINITE SERVICE LIFE

End of Year	Asset 1			Asset 2		
0	P = $9000			P = $17,000		
1		4000			2000	
2		4000	= 6536		2000	
3		4000	·		2000	= 5571
4		4000	·		2000	
5	L = 1000	4000	·		2000	·
6			6536		2000	·
7			·		2000	·
8			·	L = 3,000	2000	·
9			·			5571
10			·			·
11			·			·
12			·			·
13			·			·
14			·			·
15			·			·
16			·			·
·			·			·
·			·			·
·			·			·
∞			6536			5571

need, the values ($6536 and $5571) are the before-tax long-term effects (equivalent annual costs) based on 100% equity financing. In the case of financing by a combination of debt and equity capital with explicit tax considerations, the specific annual costs should be determined and compared. The life used in determining the annual costs should be their respective economic lives.

Finite Service Life

When the need for an asset is determined to be some definite period of time, then the cost comparison should be based on this period of time. This text believes that the specification of a finite service life is tantamount to saying that there is no need for the asset after the specified period of time. Relatively, this is an unusual case. However, it can occur in pilot-plant facilities and other short-lived operations. As an example, suppose a service life for the data in Table 10-18 is estimated to be 12 years. This would imply the situation depicted in Table 10-20 and assumes the costs of the assets recycle. This approach requires the estimation of the two salvage values, L_1 and L_2, shown in Table 10-20. Or, as an

Table 10-20
COST COMPARISON WITH FINITE SERVICE LIFE (12 YEARS)

End of Year	Asset 1	Asset 2
0	P = $9000	P = $17000
1	4000	2000
2	·	·
3	·	·
4	·	·
5	L = 1000; 4000; P = 9000	·
6	4000	·
7	·	·
8	·	L = 3000; 2000; P = 17,000
9	·	2000
10	L = 1000; 4000; P = 9000	2000
11	4000	2000
12	4000	2000
	L_1 = ? 4000	L_2 = ? 2000

alternative, other explicit estimates can be made about what will happen in years 11 and 12 for asset 1 and in years 9 through 12 for asset 2. Both approaches have the disadvantage, previously mentioned, of making estimates for into the future. Nevertheless, it sometimes must be done. However, on the practical side, mistakes in estimating (and they are sure to occur) are offset by the discounting process. That is, the further into the future that estimates are made the less effect they have on decisions because of the discounting process. Also, as stated before, that because an initial cost comparison indicates a particular asset, there is no reason it must be kept until the end of its economic life. At each point in time, there is always the possibility of replacing it with a more economical asset. The uncertainties that exist in making cost comparisons, as in all economic analysis, are a result of the attempt to predict the future—a task that must be accepted or no analysis is possible. An economic analysis based on estimates is better than no analysis at all.

Continuing with Table 10-20, assumed that L_1 = $3000 and L_2 = $6000. These values result in the equivalent annual costs (EAC) shown in Table 10-21. Values $6536 and $5571 are obtained from the previous calculations. Values $8141 and $6753 are obtained in the following manner:

$$(9000 - 3000)\,(AP\ 15{,}2) + 3000\,(0.15) + 4000 = \$8141$$
$$(17{,}000 - 6000)\,(AP\ 15{,}4) + 6000\,(0.15) + 2000 = \$6753$$

Table 10-21
EQUIVALENT ANNUAL COSTS FOR FINITE SERVICE LIFE (12 YEARS)

End of Year	Asset 1	Asset 2
0	—	—
1	\$6536	\$5571
2	6536	5571
3	6536	5571
4	6536	5571
5	6536	5571
6	6536	5571
7	6536	5571
8	6536	5571
9	6536	6753
10	6536	6753
11	8141	6753
12	8141	6753

The before-tax equivalent annual costs for the two assets are

$$\begin{aligned} AC_1 &= [(6536)\,(PA\ 15{,}10) \\ &+ (8141)\,(PA\ 15{,}2)\,(PF\ 15{,}10)]\,(AP\ 15{,}12) \\ &= \$6656 \text{ per year} \\ AC_2 &= [(5571)\,(PA\ 15{,}8) + (6753)\,(PA\ 15{,}4)\,(PF\ 15{,}8)]\,(AP\ 15{,}12) \\ &= \$5776 \text{ per year} \end{aligned}$$

Note that values inside the brackets of these calculations give the present worth of the costs for a finite service life of 12 years.

In addition to the example just worked, there are some additional cases, under the assumption of finite need, that warrant further discussion. These cases result from various relationships between the least common denominator (LCD) of the economic lives of the alternatives and the estimated service life. There are three possible relationships between the LCD of the economic lives of the alternatives and the service life: (1) LCD > service life, (2) LCD = service life, and (3) LCD < service life. The life on which to base a cost comparison for these three possibilities is presented using as an example two assets with lives of 5 and 10 years (LCD = 10 years), respectively.

The first possibility (LCD > service life) is shown in Table 10-22 and indicates two situations. With a service life of three years, the salvage values L_1'

Table 10-22
COMPARISON OF ALTERNATIVES WHEN LCD > SERVICE LIFE

End of Year	Asset 1, $				Asset 2, $		
0	P_1				P_2	P_2	
1		EAC_1				EAC_2	
2							
3	L_1'		EAC_1'		L_2'		
4				← Service life			EAC_2'
5	P_1	L_1					
6							
7		EAC_1''		Service life ↙			
8	L_1''					L''_2	
9							
10					L_2		

and L_2' would have to be estimated and the comparison would be based on these values and a three-year life for each asset (EAC_1 and EAC_2). With a service life of eight years, the salvage values L_1'' and L_2'' would have to be estimated. It would also have to be assumed (without information to the contrary) that asset number one would be replaced at the end of its life (five years) by an asset with exactly the same cost data as the original one. The equivalent annual costs could then be determined for the eight years and compared. This is the same situation worked previously that gave results of \$6656 and \$5776.

Making a cost comparison for the second possibility (LCD = service life) is a relatively simple manner *with the recycling assumption:* compute the annual costs as if the service life is infinite (5 years for asset 1 and 10 years for asset 2) and compare. This is also true if the service life is any multiple of the LCD (20, 30, . . .).

The third possibility (LCD < service life) is shown in Table 10-23 and indicates a simple basis of comparison. That is, a comparison can be based on the LCD, which implies the same numerical annual costs obtained with an infinite service life and recycling of the cost data. It may appear that this analysis conflicts with the contention that cost comparisons should be based on service life. This is not the case. The implication is that, once the LDC is reached, the most economical choice of alternatives (whatever they may be) is chosen for the remaining service period; thus, resulting in the most economical choice of alternatives over the entire service life.

In summary, all cost comparisons should be based on a time interval that is equal to the anticipated service life. In the case of infinite service, the comparison is straightforward with the recycling assumption. Also, there is no conflict

Table 10-23
LCD < SERVICE LIFE

End of Year	Asset 1, $		Asset 2, $	
0	P_1		P_2	
1				
2		EAC_1		
3				
4				
5	L_1			EAC_2
6				
7				
8				
9	LCD = 10 years			
10	↘		L_2	
.				
.				
.	Service life			
n	↘			

between the economic life and service life since the economic life is always less than an infinite service life. However, this is not necessarily true for the finite service case. It is possible to have an economic life greater than the service life. In this situation the cost comparison should be based on the service life. Implying that capital recovery costs, as well as other costs, should be based on the service life not the economic life. Of course, if the economic life is smaller than the specified finite service life and the asset is recycled as shown in Table 10-20, then the economic life should be used in determining the times at which the asset is replaced.

Study Plans

It is sometimes suggested that a "study plan" should be used when making cost comparisons. There is nothing wrong with this suggestion, provided that a "study plan" means the service life (infinite or finite). Sometimes, this is not the meaning of study plans or, if it is the meaning, it is not explicitly stated as such. Many times, study plans are based on the lives of the assets. This is incorrect. A true study plan should be the anticipated service life which is not, necessarily, related to lives of the assets. It is true that it is possible to define study plans relating to the lives of the assets that give the same cost comparison as an infinite or finite service life. However, this can lead to a misconception regarding the correct time periods on which cost comparisons should be based—the service life.

In order to further illustrate the procedures for making cost comparisons with different economic lives and also extend these procedures to include combinations of debt and equity financing, the following example is worked.

EXAMPLE 10-6

Compare the two assets shown in Table 10-24 on the basis of the following service lives: (1) infinite, (2) 20 years, (3) 35 years, (4) 5 years, and (5) 8 years.

For an infinite service life, the annual costs are determined as previously described and compared. The annual costs are shown in Tables 10-25 and 10-26 and their equivalent annual costs are

$$\begin{aligned}
\text{EAC}_1 &= [74{,}000\,(PF\ 20{,}1) + \ldots + 72{,}060\,(PF\ 20{,}5)]\,(AP\ 20{,}5) \\
&= \$73{,}506 \\
\text{EAC}_2 &= [80{,}727\,(PF\ 20{,}1) + \ldots + (57{,}558)\,(PF\ 20{,}10)]\,(AP\ 20{,}10) \\
&= \$72{,}810
\end{aligned}$$

The choice between assets 1 and 2 will depend on the relative importance of the short-term effects (annual costs) and the long-term effects (equivalent annual amount) since there is a conflict between these two effects. That is, the annual costs for asset one are less in years 1 through 4 while the equivalent annual amount of the annual costs is larger than equivalent annual amount for asset 2.

A cost comparison of the assets for a finite service life of 20 years is the same as the infinite service life since the service life of 20 years is a multiple of the LCD of the lives of the assets. Again the recycling assumption is being made.

A cost comparison of the assets for a finite service life of 35 years would also be the same as the infinite service life since the service life is greater than the LCD of the lives of

Table 10-24
COST DATA FOR EXAMPLE 10-6

	Assets 1	Assets 2
Initial costs	\$100,000	\$150,000
Annual operating costs	\$30,000	\$20,000
Tax depreciation	SYD	SYD
Life, years	5	10
Salvage values	\$10,000	\$20,000
Book depreciation	SL	SL
Life, years	5	10
Salvage values	\$10,000	\$20,000
Debt ratio	20%	20%
Cost of debt capital	10%	10%
Required equity return	22.5%	22.5%
Investment tax credit	0%	0%
Tax rate	52%	52%

Table 10-25

ANNUAL COSTS FOR ASSET 1 ($) IN EXAMPLE 10-6

Year	Book Dep.	SL Dep.	Tax Dep.	Tax Credit
1	$18,000.00	$18,000.00	$30,000.00	0.00
2	18,000.00	18,000.00	23,999.98	0.00
3	18,000.00	18,000.00	17,999.99	0.00
4	18,000.00	18,000.00	12,000.00	0.00
5	18,000.00	18,000.00	6,000.00	0.00

Year	Cap Inv.	Cap. Int.	Ret. on Debt.	Ret. on Eqty.	Tax	Cost	Annual Cost
	$100,000.00						
1	82,000.00	0.00	$2,000.00	$18,000.00	$ 6,500.00	$30,000.00	$74,500.00
2	64,000.00	0.00	1,640.00	14,760.00	9,490.01	30,000.00	73,890.00
3	46,000.00	0.00	1,280.00	11,520.00	12,480.00	30,000.00	73,280.00
4	28,000.00	0.00	920.00	8,280.00	15,470.00	30,000.00	72,669.94
5	10,000.00	0.00	560.00	5,040.00	18,460.00	30,000.00	72,059.94

Table 10-26
ANNUAL COSTS FOR ASSET 2 ($) IN EXAMPLE 10-6

Year	Book Dep.	SL Dep.	Tax Dep.	Tax Credit			
1	$13,000.00	$13,000.00	$23,636.36	0.00			
2	13,000.00	13,000.00	21,272.71	0.00			
3	13,000.00	13,000.00	18,909.08	0.00			
4	13,000.00	13,000.00	16,545.45	0.00			
5	13,000.00	13,000.00	14,181.81	0.00			
6	13,000.00	13,000.00	11,818.18	0.00			
7	13,000.00	13,000.00	9,454.54	0.00			
8	13,000.00	13,000.00	7,090.91	0.00			
9	13,000.00	13,000.00	4,727.27	0.00			
10	13,000.00	13,000.00	2,363.64	0.00			
Year	Cap. Inv.	Cap. Int.	Ret. on Debt	Ret. on Eqty.	Tax	Cost	Annual Cost
	$150,000.00						
1	137,000.00	0.00	$3000.00	$27,000.00	$17,727.27	$20,000.00	$80,727.25
2	124,000.00	0.00	2740.00	24,660.00	17,752.89	20,000.00	78,152.88
3	111,000.00	0.00	2480.00	22,320.00	17,778.49	20,000.00	75,578.44
4	98,000.00	0.00	2220.00	19,980.00	17,804.09	20,000.00	73,004.06
5	85,000.00	0.00	1960.00	17,640.00	17,829.70	20,000.00	70,429.69
6	72,000.00	0.00	1700.00	15,300.00	17,855.30	20,000.00	67,855.25
7	59,000.00	0.00	1440.00	12,960.00	17,880.91	20,000.00	65,280.90
8	46,000.00	0.00	1180.00	10,620.00	17,906.51	20,000.00	62,706.51
9	33,000.00	0.00	920.00	8,280.00	17,932.12	20,000.00	60,132.11
10	20,000.00	0.00	660.00	5,940.00	17,957.72	20,000.00	57,557.71

Table 10-27
COMPARISON OF THE ASSETS IN EXAMPLE 10-6 FOR A SERVICE LIFE OF 35 YEARS

End of Year	Equivalent Costs	
	Asset 1	Asset 2
0	—	—
1	$73,506	$72,810
2	73,506	72,810
3	73,506	72,810
4	.	.
.	.	.
.	.	.
.	.	.
29	73,506	72,810
30	73,506	72,810
31	—	—
.	—	—
.	—	—
.	—	—
35	—	—

the assets. What is being implied is shown in Table 10-27. That is, after the end of 30 years of service. The most economical alternative at that time is picked for the remaining five years of service.

For a service life of five years some adjustments to the annual costs of asset 2 are necessary since the service life is less than the life of the asset. The annual costs for asset one are the same as in Table 10-25 since its life equals the service life. The annual costs for asset 2 are shown in Table 10-28 and are explained in the following discussion.

The first task is to estimate a realizable salvage at the end of five years. In Table 10-28 it is estimated that this salvage value is the company's capital investment (book value) at the end of five years: $85,000. The argument for estimating the salvage value as the book value in cost comparisons (this argument is not necessarily valid in replacement studies) is that to do otherwise is tantamount to saying the *book* depreciation schedule is incorrect. Remember, there are *two* depreciation schedules. One is the tax depreciation schedule and the other is the book depreciation schedule for *the recovery of invested capital*. Also, the estimation of capital recovery is being made for the alternatives at $t = 0$. Consequently, the book values at any year should be in complete agreement with the estimated (expected) salvage values at the same points in time. Otherwise, the book depreciation schedule if being incorrectly estimated; or, the estimator is not being consistent with his or her own estimates.

With the assumption of a salvage value of $85,000, the annual costs for the first four years are the same as in Table 10-26. However in year 5 there may be a large tax effect,

Table 10-28

ANNUAL COSTS FOR ASSET 2 ($) WITH A SERVICE LIFE OF 5 YEARS (EXAMPLE 10-6)

Year	Book Dep.	SL Dep.	Tax Dep.	Tax Credit			
1	$13,000.00	$13,000.00	$23,636.36	0.00			
2	13,000.00	13,000.00	21,272.71	0.00			
3	13,000.00	13,000.00	18,909.08	0.00			
4	13,000.00	13,000.00	16,545.45	0.00			
5	13,000.00	13,000.00	14,181.81	0.00			
6	0.00	0.00	11,818.18	0.00			
7	0.00	0.00	9,454.54	0.00			
8	0.00	0.00	7,090.91	0.00			
9	0.00	0.00	4,727.27	0.00			
10	0.00	0.00	2,363,64	0.00			
Year	Cap. Inv.	Cap. Int.	Ret. on Debt	Ret. on Eqty.	Tax	Cost	Annual Cost
0	$150,000.00						
1	137,000.00	0.00	$3000.00	$27,000.00	$17,727.27	$20,000.00	$ 80,727.25
2	124,000.00	0.00	2740.00	24,660.00	17,752.89	20,000.00	78,152.88
3	111,000.00	0.00	2480.00	22,320.00	17,778.49	20,000.00	75,578.44
4	98,000.00	0.00	2220.00	19,980.00	17,804.09	20,000.00	73,004.06
5	85,000.00	0.00	1960.00	17,640.00	49,837.11	20,000.00	102,437.00
6	0.00	0.00	0.00	0.00	0.00	0.00	0.00
7	0.00	0.00	0.00	0.00	0.00	0.00	0.00
8	0.00	0.00	0.00	0.00	0.00	0.00	0.00
9	0.00	0.00	0.00	0.00	0.00	0.00	0.00
10	0.00	0.00	0.00	0.00	0.00	0.00	0.00

depending on the classification (ordinary or extraordinary) of the retirement, if the choice is asset 2. This is a result of the difference between the salvage value and the *tax* book value. The tax book value at the end of five years is

$$\begin{aligned} BV_5 &= 150{,}000 - \frac{40}{50}(150{,}000 - 20{,}000) \\ &= \$55{,}455 \end{aligned}$$

resulting in a gain of

$$85{,}000 - 55{,}455 = \$29{,}545$$

If the retirement is classified as extraordinary, there is a tax increase of

$$29{,}545\,(0.52) = \$15{,}363$$

in year 5 and the taxes, from the standpoint of determining the annual cost, are

$$\begin{aligned} t_5 &= (13{,}000 + 17{,}640 - 14{,}181)\left(\frac{0.52}{0.48}\right) + \frac{15{,}363}{1 - 0.52} \\ &= 17{,}830 + 32{,}007 = \$49{,}837 \end{aligned}$$

Therefore, the annual cost in year 5 is

$$R_5 = 13{,}000 + 1960 + 17{,}640 + 20{,}000 + 49{,}837 = \$102{,}437$$

As a check on the taxes, the following calculation is made

$$\begin{aligned} t_5 &= (102{,}437 - 1960 = 20{,}000 - 14{,}181)\,(0.52) + 15{,}363 \\ &= \$49{,}837 \end{aligned}$$

The equivalent annual cost for the two assets are

$$\begin{aligned} EAC_1 &= \$73{,}506 \\ EAC_2 &= [80{,}727\,(PF\ 20{,}1) + \ldots + 102{,}437\,(PF\ 20{,}5)]\,(AP\ 20{,}5) \\ &= \$80{,}812 \end{aligned}$$

and indicates that asset 1 should be chosen since its annual amounts and equivalent annual cost are less than asset 2.

If the retirement is classified as an ordinary retirement, then the annual costs would be the same as the first five years in Table 10-24 since there are not tax considerations on the gain (or loss). The equivalent annual cost is

$$\begin{aligned} EAC_2 &= [80{,}727\,(PF\ 20{,}1) + \ldots + 70{,}430\,(PF\ 20{,}5)]\,(AP\ 20{,}5) \\ &= \$76{,}511 \end{aligned}$$

The decision in this case is clearly asset 1 since both its annual costs and equivalent annual cost are less than asset 2.

The calculations for a service life of eight years are similar to the calculations for a service life of five years except a recycling assumption is necessary for asset 1. The annual costs for a service life of eight years are shown in Table 10-29 and explained in the following discussion.

For asset 1, the annual costs for the first five years are the same as in Table 10-25. At the end of five years it is assumed that the original asset is replaced by an asset having the same cost data as the original one. Consequently, the annual costs for years 6 and 7 are the

Table 10-29
ANNUAL COSTS FOR A SERVICE LIFE OF EIGHT YEARS

End of Year	Asset 1	Asset 2
0	—	—
1	\$74,500	\$80,727
2	73,890	78,152
3	73,280	75,578
4	72,670	73,004
5	72,060	70,430
6	74,500	67,855
7	73,890	65,280
8	92,780 (73,280)[a]	83,212 (62,707)[a]

[a]These values are for an ordinary retirement.

same as years 1 and 2. However in year 8, there are tax considerations if the retirements are classified as extraordinary. These tax effects are calculated in the same manner as previously discussed. For alternative 1, the tax book value of the asset one after three years is

$$BV_3 = 100{,}000 - \frac{12}{15}(90{,}000) = \$28{,}000$$

This tax book value actually occurs in year 8 because of the recycling assumption. The salvage value at this same point (year 8) is taken as the capital investment (capital unrecovered) on the basis of the arguments previously presented regarding the estimation of the recovery of capital (book depreciation schedule). This salvage value is \$46,000 (see Table 10-25). The disposal of this asset (the one purchased in year 5) will increase taxes in year 8 by the amount

$$(46{,}000 - 28{,}000)\,(0.52) = \$9360$$

Therefore, the taxes in year 8 from the standpoint of annual costs, are (see Table 10-25)

$$\begin{aligned} t &= (18{,}000 + 11{,}520 - 18{,}000)\left(\frac{0.52}{0.48}\right) + \frac{9360}{1 - 0.52} \\ &= 12{,}480 + 19{,}500 = \$31{,}980 \end{aligned}$$

and the annual cost is

$$R = 18{,}000 + 11{,}520 + 1280 + 30{,}000 + 31{,}980 = \$92{,}780$$

In the case of asset 2, the tax book value at the end of eight years is

$$BV_8 = 150{,}000 - \frac{52}{55}(130{,}000) = 27{,}072$$

and the tax effect is an increase in taxes of

$$(46{,}000 - 27{,}072)\,(0.52) = \$9843$$

The $46,000 value is the capital investment (taken as the salvage value) for asset 2 at the end of the eighth year (see Table 10-26). No significance should be attributed to this value ($46,000) being the same as in asset 1; it is merely a coincidence.

The taxes in year 8 are (see Table 10-26)

$$t = (13{,}000 + 10{,}620 - 7091)\left(\frac{0.52}{0.48}\right) - \frac{9843}{1 - 0.48} = \$38{,}412$$

and the annual cost is

$$R = 13{,}000 + 10{,}620 + 1180 + 20{,}000 + 38{,}412 = \$83{,}212$$

If the retirements are considered ordinary, there are no tax effects and the annual costs in the eight years are the same as year 3 in Table 10-25 for asset 1 and year 8 in Table 10-26 for asset 2. These values are shown in parenthesis in Table 10-29.

The equivalent annual costs for a service life of eight years and extraordinary retirement are

$$\begin{aligned}\mathrm{EAC}_1 &= [74{,}500\,(PF\ 20{,}1) + \ldots + 92{,}780\,(PF\ 20{,}8)]\,(AP\ 20{,}8)\\ &= \$74{,}784\\ \mathrm{EAC}_2 &= [80{,}727\,(PF\ 20{,}1) + \ldots + 83{,}212\,(PF\ 20{,}8)]\,(AP\ 20{,}8)\\ &= \$78{,}692\end{aligned}$$

For an ordinary retirement, they are

$$\begin{aligned}\mathrm{EAC}_1 &= [74{,}500\,(PF\ 20{,}1) + \ldots + 73{,}280\,(PF\ 20{,}8)]\,(AP\ 20{,}8)\\ &= \$73{,}602\\ \mathrm{EAC}_2 &= [80{,}727\,(PF\ 20{,}1) + \ldots + 62{,}707\,(PF\ 20{,}8)]\,(AP\ 20{,}8)\\ &= \$77{,}449\end{aligned}$$

From a decision standpoint, the choice of an asset will be a balance between the short-term effects (annual costs) and the long-term effects (equivalent annual costs) since there is a conflict between these two effects.

LEASE VERSUS PURCHASE

Many times companies must decide between the alternatives of leasing or buying equipment. To some, leasing has an intuitive appeal because it is believed to have a tax advantage and can provide needed working capital by avoiding long-term capital commitments. This latter belief is true. However, the tax advantage may or may not be true because the existence of a tax advantage depends on the depreciation amounts and operating costs of the purchase alternative.

It is true that leasing is often more desirable than purchasing. However, all lease-buy decisions should be determined by an actual cost comparison. The following examples shows the procedures that may be used for making lease-buy decisions.

EXAMPLE 10-7

In this example the assumption of 100% equity financing is made and two cases are considered: a before-tax and after-tax cost comparison. These comparisons assume infinite need.

Suppose it is possible to lease certain equipment to satisfy some particular need at an annual cost of \$7000. Or, equipment can be purchased, with the cost data shown in Table 10-30, that provides the same service.

Before-Tax Comparison: The first step in making a before-tax cost comparison is to correct the after-tax required return on equity to a before-tax return using Eq. (10-1):

$$(\text{MARR})_b = \frac{0.15}{(1 - 0.50)} = 0.30 = 30\%$$

The equivalent annual before-tax cost of the equipment is

$$(\text{AC}) = (20{,}000 - 2000)\,(AP\ 30{,}8) + 2000\,(0.3) + 500 = \$7254$$

Comparison of this value and the annual leasing cost of \$7000 indicates that leasing the equipment is the better decision.

After-Tax Comparison: In making an after-tax cost comparison, explicit consideration is given to the difference in taxes resulting from the difference in tax deductions: namely,

$$[7000 - (2250 + 500)](0.50) = \$2125 \text{ per year}$$

The basis of this calculation is discussed in Example 10-2.

Therefore, the equivalent annual cost of the equipment is

$$\begin{aligned} \text{AC} &= (20{,}000 - 2000)\,(AP\ 15{,}8) + 2000\,(0.15) + 500 + 2125 \\ &= \$6937 \end{aligned}$$

and indicates that purchasing the equipment is the better decision when compared to the annual leasing cost of \$7000. It also indicates that there is a conflict between the before-tax and after-tax comparison. The correct decision in this case is to purchase the equipment because taxes have been explicitly considered in the second analysis. This conflict also points out one of the major contentions of this text. That is, cost comparisons that do not *explicitly* consider taxes can lead to incorrect decisions.

In Example 10-7 the assumption of 100% equity financing is made. In the next example this assumtion is relaxed and also short-term effects of purchasing equipment are considered.

Table 10-30
COST DATA FOR EXAMPLE 10-7

Initial cost	\$20,000
Economic life	8 years
Salvage value	\$2,000
Annual operating cost	\$500
Tax rate	50%
Required after tax return on equity	15%

EXAMPLE 10-8

Suppose it is possible to lease certain equipment at an annual cost of $3500 and, as an alternative, the equipment described as alternative 2 in Table 10-8 can be purchased. The service life is assumed to be infinite.

In order to consider the debt and equity financing and the short-term effects, the annual cost are generated. These cost are shown in Table 10-10. A comparison of the annual costs in Table 10-10 and the annual leasing cost of $3500 indicates that in some years the leasing cost is higher and in other years it is lower. If the equivalent annual cost (long-term effect) is computed for the annual costs given in Table 10-10, the result is $3954 (see Example 10-3). From the standpoint of deciding which alternative should be chosen, both the short-term and long-term effects must be considered. It is entirely possible that a decision might be to purchase the equipment because of the lower annual costs in the first two years even though the long-term effects are higher. In the final analysis, the decision of lease or buy would depend on the relative merits of the short-term and long-term effects.

Summary

In this chapter a very important procedure was presented for making cost comparisons—generating the annual costs. This procedure has the distinct advantage of being completely general. That is, it allows for explicit tax and short-term considerations as well as debt and equity financing to be included in cost comparisons.

Problems

10-1. A company is considering the installation of some equipment. The two alternatives under consideration are shown in the following table:

	Alternative	
	1	2
Initial cost	$100,000	$130,000
Salvage	$15,000	$10,000
Life	10 years	10 years
Annual Costs	$10,000	$5,000
Required equity return (after tax)	12%	12%
Tax rate	52%	52%

Assuming the need for the service provided by this equipment is at least 10 years, determine the following:

(a) A before-tax equivalent annual cost comparison.

(b) A before-tax present worth comparison.

(c) A before-tax capitalized cost comparison.
(d) What assumption is implied in part (c).

10-2. Repeat parts (a) and (b) of Problem 10-1 under the following conditions:
(a) The need for the service provided by the equipment is only 5 years and the estimated realizable salvage values, at that time, are L_1 = \$30,000 and L_2 = \$50,000.
(b) The need for the service provided by the equipment is for only 15 years.

10-3. For the data given in Problem 10-1, determine the equivalent annual costs for the two alternatives taking into account taxes explicitly. Assume the tax depreciation model is straight line and the tax life and salvage values are the same as given in Problem 10-1.

10-4. Suppose the equipment in Problem 10-1 is needed for only 5 years and qualifies as Sec. 1245 property. Also, the net realizable salvage values (what can actually be obtained) for the two alternatives are L_1 = \$30,000 and L_2 = \$50,000 at the end of 5 years. Assuming straight line depreciation, a tax life of 10 years, and a tax salvage value of L_1 = \$15,000 and L_2 = \$10,000, determine an after-tax equivalent annual cost comparison under the following conditions:
(a) Ordinary retirements and no investment tax credit.
(b) Extraordinary retirements and no investment tax credit.
(c) Repeat parts (a) and (b) with an investment tax credit of 10%.

10-5. For the data given in the following table, determine the annual costs, using the flow-through method, for each alternative and the alternative which should be chosen. Assume the alternatives provide equal service and the tax and book depreciation models are the same except book depreciation uses a straight-line model.

	Alternative	
	1	2
First cost	\$50,000	\$70,000
Annual operating & maintenance	\$10,000	\$6,000
Tax depreciation	SYD	SYD
Life	10 years	10 years
Salvage	\$5,000	\$7,000
Debt ratio	25%	25%
Cost of debt capital	10%	10%
Required return on equity	30%	30%
Investment tax credit	10%	10%
Tax rate	52%	52%

10-6. Listed in the following table are three mutually exclusive projects that provide equal service. Determine the annual costs using the flow-through method for each project and decide which one should be chosen assuming the need for the service is infinite. What assumption is implied in your analysis.

	Projects		
	1	2	3
Initial cost	$200,000	$300,000	$500,000
Annual costs	$50,000	$30,000	$20,000
Tax depreciation	SYD	SYD	SYD
Life	10 years	10 years	15 years
Salvage	0	$20,000	$30,000
Book depreciation	SL	SL	SL
Life	15 years	15 years	20 years
Salvage	0	$20,000	$30,000
Debt Ratio	20%	20%	20%
Required debt return	10%	10%	10%
Required equity return	22.5%	22.5%	22.5%
Investment tax credit	10%	10%	10%

10-7. Suppose in Problem 10-6 the investment tax credit is zero. Which project should be chosen?

10-8. Using the flow-through method, compare the two assets shown in the following table on the basis of the following service lives: (1) infinite, (2) 30 years, (3) 40 years, and (4) 20 years.

	Asset	
	A	B
Initial cost	$150,000	$200,000
Annual costs	$40,000	$30,000
Tax depreciation	SYD	SYD
Life	10 years	15 years
Salvage value	$10,000	$20,000
Book depreciation	SL	SL
Life	10 years	15 years
Salvage value	$10,000	$20,000
Debt ratio	40%	40%
Cost of debt capital	9%	9%
Required return on equity	19%	19%
Investment tax credit	10%	10%
Tax rate	52%	52%

10-9. A company is considering the installation of either a fully automated or semiautomated materials-handling system. The automated system has a higher initial cost but requires less maintenance and operating costs. While, in the case of the semiautomated system, the initial cost is less but annual operating and maintenance costs are higher. Given the data in the following table for the two systems and assuming both systems provide equal service, determine which system should be chosen using the flow-through methods and assuming infinite need.

	Semiautomated	Automated
Initial costs	$1,000,000	$1,500,000
Annual costs	$300,000	$100,000
Tax depreciation	DDB	DDB
Life	10 years	10 years
Salvage	$100,000	$150,000
Book depreciation	SL	SL
Life	20 years	20 years
Salvage	$100,000	$150,000
Debt ratio	25%	25%
Required debt return	10%	10%
Required equity return	30%	30%
Investment tax credit	10%	10%
Tax rate	52%	52%

10-10. An electric utility company must expand their distribution system in order to supply electrical service to a shopping center and a large housing development. The company has a choice between installing some of the necessary facilities underground or above ground. Given the etimates in the following table, determine which alternative should be chosen using the normalizing method.

	Aboveground	Underground
Initial cost	$1,500,000	$3,000,000
Annual maintenance	$300,000	$100,000
Tax depreciation	SYD	SYD
Life	20 years	20 years
Salvage	$200,000	$400,000
Book depreciation	SL	SL
Life	20 years	20 years
Salvage	$200,000	$400,000
Investment tax credit	4%	4%
Capitalized interest	8%	8%
Required return on debt	10%	10%
Required return on equity	15%	15%
Debt ratio	60%	60%
Tax rate	52%	52%

10-11. Suppose in Problem 10-10, the owners of the shopping center and the housing development are opposed to above-ground electrical facilities on the basis that they (poles, lines, etc.) do not enhance the appearance of their developments and want to know what additional amount (at the present time) would be required for the installation of underground facilities. What would be an equitable amount to charge?

10-12. Using the data given below, determine the economic life of the asset assuming the asset is Sec. 1245 property and ordinary retirements.

Initial cost	$150,000
Tax depreciation	SYD
Life	10 years
Salvage	$20,000
Debt ratio	20%
Required return on equity	22.5%
Debt capital	10%
Investment tax credit	0
Tax rate	52%

End of Year	Salvage	Oper. & Maint.
1	$120,000	$20,000
2	100,000	25,000
3	80,000	30,000
4	60,000	35,000
5	50,000	40,000
6	40,000	45,000
7	35,000	50,000
8	30,000	55,000
9	25,000	60,000
10	20,000	65,000

10-13. Using the flow-through method and the data given in the following table, determine the economic life of the asset assuming Sec. 1245 property and ordinary retirements.

Initial cost	$150,000
Tax depreciation	SYD
Life	8 years
Salvage	$20,000
Debt ratio	20%
Required return on debt	10%
Required return on equity	22.5%

Investment tax credit		0%
Tax rate		52%

End of Year	Salvage	Maint. & Operating Costs
1	$110,000	2,000
2	80,000	3,000
3	70,000	5,000
4	60,000	10,000
5	50,000	15,000
6	40,000	20,000
7	30,000	25,000
8	20,000	30,000

10-14. Using the data given in the following table and the flow-through method, determine the economic life of the asset assuming Sec. 1245 property and ordinary retirements.

Initial cost	$100,000
Tax depreciation	SYD
Life (years)	10
Salvage	$10,000
Debt ratio	20%
Required return on debt	10%
Required return on equity	22.5%
Investment tax credit	0%
Tax rate	52%

End of Year	Salvage	Oper. & Maint.
1	$91,000	$10,000
2	82,000	10,000
3	73,000	10,000
4	64,000	10,000
5	55,000	10,000
6	46,000	15,000
7	37,000	20,000
8	28,000	25,000
9	19,000	30,000
10	10,000	40,000

10-15. Repeat Problem 10-14 with the assumption that an investment tax credit of 10% is applicable. Remember in working this problem that investment tax penalties can occur with certain useful lives.

10-16. A company is in need of additional storage space. It can lease the space or expand their present storage space. Cost estimates associated with an expansion of existing facilities are given below:

Initial cost	$1,000,000
Annual costs	$200,000
Tax depreciation	straight line
Life (years)	30
Salvage	$100,000
Book depreciation	straight line
Life (years)	30
Salvage	$100,000
Debt ratio	20%
Required debt return	10%
Required equity return	22.5%
Investment tax credit	0%
Tax rate	52%

If the annual cost of renting the storage space is (a) $650,000, (b) $250,000, (c) $500,000, and (d) $600,000 should the space be leased or built?

10-17. Repeat Problem 10-15 assuming extraordinary retirements.

CHAPTER 11
REPLACEMENT ANALYSIS

A frequent question in economic analysis is, "Should a particular piece of equipment (an asset) be replaced?" This question usually results from some inadequacy with the existing equipment or from the availability of more efficient equipment. Whatever the reason, the replacement decision should be based on a sound economic analysis of the alternatives. This chapter presents various approaches for making replacement studies. As in Chapter 10, this chapter considers, basically, two approaches in making replacement studies. That is, before- and after-tax replacement studies assuming 100% equity financing and after-tax replacement studies assuming a combination of debt and equity financing.

EXAMPLE 11-1

Four years ago, a compressor was purchased for \$50,000 and at that time was estimated to have a life of 10 years with a salvage of \$5000. In the past, operation and maintenance costs have amounted to \$7000 a year for this compressor and are expected to continue at this rate. The present realizable salvage value for this existing compressor is \$25,000 (this could be the "trade-in" value). As a result of past experience with this compressor, it is now estimated that it will last only five more years and will have a salvage of \$4000 at that time.

A new compressor is now available that has an estimated annual operating and maintenance cost of \$2000. It costs \$60,000, has an estimated salvage of \$6000, and life of 10 years. The required return on equity before-tax is 25%.

Before-Tax Comparison: It is highly desirable to portray the data for replacement studies as shown in Table 11-1. A simple equivalent annual cost for the two assets shown in Table 11-1 provides a before-tax comparison as follows:

$$\text{EAC}_e = (25{,}000 - 4000)\,(AP\ 25{,}5) + 4000\,(0.25) + 7000 = \$15{,}810$$
$$\text{EAC}_n = (60{,}000 - 6000)\,(AP\ 25{,}10) + 6000\,(0.25) + 2000 = \$18{,}625$$

Which indicates the existing asset should not be replaced.

Table 11-1
COST DATA FOR EXAMPLE 11-1

End of Year	Existing Asset			New Asset	
−4	P = \$50,000; n = 10; L = \$5000				
−3					
−2					
−1					
0	S = 25,000		0	P = \$60,000	
1		7000	1		2000
2		7000	2		2000
3		7000	3		2000
4		7000	4		2000
5	L = 4,000	7000	5		2000
			.		.
			.		.
			.		.
					.
			10	L = 6,000	2000

A Comment: Some comments are necessary in regard to Example 11-1. First, as in most replacement studies, the life of the new asset is longer than the existing asset. The approach to differences in the lives used in a replacement study is essentially the same as discussed in the chapter on cost comparisons (Chapter 10); that is, determine if the expected need for the service provided is infinite or finite. In this example, the answers \$15,810 and \$18,625 are based on infinite need and an assumption of recycling. There are two ways of looking at recycling in replacement studies. Either way leads to the same numerical values for comparison purposes *as long as the need is assumed to be infinite.* The reason both ways are mentioned is because one is more plausible than the other. The first way is to assume both alternatives in Table 11-1 recycle out to infinity. That is, the existing asset is replaced every five years with an asset having the same cost data. While the new asset is replaced every 10 years with an asset having the same cost data. The second way is to assume that the new asset is the best possible replacement at the present time and it will replace the existing asset after five years, as shown in Table 11-2. It is the contention of this text that the second recycling assumption is more plausible than the first for replacement analysis. Furthermore, it is assumed that this is the case when the need is finite. For example, suppose the need is determined to be 12 years (a finite need). Using the second recycling assumption, the situation depicted in Table 11-3 would be the basis of the replacement comparison. The salvage values L_1 and L_2 would have to be estimated and then the equivalent annual costs for the two alternatives for the twelve year period would be computed and compared. It may seem that "buying" an asset at the end of year 10 for the new asset is not very realistic. However, due to lack of information to the contrary, it is as good as assumption as any. From a practical standpoint, the arguments presented in Chapter 10 should be remembered; that is, once a particular comparison is

Table 11-2
EQUIVALENT ANNUAL COSTS FOR RECYCLING ASSUMPTION 2

End of Year	Existing Asset	New Asset
0		
1	\$15,810	\$18,625
2	15,810	·
3	15,810	·
4	15,810	·
5	15,810	·
6	18,625	·
7	·	·
8	·	·
9	·	·
10	18,625	·
·	·	·
·	·	·
·	·	·
∞	18,625	18,625

Table 11-3
BASIS OF COST COMPARISON FOR EXAMPLE 11-1 WITH A FINITE NEED OF 12 YEARS

End of Year	Existing Asset			New Asset		
0	S = \$25,000			P = \$60,000		
1		7000			2000	
2		7000			·	
3		7000			·	
4		7000			·	
5	L = 4,000	7000 P = 60,000			·	
6			2000		·	
7			2000		·	
8			2000		·	
9			·		·	
10			·	L = 6,000	·	P = 60,000
11			·		·	
12		L_1;	2000		·	
					2000; L_2	

made and the best alternative is undertaken, this is not a "legal commitment" to continue with it to the end of its needed life. It is always possible to change (theoretically at least) to another asset or an alternative course of action (repair or overhaul) if a later economic evaluation indicates it is advantageous.

Another question that often arises in replacement analysis is in regard to *sunk costs*. In general, a sunk cost is a cost that is a result of past decisions (or estimates) and cannot be changed by future decisions. It can be defined in several ways. This text chooses to define sunk cost as the difference between the present book value and the realizable salvage value. For example, if straight-line depreciation is assumed to be the depreciation model for the existing asset in Example 11-1, the present book value is

$$50{,}000 - \frac{4}{10}(50{,}000 - 5000) = \$32{,}000$$

Therefore, the sunk cost is

$$32{,}000 - 25{,}000 = \$7000$$

Since this value ($7000) is a result of past decisions, it has no place in a *before-tax* replacement analysis. However, it can have an effect in *after-tax* replacement analysis as shown in later examples.

EXAMPLE 11-2

This example is an extension of Example 11-1. In replacement analysis there often occurs a third alternative in addition to the basic two alternatives of keeping an existing asset or replacing it, that is, the alternative of repairing or overhauling an existing asset. Suppose in Example 11-1 that the alternative of overhauling the existing asset at a cost of $5000 is possible in addition to the two alternatives given in Example 11-1. Furthermore, it is estimated that this overhaul will extend the life of the existing asset to seven years, reduce annual operations and maintenance costs to $5,000, and the salvage value will be $3,000 at the end of the seven years.

Under the assumption of infinite need, the equivalent annual cost for this alternative is

$$\begin{aligned} EAC_r &= (25{,}000 + 5000 - 3000)\,(AP\ 25{,}7) + 3000\,(0.25) + 5000 \\ &= \$14{,}293 \end{aligned}$$

A comparison of this result with the two equivalent annual costs in Example 11-1 indicates that this alternative has the greatest economic advantage (smallest equivalent annual cost).

EXAMPLE 11-3

In this example explicit tax considerations are included in Example 11-1. It is pointed out again that the analysis in this example is still based on an assumption of 100% equity financing.

The first step is to convert the 25% before-tax return to an after-tax return. Assuming a tax rate of 52%, the after tax required return on equity is

$$MARR_a = 25(1 - 0.52) = 12\%$$

The next step is to consider the "cost" of keeping the existing asset. This cost is made up of two parts: (1) the net realizable salvage value and (2) possible tax considerations resulting from the disposal of the existing asset. The first part is the same as in Example 11-1, that is, \$25,000. The second part of this cost is based on the difference between *tax* book values and actual realizable salvage values. The tax treatment of these differences is determined by the answers to the following questions:

1. Is the asset Sec. 1245 or 1250 property?
2. Is the depreciation accounting for the asset a single account or is it included under the ADR system (see Chapter 4)?
3. If the asset is under the ADR system, is the retirement of the asset an ordinary or extraordinary retirement?

Initially, it will be assumed that the existing asset is Sec. 1245 property under the ADR system and the retirement is considered to be extraordinary. Other possibilities are considered later.

At this point it is necessary to determine if a loss or gain results from the disposal of the existing asset. This is accomplished by determining the difference between the *tax book value* and the present net realizable salvage value. Assuming, for the sake of simplicity, that tax depreciation is based on a straight-line model, there is a loss of \$7000 (previously calculated). Under the previously stated tax assumptions, this loss can result in a tax reduction. In actuality, this loss is combined with other losses and used to offset possible gains as discussed in Chapter 4. In replacement analysis it is often difficult (if not impossible) to determine what other losses or gains may occur. Consequently the tax effects resulting from the sale of a particular asset are often *approximated* on the basis that there are no gains or losses from other depreciable assets. This is the approach taken in this text. True, it is an approximation. However, an approximation of the tax effects is better than ignoring the tax effects.

On the basis of no other offsetting gains, the loss of \$7000 can be considered as a direct reduction in taxable income or a net reduction in taxes of 7000(0.52) = \$3640. Now, this reduction in taxes is only realized if the existing asset is sold. Therefore, the total "cost" of continuing with the existing asset is the sum of the net realizable salvage value and the tax effect, a total of

$$25{,}000 + 3640 = \$28{,}640$$

In addition to approximating the tax effect on the present net realizable salvage, a similar consideration is required for the future salvage value of the existing asset. The tax book value for the existing asset at the end of an additional five years (a total of nine years) of service is

$$50{,}000 - \frac{9}{10}(50{,}000 - 5000) = \$9500$$

Meaning a loss of

$$9500 - 4000 = \$5500$$

if the asset is sold for \$4000 at that time. Therefore, the salvage value at the end of five years and its tax effect are

$$4000 + (5500)(0.52) = \$6860$$

The next tax effect that must be considered is the difference between operation and maintenance costs and depreciation amounts for the two alternatives. This tax effect is discussed in Chapter 10. For this example, assuming straight-line depreciation, this annual tax effect is

$$[(7000 + 4500) - (2000 + 5400)](0.52) = \$2132$$

Finally, the equivalent annual costs are determined:

$$\text{EAC}_e = (28{,}640 - 6860)\,(AP\ 12{,}5) + 6860\,(0.12) + 7000 = \$13{,}865$$
$$\text{EAC}_n = (60{,}000 - 6000)\,(AP\ 12{,}10) + 6000\,(0.12) + 2000 + 2132 = \$14{,}410$$

This comparison indicates the existing asset should not be replaced. This is the same decision indicated by the before-tax comparison in Example 11-1. However, a before-tax and an after-tax comparison will not always be in agreement. When there is disagreement between these two methods, the decision should be based on the after-tax analysis, implying, again, that an after-tax analysis should always be made when possible.

Another method that is used for determining the equivalent annual costs is to directly include the tax effects of the costs and depreciation as shown below:

$$\begin{aligned}\text{EAC}_e &= (28{,}640 - 6860)\,(AP\ 12{,}5) + (6860)\,(0.12)\\ &\quad + 7000 - 7000\,(0.52) - 4500\,(0.52) = \$7885\\ \text{EAC}_n &= (60{,}000 - 6000)\,(AP\ 12{,}10) + 6000\,(0.12)\\ &\quad + 2000 - 2000\,(0.52) - 5400\,(0.52) = \$8430\end{aligned}$$

This procedure gives different answers than the preceeding method. However the differences between the equivalent annual costs ($545) are the same for both methods. This latter method does have some computational advantages when more than two alternatives are involved in the replacement analysis. Basically, it is a matter of preference since either method will lead to the same decision because the differences will be equal.

EXAMPLE 11-4

As an extension of Example 11-3, consider the overhaul alternative described in Example 11-2. The first thing that must be determined is whether the $5000 overhaul is an expense or a capital expenditure. If it can be treated as an expense, the effective after-tax cost of the overhaul is (5000)(1 − 0.52) = $2400. Therefore, the initial cost for the alternative of overhauling the existing asset is

$$25{,}000 + 3640 + 2400 = \$31{,}040$$

The salvage value, with tax adjustment, for this alternative is

$$3000 + (5000 - 3000)(0.52) = \$4040$$

The tax book value for this alternative is taken as the salvage value for tax purposes ($5,000) because the overhaul is estimated to extend the life of the existing asset past its tax depreciation life of 10 years.

Under the assumption of infinite need, the equivalent annual cost is

$$\begin{aligned}\text{EAC}_r &= (31{,}040 - 4040)\,(AP\ 12{,}7) + 4040\,(0.12)\\ &\quad + 5000 - 5000\,(0.52) - 4500(0.52)\,(PA\ 12{,}6)\,(AP\ 12{,}7) = \$6693\end{aligned}$$

where the expression 4500(0.52) (*PA* 12,6)(*AP* 12,7) is the tax effect of the depreciation for six years spread over a seven-year period. Remember, the basic assumption is that the new asset is the best available asset on the market today and whatever alternative is chosen, this new asset will replace the chosen alternative at the end of its life and this new asset then cycles out to infinity.

If this third alternative (overhaul) is included in Example 11-3 and compared with the values $7885 and $8430, it is the choice since it has the least equivalent annual cost.

EXAMPLE 11-5

In Example 11-4, the assumption is made that the overhaul cost ($5000) can be expensed for tax purposes. This example considers the case where part of the overhaul cost ($3000) is depreciated and the other part ($2000) is expensed. From the standpoint of the tax laws, the "acid test" is whether or not the overhaul extends the life of the existing asset. If it does, the overhaul is considered as a depreciable asset.

Suppose, the depreciable part of the overhaul expense is depreciated on the basis of a life of seven years a salvage value of $200 for tax purposes and a straight-line model. Meaning that the depreciation deduction for tax purposes, resulting from the $3000, is

$$\frac{3000 - 200}{7} = \$400 \text{ per year}$$

The initial cost for purposes of analysis is

$$25{,}000 + 3640 + 2000(1 - 0.52) + 3000 = \$32{,}600$$

This value is based on the same considerations as in Example 11-4. The $3000 value is the depreciable amount and the $2000 value is the expensed amount. The salvage value is

$$2800 + (5000 - 2800)(0.52) + 200 = \$4144$$

This salvage value is based on the asset, with overhaul, having a total realizable salvage value of $3000 at the end of seven years. Part of this total value ($200) is due to the overhaul while the remaining amount ($2800) is due to the original asset. The difference between the tax book value at the end of even years ($5000) and the $2800 is a loss. There are no tax effects on the $200 because of the assumption that this part of the salvage value, attributed to the overhaul, is equal to the book value (salvage) of the overhaul. In actuality, the portion of the realizable salvage attributed to the overhaul makes no difference in the salvage value. For example, if it is assumed that $500 of the $3000 is attributed to the overhaul, the salvage value is

$$2500 + (5000 - 2500)(0.52) + 500 - (500 - 200)(0.52) = \$4144$$

which gives the same result as before. The (500 − 200) (0.52) term is the tax effect on a gain of $300 and is negative because it would be taxed as ordinary income (Sec. 1245 property).

The equivalent annual cost for this alternative is

$$\begin{aligned} \text{EAC}_r &= (32{,}600 - 4144)\,(AP\ 12{,}7) + 4144\,(0.12) \\ &\quad + 5000 - (5000)\,(0.52) - 4500\,(PA\ 12{,}6)\,(AP\ 12{,}7)\,(0.52) \\ &\quad - 400\,(0.52) = \$6816 \end{aligned}$$

OTHER TAX CONSIDERATIONS

In addition to the tax effects included in the preceding examples, other tax considerations can be involved in replacement analysis. It is the purpose of this section to discuss some of these tax considerations.

Sec. 1245 Property Gains

If the net realizable salvage is greater than the book value and the retirement is classified as extraordinary, the gain (the difference) is subject to tax considerations. For example, suppose the net realizable salvage value for an asset is \$25,000, the tax book value is \$15,000, and the original purchase price of the asset if \$40,000. In this case, the initial cost of the existing asset used in the replacement analysis, assuming $T = 52\%$, is (see Chapter 4)

$$25{,}000 - (25{,}000 - 15{,}000)(0.52) = \$19{,}800$$

This value is based on the assumption that there are no offsetting losses from the sale of other Sec. 1245 or 1250 property.

If the net realizable salvage is higher than the original purchase price (an unusual case), say for example \$50,000, then an approximation for the initial cost is

$$\$50{,}000 - (40{,}000 - 15{,}000)\,(0.52) - (50{,}000 - 40{,}000)\,(0.30) = \$34{,}000$$

Which assumes a tax rate on the capital gain (\$10,000) of 30%. The basis of this approximation is shown in Figure 4-1.

The same type of tax effects should also be included in future estimated salvage values when applicable.

Ordinary Retirements

In Example 11-3 it is assumed that the retirement of the existing asset is extraordinary and the initial cost (\$28,640) used in the analysis reflected the tax effect. However, if the retirement is considered an ordinary retirement (not an unusual case), there are no tax effects on gains or losses. Consequently, the net realizable value of \$25,000 would be the initial cost. This would also be true for the salvage value at the end of five years. Therefore, the equivalent annual cost for the existing asset is

$$\text{EAC}_e = (25{,}000 - 4000)\,(AP\ 12{,}5) + 4000\,(0.12) + 7000 = \$13{,}305$$

While the equivalent annual cost for the new asset would be the same as calculated in Example 11-3 (\$14,410).

Single Asset Depreciation Accounting

If a single account is used for a particular asset, then the retirement of the asset is, in effect, an extraordinary retirement and the tax effects shown in Example 11-3 should be included in the analysis.

Sec. 1250 Property

All retirements of Sec. 1250 Property are classified as extraordinary under present tax laws. Consequently, there will always be some tax effects unless the book value for tax purposes is exactly equal to the net realizable salvage value. For losses resulting from the sale of Sec. 1250 property, the tax effect approximation is the same as Sec. 1245 property. When there is a gain, the tax effect depends on the method of depreciation used for tax purposes. If straight-line depreciation has been used for tax purposes, the entire gain is a long-term gain (see Example 4-3) and a tax rate of 30% should be used in approximating the tax effect. If an accelerated depreciation model has been used, the tax effect is a combination of ordinary income and possible long-term capital gains. As an example, consider the asset in Example 4-3 and a selling price of $950,000. In this case, the initial cost for replacement analysis is

$$950{,}000 - (100{,}646)\,(0.52) - 70{,}000\,(0.3) = \$876{,}664$$

Investment Tax Credit

For an existing asset, the original investment tax credit has no bearing on a replacement analysis since it has been realized in the past. However, there may be an investment tax penalty. Present tax laws state that if Sec. 38 property is disposed of before the end of its estimated useful life, the tax for the year of disposal is increased by the difference between the credit originally allowed and the credit that would have been allowed if the computation had been based on the shorter useful life. That is, if the asset is kept at least seven years there is no investment tax penalty. As an example, assume that investment tax considerations are to be included in Example 11-5. First, if the existing asset is sold, there is an investment tax penalty of

$$50{,}000\,(0.10) - \frac{1}{3}\,(0.10)\,(50{,}000) = \$3333$$

because the useful life is four years. Second, the overhaul expense of $3000 qualifies for the full investment tax credit since the estimated life is seven years, that is,

$$3000\,(0.10) = \$300$$

Finally, if the new asset is purchased, it will qualify for the full investment tax credit:

$$60{,}000\,(0.10) = \$6000$$

Now, the penalty of $3333 and credit of $6000 are only experienced if the new asset is purchased. Therefore, the net (a credit) of $2667 should be introduced into the equivalent annual cost of the new asset. Furthermore, the $300 credit is only experienced if the existing asset is kept and overhauled and, therefore, should be included in the equivalent annual cost of the existing asset.

As mentioned in Chapter 5, there is some slight controversy in regard to the year in which investment tax credits (or penalties) should be taken; some suggest year zero, others suggest the first year. This text believes that they should be taken in the first year. Actually, as long as there is consistency, the decision will not be altered.

Adjustments for these investment tax considerations are included in the following manner:

$$\begin{aligned}\text{EAC}_r &= (32{,}600 - 4144)\,(AP\ 12{,}7) + 4144\,(0.12) + 5000 \\ &\quad - 5000(0.52) - 4500(PA\,12{,}6)\,(AP\,12{,}7)\,(0.52) \\ &\quad - 400(0.52) - 300(PF\ 12{,}1)\,(AP\ 12{,}7) = \$6757 \\ \text{EAC}_n &= (60{,}000 - 6000)\,(AP\ 12{,}10) + 6000\,(0.12) + 2000 \\ &\quad - 2000\,(0.52) - 5400\,(0.52) - 2667\,(PF\ 12{,}1)\,(AP\ 12{,}10) = \$8008\end{aligned}$$

Comparing these results with EAC_e = \$7885, the decision in the overhaul alternative.

REPLACEMENT ANALYSIS EXTENSIONS

The purpose of this section is to extend replacement analysis to include combinations of debt and equity financing and the determination of specific yearly costs. These extensions are possible through the use of the concept of minimum annual revenue requirements (annual costs) as shown in the following example.

EXAMPLE 11-6

Three years ago the asset with the estimated data shown in Table 11-4 was installed. The original estimate of maintenance and operation costs (\$1700) has been found to be low. They have been \$3000 per year and are expected to continue at this amount if this asset is

Table 11-4
ORIGINAL DATA FOR EXISTING ASSET IN EXAMPLE 11-6

First cost	\$12,000
Maintenance and operating cost, \$/yr	1,700
Tax depreciation	SYD
Life	5 years
Salvage value	\$2,000
Book depreciation	Straight line
Life	8 years
Salvage value	\$2,000
Debt ratio	50%
Cost of debt capital	8%
Cost of equity capital	12%
Investment tax credit	0%
Tax rate	52%

not replaced. Consequently, this existing asset is being considered for replacement by the asset shown in Table 11-5.

If the existing asset is kept, it is estimated to have a life of only four more years and a salvage value of $1,000 at that time. The present realizable salvage value of the existing asset is $3000.

Should the existing asset be replaced? In order to answer this question several steps are required. The first step is to generate the annual costs (minimum annual revenue requirements) for the data given in Table 11-4. This step is done, using the flow-through method, and the results are shown in Table 11-6.

The next step is to generate the annual costs for the existing asset for the next *four* years (its estimated remaining life). These annual costs are shown in Table 11-7 and are explained in the following sample calculations.

The tax depreciation and interest in Table 11-7 are obtained from Table 11-6. What is being implied is that, from a tax standpoint, these deductions will remain the same if the decision is to keep the existing asset and they are attributed to the original cost ($12,000) of the existing asset.

In this example both the future salvage value and life have been reestimated to have, respectively, values of $1000 and four years. This point was made earlier in this chapter, that is, it is completely correct to revise estimates of life, salvage, and costs for an existing asset. There is no requirement that original estimates for the existing asset have to be used in the replacement analysis unless, of course, they are still considered to be correct

In this example, it is assumed that the existing asset is Sec. 1245 property and an extraordinary retirement. Consequently, the chargeable investment in year zero is

$$K_0 = 3000 + (4000 - 3000)(0.52) = \$3520$$

The $4000 value is the present tax book value. The basis of this calculation is discussed previously in this chapter. In addition to that discussion, another way of looking at the $3520 value is that the existing asset can return to the company $3000 at this time as a result of its net realizable salvage value. Also, it can provide a reduction in taxes of $520.

Table 11-5

DATA FOR NEW ASSET IN EXAMPLE 11-6

First cost	$20,000
Maintenance and operating cost, $/yr	500
Tax depreciation	SYD
Life	5 years
Salvage value	$2,000
Book depreciation	Straight line
Life	8 years
Salvage value	$2,000
Debt ratio	50%
Cost of debt capital	8%
Cost of equity capital	12%
Investment tax credit	10%
Tax rate	52%

Table 11-6

ORIGINAL ANNUAL COSTS FOR EXISTING ASSET IN EXAMPLE 11-6

End of Year	Book Depreciation	Tax Depreciation	Chargeable Investment	Return on Equity	Debt Interest	Tax	Maint. and Oper. Cost	Annual Cost
0	—	—	$12,000	—	—	—	—	—
1	$1250	$3333	10,750	$720	$480	−$1477	$1700	$2673
2	1250	2667	9,500	645	430	−836	1700	3189
3	1250	2000	8,250	570	380	−195	1700	3705
4	1250	1333	7,000	495	330	446	1700	4221
5	1250	667	5,750	420	280	1087	1700	4737
6	1250	—	4,500	345	230	1728	1700	5253
7	1250	—	3,250	270	180	1647	1700	5047
8	1250	—	2,000	195	130	1565	1700	4840

Table 11-7

ANNUAL COSTS FOR THE EXISTING ASSET IN EXAMPLE 11-6

End of Year	Tax Decreciation, D_t	Tax Debt Interest, I_t	Tax Book Value, B_x	Capital Rec. Costs, D_b	Chargeable Investment, K_x	Return, f_x	Oper. and Maint., C	Tax, t_x	Annual Cost, R_x
0	—	—	$4000	—	$3520	—	—	—	—
1	$1333	$330	2667	$630	2890	$352	$3000	−$738	$3244
2	667	280	2000	630	2260	289	3000	−30	3889
3	—	230	2000	630	1630	226	3000	678	4534
4	—	180	2000	630	1000	163	3000	−419	3374

Thus, if the decision is made to keep the existing asset, the company is, in effect, "loaning" \$3520 to the existing asset. This "loan," minus the salvage value, must be recovered through the capital recovery costs. The capital recovery costs are assumed in this example to be on a straight-line basis:

$$\frac{3520 - 1000}{4} = \$630 \text{ per year}$$

There is no requirement that these costs be determined on a straight-line basis. They could be determined on the basis of any depreciation model.

The chargeable investment for each year is given by the equation

$$K_x = K_{x-1} - D_b \qquad (11\text{-}1)$$

For example,

$$\begin{aligned} K_0 &= \$3520 \\ K_1 &= 3520 - 630 = \$2890 \\ K_2 &= 2890 - 630 = \$2260 \end{aligned}$$

In addition to the capital recovery costs, the asset must recover a return each year on the unrecovered amount (the chargeable investment). The rate of this return is the incremental weighted cost of capital that is estimated to be applicable in the future. This value may be different than that used initially for the existing asset. In this example, it is assumed there has been no change. Therefore, the required rate of return is

$$0.5(8) + (1 - 0.5)(12) = 10\%$$

and the returns for the first two years are

$$\begin{aligned} f_1 &= 3520\,(0.10) = \$352 \\ f_2 &= 2890\,(0.10) = \$289 \end{aligned}$$

Note that in this context of return both debt and equity returns are included in these values.

Taxes are based on the formula presented in Chapter 7 for the flow through:

$$t = \left[(D_b + f_e - D_t)\left(\frac{T}{1 - T}\right)\right] - \frac{C_t}{1 - T} \qquad (11\text{-}2)$$

Equation (11-2) must be modified when used for an *existing asset* since the return of debt interest included in f is different than the debt interest for tax purposes (I_t). That is, the interest deduction for tax purpose is based on money available at the time of the existing asset was originally purchased. This money is, in part (some is equity), a result of some debt obligation incurred (or in existence) at the time of purchase. Consequently, the interest deduction for tax purposes does not change if it is assumed that the basic debt obligation does not change. This is assumed in Eq. (11-3). The debt return included in the total return, f, is based on the net present realizable salvage value (with tax adjustments if appropriate) since this is the amount of money being "invested" in the existing asset. With these consideratons, Eq. (11-2) can be written as

$$t = \left[(D_b + f - D_t - I_t)\left(\frac{T}{1 - T}\right)\right] - \frac{C_t}{1 - T} \qquad (11\text{-}3)$$

In Eq. (11-3), C_t is the investment tax credit, it can be a tax credit (positive) or an additional tax (negative) resulting, in a particular year, from the sale of the asset. In this example, there is a tax credit in the fourth year for the existing asset since the tax book value ($2000) is greater than the estimated net realizable salvage (estimated book depreciation salvage value) of $1000:

$$(2000 - 1000)(0.52) = \$520$$

Therefore, using Eq. (11-3), the taxes for the fourth year are

$$t_4 = (630 + 163 - 0 - 180)\left(\frac{0.52}{0.48}\right) - \left(\frac{520}{0.48}\right) = \$-419$$

Since there are no investment tax considerations for the existing asset, the taxes for the first two years are

$$t_1 = (630 + 352 - 1333 - 330)\left(\frac{0.52}{0.48}\right) = \$-738$$

$$t_2 = (630 + 289 - 667 - 280)\left(\frac{0.52}{0.48}\right) = \$-30$$

The annual costs for an *existing asset* are determined using the equation

$$R_x = D_{bx} + f_x + C_x + t_x \tag{11-4}$$

For the first two years and fourth year, the annual costs are

$$R_1 = 630 + 352 + 3000 - 738 = \$3244$$
$$R_2 = 630 + 298 + 3000 - 30 = \$3889$$
$$R_4 = 630 + 163 + 3000 - 419 = \$3374$$

As a check, the taxes are calculated using the annual costs as gross incomes.

$$t_2 = (3244 - 3000 - 1333 - 330)(0.52) = \$-738$$
$$t_2 = (3889 - 3000 - 667 - 280)(0.52) = \$-30$$
$$t_4 = (3374 - 3000 - 0 - 180)(0.52) - 520 = \$-419$$

The final step is to generate the annual costs for the replacement asset shown in Table 11-8. These annual costs are calculated in the same manner as discussed in Chapter 7 for the flow-through method.

If the equivalent annual costs for both assets are determined, the results are

$$\text{EAC}_0 = [3244(PF\ 10,1) + 3889\,(PF\ 10,2) + 4534\,(PF\ 10,3) + 3374\,(PF\ 10,4)]\,(AP\ 10,4) = \$3746$$
$$\text{EAC}_n = [-789\,(PF\ 10,1) + 2916\,(PF\ 10,2) + \ldots + 5889\,(PF\ 10,8)]\,(AP,\ 10,8) = \$3974$$

These values indicate that the existing asset should not be replaced based on the equivalent annual cost (long-term effects). However a comparison of the annual costs for the first two years shows that the new asset has a decided advantage. It may be possible that these short-term effects out-weight the long-term effects.

Table 11-8
ANNUAL COSTS FOR NEW ASSET IN EXAMPLE 11-6

End of Year	Book Depreciation, D_b	Tax Depreciation, D_t	Invest. Tax Credit, C_t	Chargeable Investment, K_x	Return on Equity, f_{ex}	Debt Interest, I_x	Tax, t_x	Maint. and Oper. Cost, C	Annual Cost, R_x
0	—	—	—	$20,000	—	—	—	—	—
1	$2250	$6000	$1333	17,750	$1200	$800	−$5539	$500	−$789
2	2250	4800	—	15,500	1065	710	−1609	500	2916
3	2250	3600	—	13,250	930	620	−455	500	3845
4	2250	2400	—	11,000	795	530	699	500	4774
5	2250	1200	—	8,750	660	440	1852	500	5702
6	2250	—	—	6,500	525	350	3006	500	6631
7	2250	—	—	4,250	390	260	2860	500	6260
8	2250	—	—	2,000	255	170	2714	500	5889

TAX CONSIDERATIONS IN THE DETERMINATION OF ANNUAL COSTS

In determining annual costs, the tax effects described earlier in this chapter may be applicable, such as (1) Sec. 1245 gains, (2) ordinary retirements, (3) Sec. 1250 property, and (4) investment tax credits and penalties. Including the tax effects of the first three items into the annual costs is a rather simple matter. All that is necessary is to use for the initial chargeable investment, K_0, the tax adjusted net realizable salvage values as described earlier. Tax effects for future salvage values are included as shown in Example 11-6.

In the case of investment tax credits and penalties there are some slight computational differences. In order to show how investment tax considerations can be included in the annual costs and also provide an extension to Example 11-6, the following example is worked.

EXAMPLE 11-7

Suppose that in addition to the two alternatives described in Example 11-6, there is the possibility of overhauling the existing asset at a cost of \$1200. As a result of this overhaul, it is estimated that the life of the existing asset will be six years and have a total (overhaul plus existing asset) salvage value of \$1420 at that time. Furthermore, the overhaul is expected to reduce annual operation and maintenance costs to \$2000. It is determined that \$300 of the overhaul cost can be expensed and the remaining \$900 must be capitalized. For tax purposes, the \$900 is depreciated on the basis of a \$60 salvage value, six years, and a sum of the years depreciation model.

The annual costs for the overhaul alternative in this example are shown in Table 11-9 and explained in the following sample calculations. Actually, many of the columns in Table 11-9 can be combined. However, it is believed that the format used in this table is helpful from the standpoint of understanding the procedure.

The original tax depreciation (D_t) and tax debt interest (I_t) are the same as given in Table 11-6. As before, these values are based on the original cost of the existing asset.

The tax depreciation amounts for the \$900 depreciable portion are determined by the sum-of-the-years depreciation model. For example, the depreciation amounts for the first two years are

$$D'_{t1} = \frac{6}{21}(900 - 60) = 240$$

$$D'_{t2} = \frac{5}{21}(900 - 60) = 200$$

The capital recovery costs are assumed to be on the basis of a straight line depreciation model; namely

$$D'_d = \frac{900 - 60}{6} = \$140 \text{ per year}$$

The chargeable investment of the overhaul is given by the equations

$$K'_x = K'_{x-1} - D'_{bx} \tag{11-6}$$

and for the first three years is

$$K'_0 = 900$$
$$K'_1 = 900 - 140 = 760$$
$$K'_2 = 760 - 140 = 620$$

Table 11-9

ANNUAL COSTS FOR OVERHAUL ALTERNATIVE IN EXAMPLE 11-7

End of Year	Original Tax Dep., D_t	Tax Dep. on Overhaul, D'_t	Original Tax Debt Interest, I_t	Capital Rec. Cost for Overhaul, D'_b	Chargeable Invest. for Overhaul, K'_x	Debt Int. on Overhaul, f'_d	Return on Overhaul, f'_e
0	—	—	—	—	$900	—	—
1	$1333	$240	$330	$140	760	$36	$54
2	667	200	280	140	620	30	46
3	—	160	230	140	480	25	37
4	—	120	180	140	340	19	29
5	—	80	130	140	200	14	20
6	—	40	—	140	60	8	12

Table 11-9 *(Con't)*

End of Year	Capital Rec. on Original, D_b	Chargeable Invest. for Original, K_x	Return on Original f	Maint. and Oper., C	Repair, C'	Invest. Tax Credit, C_t	Tax Book Value—Orig. B_x	Tax Book Value—Overhaul, B'_x	Tax, t_x	Annual Costs R_x
0	—	$3520	—	—	—	—	$4000	900	—	—
1	$360	3160	$352	$2000	$300	$ 60[a]	2667	660	−$1205	$2037
2	360	2800	316	2000	—	—	2000	460	−309	2583
3	360	2440	280	2000	—	—	2000	300	463	3305
4	360	2080	244	2000	—	—	2000	180	512	3304
5	360	1720	208	2000	—	—	2000	120	561	3303
6	360	1360	172	2000	—	333[b]	2000	60	4	2696

[a]Investment tax credit.

[b]Tax credit from sale of asset.

The debt interest on the overhaul is given by the equation

$$\begin{aligned} f'_{dx} &= (c)(K'_{x-1})i_d \\ f'_{d1} &= 0.5\ (900)\ (0.08) = 36 \\ f'_{d2} &= 0.5\ (760)\ (0.08) = 30 \end{aligned} \tag{11-7}$$

The equity return on the overhaul is given by the equation

$$\begin{aligned} f'_{ex} &= (1 - c)(K'_{x-1})i_e \\ f'_{e1} &= (1 - 0.5)\ (900)\ (0.12) = 54 \\ f'_{e2} &= (1 - 0.5)\ (760)\ (0.12) = 46 \end{aligned}$$

The initial chargeable investment for the original investment is the present net realizable salvage value plus the tax effect, that is,

$$K_0 = 3000 + (4000 - 3000)\ (0.52) = \$3520$$

The salvage value at the end of the six years for the existing asset without overhaul is

$$1420 - 60 = \$1360$$

The capital recovery costs, based on a straight-line model, for the existing asset are

$$D_b = \frac{3520 - 1360}{6} = \$360 \text{ per year}$$

The chargeable investment for the existing asset is given by the equation

$$\begin{aligned} K_x &= K_{x-1} - D_b \\ K_0 &= 3520 \\ K_1 &= 3520 - 360 = \$3160 \\ K_2 &= 3214 - 360 = \$2800 \end{aligned}$$

The return on the original chargeable investent is given by the equation

$$f_x = (i)K_{x-1} \tag{11-8}$$

where i is the weighted cost of capital and for this example is

$$i = 0.5(8) + (1 - 0.5)\ (12) = 10\%$$

For the first two years, this return is

$$\begin{aligned} f_1 &= (0.10)\ (3520) = \$352 \\ f_2 &= (0.10)\ (3160) = \$316 \end{aligned}$$

The maintenance and operation costs and also the repair cost are given in the statement of this example. The tax book value for the original investment and for the overhaul are included for the sake of completeness and the need for the values of B_0 and B_6 in determining tax effects.

The investment tax credit is

$$C'_t = \frac{2}{3}(0.10)(900) = \$60$$

Since the life is six years, only two thirds of the full investment tax credit (\$90) is allowed.

The taxes are calculated by modifying Eq. (11-3) to the following:

$$t_x = (D_b + D'_b + f_e + f_d + f - D_t - D'_t - I_t - f_d)_x \left(\frac{T}{1-T}\right) - \frac{C'_t}{1-t} \qquad (11\text{-}9)$$

As examples, the taxes for the first two years are

$$t_1 = (360 + 140 + 54 + 36 + 352 - 1333 - 240$$
$$- 330 - 36)\left(\frac{0.52}{0.48}\right) - \frac{60}{0.48} = \$-1205$$
$$t_2 = (360 + 140 + 46 + 30 + 316 - 667 - 200$$
$$- 280 - 30)\left(\frac{0.52}{0.48}\right) = \$-309$$

The tax credit of \$333 shown in year 6 is based on the consideration that the total tax book value at this point is

$$2000 + 60 = \$2060$$

Therefore, with a total salvage of \$1420, the tax credit is

$$(2060 - 1420)(0.52) = \$333$$

and the taxes for the sixth year are

$$t_6 = (360 + 140 + 12 + 8 + 172 - 0 - 40 - 0 - 8)\left(\frac{0.52}{0.48}\right) - \frac{333}{0.48}$$
$$= 698 - 694 = \$4$$

The annual costs are determined using the equation

$$R_x = (D_b + D'_b + f + f'_e + f'_d + C + C' + t)_x \qquad (11\text{-}10)$$

For the first two years and sixth year, the annual costs are

$$R_1 = (360 + 140 + 352 + 54 + 36 + 2000 + 300 - 1205) = \$2037$$
$$R_2 = (360 + 140 + 316 + 46 + 30 + 2000 + 0 - 309) = \$2583$$
$$R_6 = (360 + 140 + 172 + 12 + 8 + 2000 + 0 + 4) = \$2696$$

As a check on the taxes, the taxes are calculated using the annual costs as the gross incomes.

$$t_1 = (2037 - 2000 - 300 - 1333 - 240 - 330 - 36)(0.52) - 60$$
$$= \$-1205$$
$$t_2 = (2583 - 2000 - 0 - 667 - 200 - 280 - 30)(0.52)$$
$$= \$-309$$
$$t_6 = (2696 - 2000 - 0 - 0 - 40 - 0 - 8)(0.52) - 333$$
$$= \$4$$

Note that in these calculations the expensed portion of the overhaul cost (\$300) is inculded in year 1. This is slightly different than in Example 11-4 and 11-5 where the expensed amounts are taken in year 0. This is a matter of preference. That is, it is felt, in determining specific yearly costs (as opposed to equivalent annual costs), that including the expense portion in this matter clearly shows its effect on the yearly costs.

Using the annual costs in Table 11-9, the equivalent annual cost is

$$[2037(PF\ 10,1) + 2582(PF\ 10,2) + \ldots + 2697(PF\ 10,6)](AP\ 10,6) = \$2824$$

This value of $2824 indicates that the overhaul alternative has a decided long-term advantage over the other two alternatives described in Example 11-6. However, there is still a decided short-term advantage for the new asset. The final choice of an alternative would have to be a balance between the short-term and long-term effects.

In Example 11-7 there is no investment tax penalty consideration because the life is seven years. If there is an investment tax penalty, it can be included in the replacement analysis by adjusting the first-year annual cost of the new asset. For example, suppose there is an investment tax penalty of $800 resulting from an early disposal of an existing asset. The calculation of this penalty is shown in Chapter 4. Further suppose that the new asset has an investment tax credit of $2,000. The net of these two values ($1200) is used for C_t in the formula for calculating the taxes of the *new asset* in year one. The remaining calculations are the same.

FINITE NEED

In Examples 11-6 and 11-7 the comparisons of the three alternatives are based on an infinite need and the recycling assumptions discussed in Example 11-1. The next example discusses the comparison of alternatives when a finite need is specified.

EXAMPLE 11-8

In this example a comparison of the alternatives described in Example 11-7 and 11-8 is made for a finite need of eight years.

In order to make a comparison, it is necessary to determine the annual cost for each alternative. The annual cost for the three alternatives over an eight-year (the need) period are shown in Table 11-10. The basis of these costs are explained by the following calculations and comments.

Table 11-10
ANNUAL COSTS FOR EXAMPLE 11-8

End of Year	Existing Asset	Existing Asset with Overhaul	New Asset
0	—	—	—
1	$ 3,244	$ 2,037	−$789
2	3,889	2,583	2916
3	4,534	3,305	3845
4	3,374	3,304	4774
5	−789	3,444	5702
6	2,916	2,696	6631
7	3,845	−789	6260
8	14,612 (6162)[a]	12,518 (5693)[a]	5889

[a]Values in parentheses are for ordinary retirements.

The annual costs for the new asset are the same as in Table 11-8 since the need and economic life are the same. In the case of the existing asset, the annual costs for the first four years are the same as given in Table 11-7. At the end of four years it is assumed that the existing asset is replaced by the new asset. This assumption is discussed earlier in this chapter and also in Chapter 10. With this assumption the annual costs for years 5, 6, and 7 are the same as years 1, 2, and 3 for the new asset. The annual cost for year 8 must be adjusted to include possible tax effects resulting from the disposal of the new asset after four years of use. It is assumed that the net realizable salvage is the company's book value after four years, that is,

$$BV_4 = 20{,}000 - 4\,(2250) = \$11{,}000$$

The basis of this assumption is discussed in Example 10-6. The tax book value after four years is

$$BV_4 = 20{,}000 - (6000 + 4800 + 3600 + 2400) = \$3200$$

Consequently, assuming Sec. 1245 property and an extraordinary retirement, there is a tax increase of

$$(11{,}000 - 3{,}200)(0.52) = \$4056$$

Adjusting the fourth year annual cost gives

$$4774 + \frac{4056}{1 - 0.52} = \$13{,}224$$

In addition to this adjustment, there is an investment tax adjustment. The annual cost in year four (\$−789) is based on two-thirds of the allowable investment tax credit since the tax life is five years. However, if the asset is disposed of at the end of year eight (four years of use), there is an investment tax penalty of

$$\frac{2}{3}(20{,}000)(0.10) - \frac{1}{3}(200{,}000)(0.10) = \$666$$

This penalty is a result of only four years of use. Including this investment tax penalty, the annual cost in year 8 is

$$13224 + \frac{666}{1 - 0.52} = \$14{,}612$$

If the retirement is considered as ordinary, the annual cost in year 8 is

$$4774 + \frac{666}{1 - 0.52} = \$6162$$

The annual costs for the overhaul alternatives are determined in a very similar manner. The annual costs for the first six years are the same as those given in Table 11-9. The annual cost in year 7 is the same as year 1 with the replacement assumption mentioned earlier. The annual cost in year 8 must also consider the tax adjustments for the difference in tax book value and selling price (company book value) and the investment tax penalty. The salvage value at the end of four years is

$$BV_2 = 20{,}000 - 2(2250) = \$15{,}500$$

and the tax book value is

$$BV_2 = 20{,}000 - (600 + 4800) = \$9200$$

Therefore, the tax increase as a result of the disposal is

$$(15{,}500 - 9200)(0.52) = \$3276$$

The entire investment tax credit of $1333 is a penalty since the asset is used for only two years (years 7 and 8). As a result of these considerations, the annual cost in year 8 for the overhaul alternative is

$$2916 + \frac{3276}{1 - 0.52} + \frac{1333}{1 - 0.52} = \$12{,}518$$

If the retirement is considered ordinary, the annual cost is

$$2916 + \frac{1333}{1 - 0.52} = \$5693$$

The equivalent annual costs for the three alternatives with extraordinary and ordinary retirement (in parentheses) are

$$\begin{aligned} EAC_1 &= [3244(PF\ 10,1) + \ldots + 14{,}612(PF\ 10,8)](AP\ 10,8) \\ &= \$4091\ (\$3353) \\ EAC_2 &= [2037(PF\ 10,1) + \ldots + 12{,}518(PF\ 10,8)](AP\ 10,8) \\ &= \$3342\ (\$2745) \\ EAC_3 &= [-789(PF\ 10,1) + \ldots + 5889(PF\ 10,8)](AP\ 10,8) \\ &= \$3974 (\$3974) \end{aligned}$$

The choice of an alternative would be a balance between the short-term and long-term effects.

Problems

11-1. A company is considering replacing certain machinery that has been in service for 5 years. This machinery originally cost $100,000, and had an estimated salvage value of $15,000 at the end of 15 years. Operations and maintenance costs have amounted to $25,000 per year. If this existing equipment is kept, it is believed the annual operating and maintenance costs will be the same for future years. However, based on the company's past experience with the existing machinery, it is now estimated that it will last only 5 more years and have a salvage value of $20,000 at that time. The present realizable salvage value for the existing machinery is $50,000.

An evaluation of machinery presently available that provides equal service has revealed a possible economic advantage. There is now available machinery that will reduce annual operating and maintenance costs to $10,000 per year. This new machinery will cost $115,000, last 15 years, and have a salvage value of $10,000.

Assuming 100% equity financing and a required before-tax return on equity of 25%, determine a before-tax equivalent annual cost comparison under the following conditions:

(a) The need for the service provided by the equipment is infinite.

(b) The need is for only 5 years.
(c) The need is for 10 years.
(d) The need is for 30 years.

It has been estimated that the salvage values of the new equipment are $70,000 and $30,000 at the end of 5 and 10 years of service, respectively.

11-2. In Problem 11-1, what is the sunk cost if the existing machinery is sold? Assume straight-line depreciation is used for the existing machinery.

11-3. Make an after-tax comparison of the alternatives described in Problem 11-1 under the following conditions:

(a) The machinery qualifies as Sec. 1245 property, retirements are extraordinary, the need is infinite, and there is no investment tax credit.
(b) Same as part (a), only the retirements are ordinary.

Assume in parts (a) and (b) that straight-line depreciation is used for tax purposes and the tax rate is 52%.

11-4. Repeat Problem 11-3 for extraordinary retirements and the following needs:

(a) 5 years
(b) 10 years
(c) 30 years

Assume that the net realizable salvage values for new equipment are respectively, $70,000 and $30,000, at the end of 5 and 10 years of service.

11-5. Determine the equivalent annual cost of an overhaul alternative for Problems 11-1 and 11-3 if the overhaul cost is $15,000 and will reduce annual costs to $12,000. Assume, for tax purposes, that $5000 of the overhaul cost qualifies as a repair expense and the remaining amount ($10,000) is depreciated. The overhaul is to be depreciated, for tax purposes, on the basis of a straight-line model and a salvage value of $1000. It is estimated that this overhaul will extend the life of the existing asset to 10 years and the total salvage at that time will also be $7000, the total salvage value at the end of five years of service is $35,000.

11-6. Five years ago an asset with the data shown below was installed. The original estimate of the annual costs ($5000) has been found to be low. These costs have amounted to $10,000 per year and are expected to continue at this amount if the asset is not replaced. Consequently, this asset is being considered for replacement by the asset shown below:

Initial cost	$95,000
Annual costs	$5,000
Tax depreciation	SYD
Life	15 years
Salvage	$5,000
Book depreciation	SL
Life	15 years
Salvage	$5,000
Debt ratio	40%
Required debt return	9%
Required equity return	19%
Investment tax credit	10%
Tax rate	52%

Initial cost	$100,000
Annual costs	$5,000
Tax depreciation	SYD
Life	10 years
Salvage	$10,000
Book depreciation	SL
Life	10 years
Salvage	$10,000
Debt ratio	40%
Required debt return	9%
Required equity return	19%
Investment tax credit	10%
Tax rate	52%

It is now estimated that the existing asset will last only five more years and will have a zero salvage value at that time. It is believed the existing asset can be sold for $30,000. Under the following conditions, determine if the existing asset should be replaced.

(a) The assets are Sec. 1245 property and retirements are classified as extraordinary.
(b) Capital recovery costs for the existing asset are assumed to be on the basis of a straight-line depreciation model.
(c) Infinite need.

11-7. Repeat Problem 11-6 for ordinary retirements.

11-8. Repeat Problems 11-6 and 11-7 for a finite need of 15 years.

11-9. Determine the annual costs for an overhaul alternative for Problem 11-6 using the following data:

Total cost of overhaul	$10,000
Repair portion	$2,000
Depreciable portion	$8,000
Tax method for overhaul	SYD
Life	10 years
Salvage	$1,000
Book depreciation	SL
Life	10 years
Salvage	$1,000
Total salvage (overhaul and original asset) =	$7,000
Annual costs	$7,000

CHAPTER 12
SOME RISK CONSIDERATIONS IN CASH FLOW MODELS

In the previous chapters, the models presented for the economic analysis of capital expenditures are based on single estimates of the variables involved in a particular model. That is, the existence of variability in these estimates is not recognized. This chapter recognizes this variability by explicitly including it in the models considered. In considering these models, a knowledge of probability concepts and statistics is necessary. Consequently, this chapter presupposes some knowledge in these areas. The usual beginning course in probability and statistics is considered to be sufficient for the material presented in this chapter.

The explicit inclusion of variability in cash flows models can be approached in two general ways—analytical and simulation approaches.

ANALYTICAL APPROACHES

In this section several basic analytical approaches to probabilistic cash flow models are presented. These approaches are based on the fundamental relationships regarding the combination of random variables given below:

Sums of Random Variables

If $x_1, x_2, \ldots, x_n$ are random variables and μ_i is the mean of the ith random variable, the expected value of the sum of these random variables, μ_s, is

$$E(x_1, x_2, \ldots, x_n) = \mu_s = \sum_{i=1}^{n} \mu_i \tag{12-1}$$

and the variance of the sum, σ_s^2, is

$$V(x_1, x_2, \ldots, x_n) = \sigma_s^2 = \sum_{i=1}^{n} \sigma_i^2 + 2\sum_{i=1}^{n-1} \sum_{j=i+1}^{n} \sigma_{ij} \tag{12-2}$$

Where, σ_i^2 is the variance of the ith random variable and σ_{ij} are the covariance terms. A substitution often made for covariance terms is

$$\sigma_{ij} = r_{ij}\,\sigma_i\,\sigma_j \tag{12-3}$$

where r_{ij} is the correlation coefficient for the random variables x_i and x_j.

Difference of Two Random Variables

The expected value of the difference of two random variables, μ_D, is

$$E(x_i - x_j) = \mu_D = \mu_i - \mu_j \tag{12-4}$$

where x_i and x_j are the random variables and μ_i and μ_j are the means of the random variables. The variance of the difference of two random variables is

$$\sigma_D^2 = \sigma_i^2 + \sigma_j^2 - 2\sigma_{ij} \tag{12-5}$$

Multiplication of Two Independent Random Variables

If two *independent* random variables are multiplied, the expected value of the product, μ_p, is

$$E(x_i x_j) = \mu_p = (\mu_i)(\mu_j) \tag{12-6}$$

and the variance, σ_p^2, is

$$V(x_i x_j) = \sigma_p^2 = \mu_i^2\sigma_j^2 + \mu_j^2\sigma_i^2 + \sigma_j^2\sigma_i^2 \tag{12-7}$$

There is no analogous expression for the dependent case.

A Random Variable Multiplied by a Constant

When a random variable x is multiplied by a constant c to give the random variable $z = cx$, the mean, μ_z, and variance σ_z^2, of the product are

$$E(cx) = \mu_z = c\mu_x \tag{12-8}$$
$$V(cx) = \sigma_z^2 = c^2\sigma_x^2 \tag{12-9}$$

Now, the application of the relationships given by Eqs. (12-1) through (12-9) to probabilistic cash flow models is shown in the following examples.

EXAMPLE 12-1

In this example the computation of the mean and standard deviation of the net present value distribution is shown (also see Reference 1) for the situation depicted in Table 12-1. It is assumed in this example that the cash flows are independent. Also, the number of years, n, and minimum acceptable rate of return, k, are known with certainty. With these assumptions, the mean and variance of the net present value can be determined by applying some of the previously presented relationships for random variables.

Since k and n are constants, the discount factor, $(1 + k)^{-t}$, is also a constant, implying that Eqs. (12-8) and (12-9) are applicable. After discounting, Eqs. (12-1) through (12-5) are applicable since the sum and difference of random variables are involved. The

Table 12-1
MEAN AND VARIANCE OF CASH FLOWS FOR EXAMPLE 12-1

Year	Cash Flow	
	Mean	Variance
0	μ_0	σ_0^2
1	μ_1	σ_1^2
2	μ_2	σ_2^2
3	μ_3	σ_3^2
.	.	.
t	μ_t	σ_t^2
.	.	.
n	μ_n	σ_n^2

difference is involved because the initial cash flow will, in all probability, be an outflow (the investment). With these considerations, the expected net present value and variance are

$$E(\text{NPV}) = \sum_{t=0}^{n} \mu_t(PFk,t) \tag{12-10}$$

$$V(\text{NPV}) = \sum_{t=0}^{n} \sigma_t^2(PFk,2t) \tag{12-11}$$

As a numerical example of the use of Eqs. (12-10) and (12-11), consider the data given in Table 12-2. Using Eq. (12-10) and a minimum acceptable rate of return of 12%, the expected net present value is

$$E(\text{NPV}) = -1500 + 1000(PF12,1) + 1000(PF12,2) + 2000(PF12,3) = \$1614$$

and the variance is

$$V(\text{NPV}) = (600)^2 + (700)^2(PF12,2) + (700)^2(PF12,4) + (1000)^2(PF12,6) = 156.86(10)^4$$

Table 12-2
CASH FLOWS FOR EXAMPLE 12-1

End of Year	Cash Flows	
	Mean	Standard Deviation
0	−$1500	$ 600
1	1000	700
2	1000	700
3	2000	1000

The standard deviation of the net present value, $\sigma(\text{NPV})$, is

$$\sigma(\text{NPV}) = \sqrt{156.86(10)^4} = \$1252$$

EXAMPLE 12-2

In Example 12-1 it is assumed that the cash flows are independent. This assumption is not required. It is possible to determine the mean and variance of the net present value distribution when the cash flows are dependent. For the dependent case, the mean of the net present value is the same as Eq. (12-10). In order to determine the variance of the net present value, Eq. (12-11) must be modified to include the covariance terms. This modification is shown in the following equation.

$$V(\text{NPV}) = \sum_{t=0}^{n} \sigma_t^2(PFk,2t) \pm 2 \sum_{t=0}^{n-1} \sum_{j=t+1}^{n} \sigma_{t,j}(PFk,(t+j)) \tag{12-12}$$

substituting Eq. (12-3) for the covariance terms, Eq. (12-12) can be written as

$$V(\text{NPV}) = \sum_{t=0}^{n} \sigma_t^2(PFk,2t) \pm 2 \sum_{t=0}^{n-1} \sum_{j=t+1}^{n} r_{t,j}\sigma_t\sigma_j(PFk,\ (t+j)) \tag{12-13}$$

The correct sign (plus or minus) to use in Eqs. (12-12) and (12-13) is determined by the relationship (addition or subtraction) involved between the random variables.

As a numerical example, suppose the cash flows in Example 12-1 are dependent and it is believed the correlation coefficients given in Table 12-3 provide a correct representation of this dependency. For this case, the mean of the net present value distribution is the same as calculated previously, that is, \$1614. Using Eq. (12-13) the variance is

$$\begin{aligned} V(\text{NPV}) &= (600)^2 + (700)^2\,(PF12,2) + (700)^2\,(PF12,4) + (1000)^2(PF12,6) \\ &\quad - 2(.8)(600)(700)(PF12,1) - 2(.5)(600)(700)(PF12,2) \\ &\quad - 2(.3)(600)(1000)(PF12,3) + 2(.7)(700)(700)(PF12,3) \\ &\quad + 2(.5)(700)(1000)(PF12,4) + 2(.6)(700)(1000)(PF.2,5) \\ &= 156.86(10)^4 + 21.88(10)^4 = 178.84(10)^4 \end{aligned}$$

and the standard deviation is

$$\sigma(\text{NPV}) = \sqrt{178.84(10)^4} = \$1337$$

At this point it should be noted that a knowledge of the mean and variance (or

Table 12-3
CORRELATION COEFFICIENTS FOR EXAMPLE 12-2

	0	1	2	3
0	—	.8	.5	.3
1		—	.7	.5
2			—	.6
3	(Symmetrical)			—

standard deviation) of the net present value does not, by itself, provide a decision regarding the acceptability of a project (investment). They are simply an aid in the decision-making process. A subjective evaluation is still required. In the following discussion some additional uses of the mean and variance (or standard deviation) of the net present value are presented.

Probability Statements: It is beneficial in the decision-making process to be able to make probability statements regarding net present values. For example, it might be an aid in making a decision regarding the acceptability of a project to know the probability of the net present value being less than zero or within some specified interval. Without any knowledge or some assumption regarding the distribution of net present value, *exact* probability statements are not possible. However, it is possible to make *weak* probability statements using Tchebycheff's inequality. Mathematically, Tchebycheff's inequality can be expressed as

$$P(\mu - b\sigma \leq X \leq \mu + b\sigma) > 1 - \frac{1}{b^2} \qquad (12\text{-}14)$$

where

x = a random variable
μ = the expected value of x
σ = the standard deviation of x
b = a constant that is greater than 1

The application of Eq. (12-14) is shown in the following example.

EXAMPLE 12-3

Suppose that the mean and standard deviation of the net present value distribution for a particular project are, respectively, \$2000 and \$1000. Using Tchebycheff's inequality the following statement is possible for $b = 2$.

The probability is *greater than*

$$1 - \frac{1}{(2)^2} = .75 = 75\%$$

that the net present value will fall within the interval

$$2000 \pm (2)(1000) = \$0 \text{ to } \$4000$$

Or, the probability is 25% *or less* that the actual net present value will fall outside the \$0 to \$4000 interval.

The phrases "greater than" and "or less" are the reasons that the probability statements given by Tchebycheff's inequality are often referred to as *weak* probability statements.

In this example the probability statements involve both tails of the probability distribution. It is also possible to make one-tailed probability statements by using the following relationships (see Reference 4).

$$P(X \leq \mu - k\sigma) \leq \frac{1}{1 + k^2} \qquad (12\text{-}15)$$

$$P(X \geqslant \mu + k\sigma) \leqslant \frac{1}{1 + k^2} \qquad (12\text{-}16)$$

As an example of the use of Eqs. (12-15) and (12-16), suppose the mean and standard deviation of the net present value distribution are, respectively, \$2000 and \$1000. Then, the following statements are possible.

The probability is equal to or less than

$$\frac{1}{1 + (k)^2} = \frac{1}{1 + (2)^2} = .20 = 20\%$$

that the actual net present value will be below

$$\mu - k\sigma = 2000 - 2(1000) = \$0$$

The probability is equal to or less than

$$\frac{1}{1 + k^2} = \frac{1}{1 + (2)^2} = .20 = 20\%$$

that the actual net present value will be greater than

$$\mu + k\sigma = 2000 + 2(1000) = \$4000$$

There might be a tendency to combine Eqs. (12-15) and (12-16) in order to make probability statements regarding an interval. This is incorrect. Eq. (12-14) should be used for estimating interior probabilities since it establishes the *lowest* bound for interior probabilities.

Since Tchebycheff's inequality does not provide exact probability statements, a common approach is to assume that the net present value is normally distributed. In general, this assumption is hard to defend except in the case where the cash flows are known (or assumed) to be independent. In the independent case, a fairly strong argument can be presented for the assumption of normality on the basis of the central limit theorem. If it is decided (on some basis) that the net present value is normally distributed, the mean and standard deviation can be used to make exact probability statements and to calculate expected losses as shown in the following example.

EXAMPLE 12-4

In this example the net present value is assumed to be normally distributed with a mean and standard deviation, respectively, of \$2000 and \$1000. The method for calculating these values is shown in Example 12-1. Also, for illustrative purposes, assume that the minimum attractive rate of return used in determining these values is 15%.

The determination of exact probability statements is a relatively simple matter with the assumption that the net present value is normally distributed. For example, the probability that the net present value is less than zero is

$$k_\alpha = \frac{0 - 2000}{1000} = -2.00$$

$$\alpha = .0228 = 2.28\%$$

There are two other possible interpretations of this answer. First, by considering that a net present value less than zero is a loss, it is possible to interpret this answer as the probability of a loss. Second, it is stated in this example that 15% is used in calculating the mean and standard deviation of the net present value. Consequently, this answer can also be interpreted as the probability of a rate of return being less than 15%. This second interpretation can be extended. That is, the probability of other rates of return, i, being less than the minimum attractive rate of return, k, can also be determined. This is accomplished by computing the mean and standard deviation of the net present value using i and then calculating the probability of a net present value less than zero. Or, stated mathematically,

$$P(\text{NPV} < 0 \mid k) = P(i < k) \tag{12-17}$$

Expected Loss: With the condition that the net present value is normally distributed, it is possible to calculate the expected loss (see also References 1 and 5) using the following relationship.

$$\text{EL} = \sigma_{\text{NPV}}\ [L_N(D)] \tag{12-18}$$

where

EL = expected loss. A loss is defined as NPV < 0.
σ_{NPV} = standard deviation of the net present value
$L_N(D)$ = unit normal loss integral given in Appendix G as a numerical example, the EL, given $E(\text{NPV}) = \$2000$ and $\sigma(\text{NPV}) = \$1000$, is

$$\text{EL} = 1000\ [0.008491] = \$8.491$$

The value in the brackets is obtained from Appendix G in the following manner.

$$D = \frac{E(\text{NPV})}{\sigma(\text{NPV})} = \frac{2000}{1000} = 2.00$$
$$L_N(D) = 0.008491$$

The unit normal loss integral as used in the context of this example is

$$\text{EL} = \int_{-\infty}^{0} |X| f(X)\, dx \tag{12-19}$$

where

X = all possible negative net present values since a NPV < 0 is considered a loss.
$f(X)$ = probability density function. The tables given in Appendix G are based on a normal distribution with a mean of 0 and variance of 1.

As a result of the definition of the unit normal loss integral given by Eq. (12-19), it should be recognized that, in actuality, the EL is negative. However, the negative sign is usually omitted because it is implied in the definition of a loss.

It is also possible to determine the expected loss given a loss occurs as shown below

$$\text{EL} \mid \text{loss} = \frac{\text{EL}}{\alpha} \tag{12-20}$$

and for this example is

$$\text{EL} \mid \text{loss} = \frac{8.491}{.0228} = \$372.41$$

The expected gain EG, can also be calculated using the relationship

$$\text{EG} = \sigma(\text{NPV})\,[L_N(-D)] \tag{12-21}$$

$$L_N(-D) = \left[\frac{E(\text{NPV})}{\sigma(\text{NPV})} + L_N(D)\right] \tag{12-22}$$

For example,

$$L_N(-D) = \left[\frac{2000}{1000} + .008491\right] = 2.008491$$

$$\text{EG} = 1000(2.008491) = \$2008.491$$

It should be noted that (remembering the EL is negative) EG + EL = E(NPV).

The Mean and Variance of Cash Flows: Up to this point the discussion is concerned with the determination and possible uses of the mean and variance of the net present value distribution. However, before these uses are possible, an intermediate step is required: the determination of the mean and variance of the cash flows. Therefore, the next few examples are concerned with the development of analytical approaches for the determination of these quantities. Basically, these approaches are another application of the statistical theorems presented earlier in this chapter.

EXAMPLE 12-5

In this example the mean and variance of the cash flow equation

$$X_t = (G - C - I) - (G - C - I - D)\,T - k + L + B - P \tag{12-23}$$

are determined. Initially, it is assumed that all variables are independent and the tax rate, T, is a constant. Of course, it is highly probable that many of the variables in Eq. (12-23) are not applicable for a particular year. For example, it is highly likely to have capital expenditures, K, in earlier years and none in later years. Similarly, it is likely to have salvage values, L, only in later years.

Now in order to obtain the variance of Eq. (12-23) it is desirable to rewrite it as

$$X_t = G(1-T) - C(1 - T) - I(1 - T) + DT - K + L + B - P \tag{12-24}$$

because, in this form, each term is independent since T is assumed to be a constant. This is the principal reason for assuming the tax rate (T) is a constant. That is, all the terms in Eq. (12-24) are independent. If T is allowed to be a random variable, the application of the statistical theorems presented previously is not possible.

The variance of Eq. (12-24) is

$$\sigma_{x_t}^2 = \sigma_G^2\,(1 - T)^2 + \sigma_c^2\,(1 - T)^2 + \sigma_I^2\,(1 - T)^2 + \sigma_D^2\,T^2 + \sigma_K^2 + \sigma_L^2 + \sigma_B^2 + \sigma_P^2 \tag{12-25}$$

and the mean is

$$\overline{X}_t = \overline{G}(1 - T) - \overline{C}(1 - T) - \overline{I}(1 - T) + \overline{D}T - \overline{K} + \overline{L} + \overline{B} - \overline{P} \tag{12-26}$$

Applications of Eqs. (12-25) and (12-26) and some possible extensions are presented in the following examples.

EXAMPLE 12-6

This example determines the mean and variance of the net present value for the situation depicted in Table 12-4 and the following conditions:

1. The number of years, n, is known with certainty.
2. G, C, I, D, and P are all independent random variables.
3. The tax rate, T, and discount rate (minimum attractive rate of return) are constants.

The mean and variance of the net present value are the discounted sums of the various terms in Eqs. (12-25) and (12-26), that is,

$$E(\text{NPV}) = (1 - T) \sum_{t=1}^{n} \overline{G}_t (PFk,t) - (1 - T) \sum_{t=1}^{n} \overline{C}_t (PFk,t) \tag{12-27}$$

$$- (1 - T) \sum_{t=1}^{n} \overline{I}_t (PFk,t) + T \sum_{t=1}^{n} \overline{D}_t (PFk,t) \tag{12-28}$$

$$- \overline{K}_0 + \overline{B}_0 + \overline{L}(PFk,n)$$

$$V(\text{NPV}) = (1 - T)^2 \sum_{t=1}^{n} \sigma^2_{\overline{G}_t} (PFk,2t) + (1 - T)^2 \sum_{t=1}^{n} \sigma^2_{\overline{C}_t} (PFk,2t)$$

$$+ (1 - T)^2 \sum_{t=1}^{n} \sigma^2_{\overline{I}_t} (PFk,2t) + T^2 \sum_{t=1}^{n} \sigma^2_{\overline{D}_t} (PFk,2t)$$

$$+ \sigma^2_{\overline{K}_0} + \sigma^2_{\overline{B}_0} + \sigma^2_{\overline{L}_n} (PFk,2n)$$

Obviously, if numerical values are available for the quantities in Eqs. (12-27) and (12-28) they can be substituted into these equations to obtain E(NPV) and V(NPV). From a practical standpoint, the determination of numerical values for the quantities in Eqs. (12-27) and (12-28) is, perhaps, the most difficult task. In the final analysis, they must be determined by statistical techniques based on historical data or simply estimated on the basis of "expert judgement." Some of these estimation problems can be reduced by making certain simplifying assumptions and considering possible relationships between some of the variables. This approach is shown in the next example.

Table 12-4

CASH FLOW VARIABLES FOR EXAMPLE 12-6

Year							
0		K_0	B_0				
1	G_1	C_1	I_1	D_1	T	P_1	
.	.	.	.	.	.	.	
.	.	.	.	.	.	.	
.	.	.	.	.	.	.	
t	G_t	C_t	I_t	D_t	T	P_t	
.	.	.	.	.	.	.	
.	.	.	.	.	.	.	
.	.	.	.	.	.	.	
.	.	.	.	.	.	.	
n	G_n	C_n	I_n	D_n	T	P_n	L_n

EXAMPLE 12-7

The assumption and conditions made in this example are the same as in Example 12-6 with the following additional conditions and assumptions.

1. The salvage value, L_n, is a constant proportion, C_1, of the capital investment, K_0. That is,

$$L_n = C_1 K_0 \tag{12-29}$$

Therefore, the mean and variance of this salvage value are

$$\overline{L}_n = C_1\overline{K}_0 \tag{12-30}$$
$$\sigma^2_{L_n} = (C_1)^2\,\sigma^2_{K_0} \tag{12-31}$$

2. The depreciation life is n years. Consequently, the total depreciation, d, is

$$d = K_0 - L_n = K_0 - C_1K_0 = K_0(1 - C_1) \tag{12-32}$$

It is assumed that the distribution of *yearly* depreciation is the same and independent for all year, the expressions for the mean, $\bar{D}$, and variance, σ^2_D, of the yearly distributions are

$$\overline{d} = \overline{K}_0(1 - C_1) = \overline{D}_1 + \overline{D}_2 + \ldots + \overline{D}_n = n\overline{D} \tag{12-33}$$

$$\overline{D} = \frac{\overline{K}_0(1 - C_1)}{n} \tag{12-34}$$

$$\sigma^2_d = \sigma^2_{K_0}(1 - C)^2 = \sigma^2_{D_1} + \sigma^2_{D_2} + \ldots + \sigma^2_{D_n} = n\,\sigma^2_D$$

$$\sigma^2_D = \frac{\sigma^2_{K_0}(1 - C)^2}{n}$$

3. The amount of borrowed money, B_0, is a constant proportion, C_2, of the capital expenditure. That is,

$$B_0 = C_2K_0 \tag{12-35}$$

Consequently the mean and variance of the borrowed money are

$$\overline{B}_0 = C_2\overline{K}_0 \tag{12-36}$$
$$\sigma^2_{B_0} = (C_2)^2\,\sigma^2_{K_0} \tag{12-37}$$

If it is assumed that the principal payment distributions are the same and independent for each year, the mean and variance of the yearly principal payments are

$$\overline{B}_0 = C_2\overline{K}_0 = \overline{P}_1 + \overline{P}_2 + \ldots + \overline{P}_n = n\overline{P}$$

$$\overline{P} = \frac{C_2\overline{K}_0}{n} \tag{12-38}$$

$$\sigma^2_{B_0} = (C_2)^2\,\sigma^2_{K_0} = \sigma^2_{P_1} + \sigma^2_{P_2} + \ldots + \sigma^2_{P_n} = n\sigma^2_P \tag{12-39}$$

$$\sigma^2_P = \frac{(C_2)^2\,\sigma^2_{K_0}}{n}$$

4. The cost of borrowed money, k_d, is a constant that is assessed on the unpaid balance. Since the principal is paid in equal yearly installments, the interest payment, I_t, is

$$I_t = k_d\,[B_0 - (t - 1)P] \tag{12-40}$$
$$= k_d\left[B_0 - (t - 1)\,\frac{B_0}{n}\right] \tag{12-41}$$

Substituting for B_0 and rearranging gives

$$I_t = k_d C_2 K_0 \left(\frac{n - t + 1}{n}\right) \tag{12-42}$$

The mean and variance of Eq. (12-42) are

$$\overline{I}_t = k_d(C_2)\,\overline{K}_0 \left(\frac{n - t + 1}{n}\right) \tag{12-43}$$

$$\sigma^2_{\tilde{I}_t} = (k_d)^2\,(C_2)^2\,\sigma^2_{\tilde{K}_0} \left(\frac{n - t + 1}{n}\right)^2 \tag{12-44}$$

Finally, substituting the pertinent equations into Eqs. (12-27) and (12-28), the mean and variance of the net present value are

$$\begin{aligned} E(\text{NPV}) = {} & (1 - T) \sum_{t=1}^{n} \overline{G}_t\,(PFi,t) - (1 - T) \sum_{t=1}^{n} \overline{C}_t\,(PFi,t) \\ & - (1 - T)\,[(k_d)(C_2)\,(\overline{K}_0)] \sum_{t=1}^{n} \left(\frac{n - t + 1}{n}\right)(PFi,t) \\ & + T\left[\frac{\overline{K}_0(1 - C_1)}{n}\right] \sum_{t=1}^{n} (PFi,t) - \overline{K}_0 + C_2\overline{K}_0 \\ & + C_1\overline{K}_0\,(PFi,n) \end{aligned} \tag{12-45}$$

$$\begin{aligned} V(\text{NPV}) = {} & (1 - T)^2 \sum_{t=1}^{n} \sigma^2_{\tilde{G}_t}\,(PFi,2t) + (1 - T)^2 \sum_{t=1}^{n} \sigma^2_{\tilde{C}_t}\,(PFi,2t) \\ & + (1 - T)^2\,[(k_1)^2\,(C_2)^2\,(\sigma^2_{\tilde{K}_0})] \sum_{t=1}^{n} \left[\frac{n - t + 1}{n}\right]^2 (PFi,2t) \\ & + \frac{(T)^2(\sigma^2_{\tilde{K}_0})(1 - C_1)^2}{n} \sum_{t=1}^{n} (PFi,2t) + \sigma^2_{\tilde{K}_0} + C_2^2\sigma^2_{\tilde{K}_0} \\ & + C_1^2\sigma^2_{\tilde{K}_0}\,(PFi,2n) \end{aligned} \tag{12-46}$$

The conditions and assumptions used in Example 12-7 greatly reduce the estimating problems since most of the expressions are a function of one random variable—the capital investment.

Note that since debt and principal are explicitly included in Eqs. (12-45) and (12-46), the discount rate (i) is, in actuality, the cost of equity capital (see Chapter 7).

In Example 12-7 the relationships between various random variables are expressed as constant proportions and, as a result, the independency (from a statistical standpoint) between the various terms is maintained. In some instances this approach is not acceptable. Consequently, it is desirable to consider methods for expressing statistical dependency between various random variables in the cash flow streams.

There are two basic ways in which dependencies can occur. The first way is between variables in the same time period. The second way is between variables in different time periods. In both cases, expressions for the mean and variance of the net cash flow or net present value can be obtained by using the statistical theorems presented earlier in this chapter. This is shown in the following examples.

EXAMPLE 12-8

Suppose in Eq. (12-23) it is assumed that there is a dependency between the gross income, G, and the annual costs, C, and it is desired to obtain an expression for the mean and variance of the net cash flow. At this point, it is assumed that all other random variables are independent.

The mean of the cash flow is the same as Eq. (12-26). However, the variance is

$$\sigma_{X_t}^2 = \sigma_{G_t}^2 (1 - T)^2 + \sigma_{C_t}^2 (1 - T)^2 - 2r_{G,C}\, \sigma_{G_t} \sigma_{C_t} (1 - T)^2 + \sigma_I^2(1 - T)^2 + \sigma_D^2 T^2 + \sigma_K^2 + \sigma_L^2 + \sigma_B^2 + \sigma_P^2 \quad (12\text{-}47)$$

Equation (12-47) is the same as Eq. (12-25) except for the term with the correlation coefficient $r_{G,C}$. This term is based on an application of Eqs. (12-3) and (12-5).

Obviously this is not the only dependency that can be considered. There are others. For example, if it is desired to show a dependency between the interest and the borrowed money, it would be necessary to add the term

$$-2\, r_{I,B}\, \sigma_I \sigma_B (1 - T) \quad (12\text{-}48)$$

to Eq. (12-47).

Once the means and variances of the cash flows are determined, the mean and variance of the net present value can be determined as shown in Example 12-3. Provided, it is assumed that there is no dependency between cash flows in different years. The next example shows an approach to including dependency between random variables occurring in different years.

EXAMPLE 12-9

Suppose in Example 12-6 the gross incomes are dependent and all other conditions are the same. It is desired to obtain an expression for the mean and variance of the net present value.

Actually, the solution to this problem is very similar to the extension given in Example 12-3. The mean of the net present value is the same as Eq. (12-27). The variance of the net present value is the same as Eq. (12-28) with the addition of the following term:

$$+ 2(1 - T)^2 \sum_{t=1}^{n-1} \sum_{j=t+1}^{n} r_{G_t,G_j}\, \sigma_{G_t} \sigma_{G_j} (PFk,\ (t + j)) \quad (12\text{-}49)$$

Equation (12-49) is an expression for the various terms resulting from the dependencies assumed in this example.

A Comment: From a theoretical standpoint the inclusion of dependencies in the determination of the mean and variance of the net profit value is, of course, important since it is highly probable that such dependencies do exist. However, from a practical standpoint there are some difficulties in approaching dependencies between the random variables in the manner shown in Examples 12-8 and 12-9. These difficulties occur because of the large number of dependencies that might be expected in a truly representative capital investment of say, some 20 to 25 years. Also, it would be highly likely to have dependencies upon dependencies. That is, gross incomes dependent upon the initial investment (size of the plant, etc.), annual costs dependent upon initial investment and gross incomes, borrowed money dependent upon capital investment, and so on. The number of combinations is tremendous and, consequently, the estimating problem is, from a practical standpoint, impossible. In the final analysis, at some point, an assumption of independence between some of the random variables is required as a result of practical considerations.

A PROBABILISTIC *C-V-P* MODEL

In Chapter 8 some deterministic *C-V-P* models are presented. In this section a probabilistic model is presented based on the work of Jaedicke and Robichek (see Reference 3).

The basic before-tax model presented in Chapter 8 is

$$P = V(S - C) - F \tag{12-50}$$

If it is assumed that $V, S, C,$ and F in Eq. (12-50) are independent random variables (Jaedicke and Robichek), expressions for the mean and variance of the profit can be determined using the statistical theorems presented earlier in this chapter. First, Eq. (12-50) is rewritten as

$$P = VA - F \tag{12-51}$$

where

$$A = S - C \tag{12-52}$$

Applying Eq. (12-7) to Eq. (12-51), the variance of the profit is

$$\sigma_P^2 = \sigma_V^2\,\overline{A}^2 + \sigma_A^2\,\overline{V}^2 + \sigma_A^2\,\sigma_V^2 + \sigma_F^2 \tag{12-53}$$

since

$$\overline{A} = \overline{S} - \overline{C} \tag{12-54}$$
$$\sigma_A^2 = \sigma_S^2 + \sigma_C^2 \tag{12-55}$$

Equation (12-53) can be rewritten as

$$\sigma_P^2 = \sigma_V^2\,(\overline{S} - \overline{C})^2 + (\sigma_S^2 + \sigma_C^2)\,\overline{V}^2 + (\sigma_S^2 + \sigma_C^2)\,\sigma_V^2 + \sigma_F^2 \tag{12-56}$$

The mean of the profit is

$$\overline{P} = \overline{V}\,(\overline{S} - \overline{C}) - \overline{F} \tag{12-57}$$

Similar expressions can be obtained for the after tax profit, P_t, if the tax rate is assumed a constant and the tax and book depreciation are the same. The after-tax profit is

$$P_t = (1 - T)[V(S - C) - F] \tag{12-58}$$

Therefore, the mean and variance, respectively, of Eq. (12-58) are

$$\overline{P}_t = (1 - T)\,\overline{P} \tag{12-59}$$
$$\sigma^2_{P_t} = (1 - T)^2\,\sigma^2_P \tag{12-60}$$

Where $\bar{P}$ and σ^2 are given by Eqs. (12-57) and (12-56).

Note that the distribution of the profit is unknown. These expressions give only the mean and variance of the profit. If probability statements are desired, it is necessary to make some assumption about the distribution of the profit (a common assumption, although questionable, is that the profit is normally distributed) or use Tchebycheff's inequality as shown in Example 12-3.

DECISION TREES

Another procedure sometimes used for the analysis of cash flow models with random elements is to use decision trees. The words "decision tree" could be considered a misnomer because, strictly speaking, they do not provide a decision regarding the acceptability of a project. Basically, a decision tree enumerates all the possible outcomes as shown in the following example.

EXAMPLE 12-10

Suppose a project has the distributions for the initial investment and cash flows shown in Figure 12-1. Furthermore, for the sake of simplicity, the cash flows are assumed to be the same for each year and the life of the project is two years.

With the data given in Figure 12-1, the decision tree shown in Figure 12-2 can be drawn. Each branch shows a possible outcome. Also, the net present values for each branch and its probability can be calculated. For example, assuming the minimum

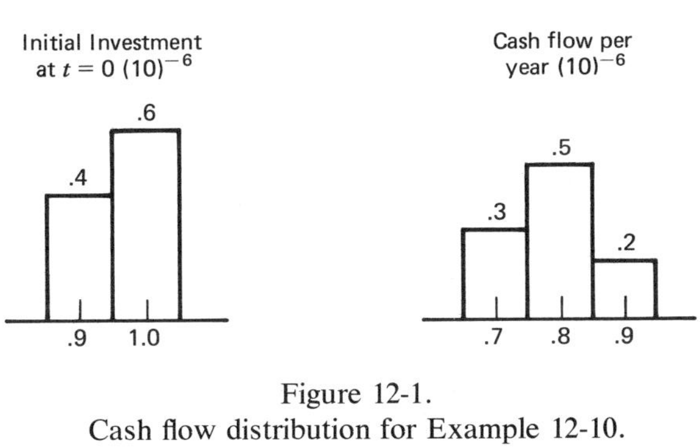

Figure 12-1.
Cash flow distribution for Example 12-10.

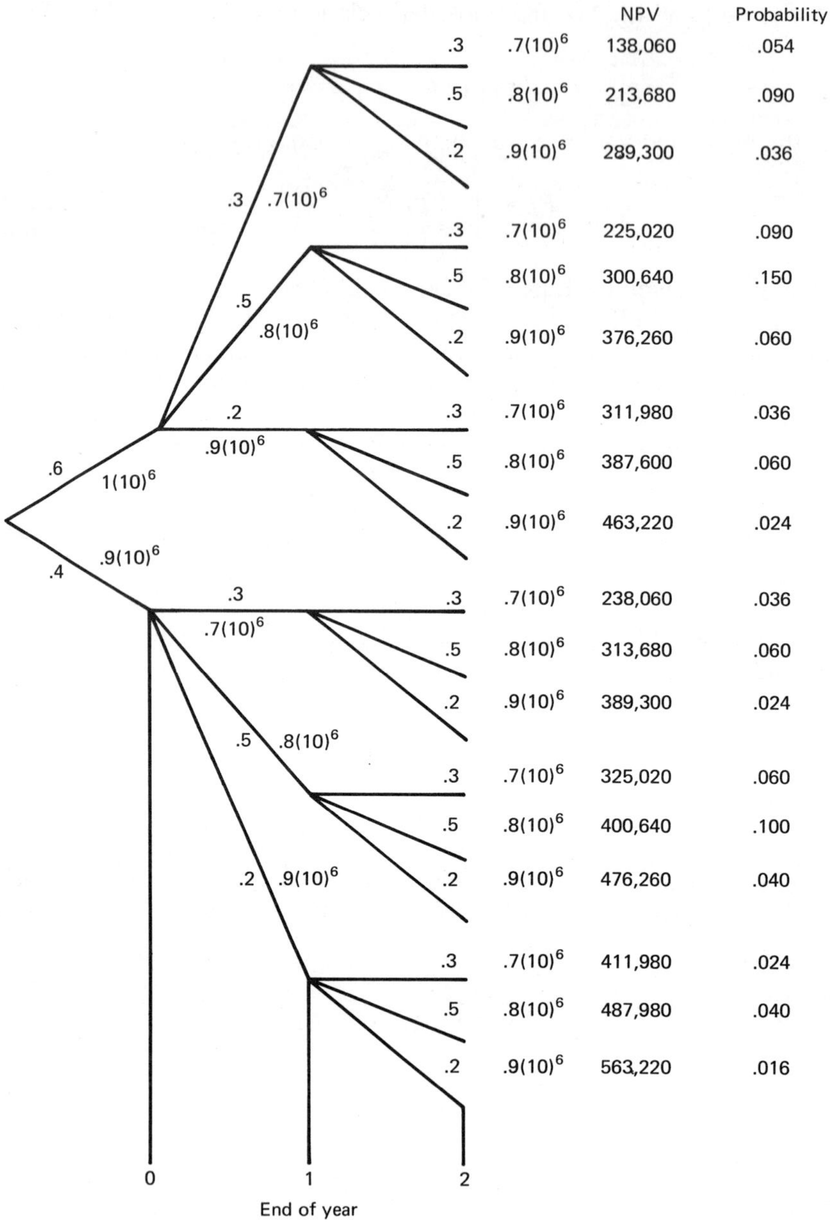

Figure 12-2.
Decision tree for Example 12-10.

acceptable rate of return is 15%, the net present value for the uppermost branch in Figure 12-2 is

$$\text{NPV} = 700{,}000(PF15{,}2) + 700{,}000\,(PF15{,}1) - 1{,}000{,}000 = 138{,}060$$

and the probability of this outcome is (.6)(.3)(.3) = .054. With the NPV for each branch and its probability, it is possible to calculate the mean, $E(\text{NPV})$, and variance of the net present value, $V(\text{NPV})$, as shown below:

$$E(\text{NPV}) = 138{,}060(0.54) + 213{,}680(.090) + \ldots + 563{,}220(.016) = \$324{,}379$$
$$V(\text{NPV}) = (138{,}060)^2\,(.054) + (213{,}680)^2\,(.090) + \ldots + (563{,}220)^2\,(.016) - (324{,}379)^2 \approx 1812.53(10)^8$$

Using these values in the decision-making process is strictly a matter of a *subjective* evaluation of the relative magnitude of the mean and variance of the net present value and the decision maker's attitude toward risk. In this particular case, there might be a tendency to consider the project acceptable since all the net present values are positive. However, in a case where there are several negative net present values, it may be more difficult to make a decision.

A Comment: A major disadvantage of using decision trees in the analysis of probabilistic cash flow models is that the number of computations can increase very rapidly. The number of end points (outcomes) on the decision tree becomes very large as the number of possible values and/or number of years increases. In general, the number of end points is equal to

$$(\text{number of possible values})^{\text{number of years}}$$

for example, if there are four possible cash flows and five years, the number of possible end points is 1024.

SIMULATION

In the solution of probabilistic cash flow models it is sometimes found that a model is not amenable to an analytical solution. In these cases, the use of Monte Carlo simulation can be a valuable tool in obtaining a solution. The applications of Monte Carlo simulation to cash flow models are discussed in this section. It is assumed that the reader has some experience in the basics of Monte Carlo simulation. References that discuss the intricacies of Monte Carlo simulation are provided at the end of this chapter. The application of Monte Carlo simulation to cash flow models is, perhaps, best shown through the use of a simple example with possible extensions discussed later.

EXAMPLE 12-11

Suppose a particular project is being considered and the variables have been estimated as shown in Table 12-5. In addition, it has been decided the following information is applicable for this project.

1. Debt ratio = 30%.
2. The debt principal is to be paid back in five equal installments. The first payment being

Table 12-5

DISTRIBUTION OF CASH FLOW VARIABLES (EXAMPLE 12-11)

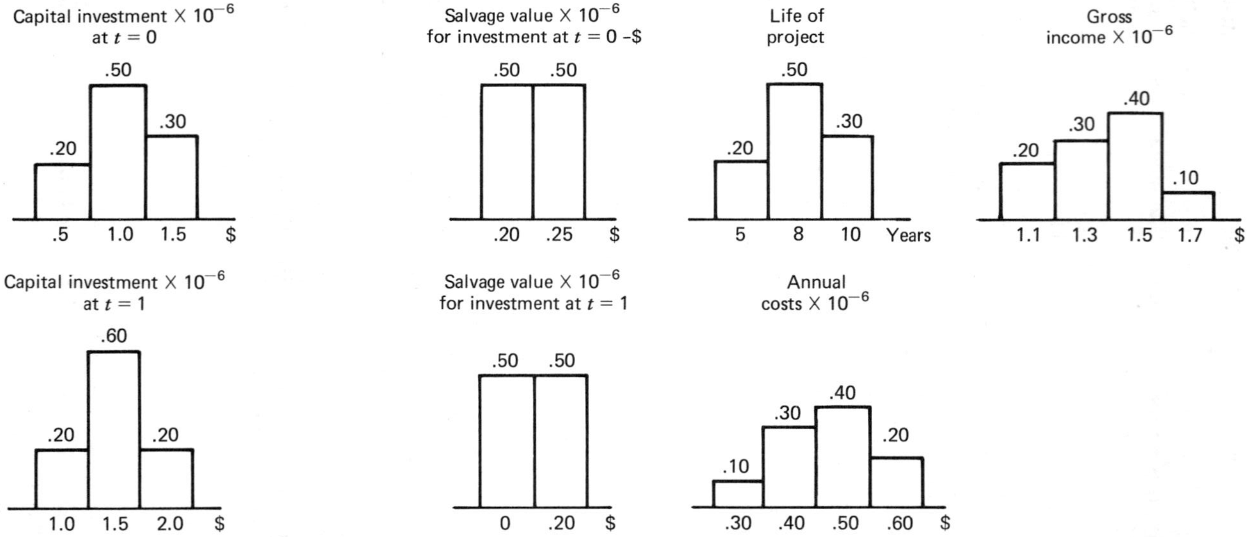

at the end of the year following the receipt of the borrowed money. The cost of debt money is 10% of the unpaid balance.

3. The tax depreciation is based on a straight-line model and a life of five years. The salvage values are obtained in the simulation.
4. The tax rate is assumed to be a constant value of 52% for each year.

Based on this data and the data given in Table 12-5, one set of possible net cash flows is shown in Table 12-6. The procedure used for obtaining these values is presented in the following discussion.

The first step is to generate three random numbers for, respectively, the life of the project, the capital investment for year zero, and the salvage value for this investment. Suppose these random numbers are 35, 54, and 43. Consequently, referring to the applicable distributions in Table 12-5, the values for the life, investment, and salvage are, respectively, eight years, \$1,000,000, and \$200,000.

The next step is to generate two random numbers for the values of the investment occurring in year 1 and its salvage. Assuming these random numbers are 68 and 22, the investment in year 1 is \$1,500,000 and its salvage value is zero. With this data it is possible to calculate the tax depreciation schedules, amount of borrowed money, principal, and interest payments. For example, the tax depreciation for the investments are

$$\frac{1{,}000{,}000 - 200{,}000}{5} = \$160{,}000 \text{ per year}$$

$$\frac{1{,}500{,}000 - 0}{5} = \$300{,}000 \text{ per year}$$

The amount of borrowed money is \$1,000,000(.3) = \$300,000, which occurs in year 0 and \$1,500,000(.3) = \$450,000, which occurs in year 1. Therefore, the principal payments in years 1 through 5 are

$$\frac{300{,}000}{5} = \$60{,}000$$

and in years 2 through 6 are

$$\frac{450{,}000}{5} = \$90{,}000$$

The interest is simply 10% of the unpaid balance.

Eight random numbers are now generated for the gross incomes and another eight random numbers for the annual costs. These values are shown in Table 12-6. After this, the net cash flows are calculated for each year using Eq. (7-15) (principal and interest payments are explicitly included). For example, the cash flows for years 0 and 1 are

$$X_0 = -\ 1{,}000{,}000 + 300{,}000 = -\ \$700{,}000$$

$$X_1 = (1{,}300{,}000 - 300{,}000 - 30{,}000) - (1{,}300{,}000 - 300{,}000 - 30{,}000 - 160{,}000)\,(.52) - \$1{,}500{,}000 + 450{,}000 - 60{,}000 = -\ \$561{,}200$$

Once the cash flows are determined, it is possible to calculate the internal rate of return, net present value, and payback period. For example, the net present value of the cash flows in Table 12-6 is \$159,072 using 25% as the required return on equity.

The procedure just shown provides only one iteration and must be repeated many times. After this procedure is repeated many times it is possible to plot a frequency

Table 12-6

SIMULATION VALUES FOR EXAMPLE 12-11

End of Year	Capital Investment and Salvage	Tax Depreciation for K_0	Tax Depreciation for K_1	Principal Payments K_0	Principal Payments K_1	Interest K_0	Interest K_1	RN G	Gross Income, G	RN C	Annual Costs, C	Cash Flow
0	−$1,000,000	—	—	—	—	—	—	—	—	—	—	−$700,000
1	−1,500,000	$160,000	—	$60,000	—	$30,000	—	32	$1,300,000	05	$300,000	−561,200
2		↑	$300,000	↑	$90,000	24,000	$45,000	56	1,500,000	89	600,000	488,080
3		|	↑	|	↑	18,000	36,000	25	1,300,000	20	400,000	495,280
4		↓	|	↓	|	12,000	27,000	70	1,500,000	45	500,000	550,480
5		160,000	↓	60,000	↓	6,000	18,000	61	1,500,000	30	400,000	605,680
6			300,000		90,000		9,000	15	1,100,000	16	400,000	397,680
7								28	1,300,000	30	400,000	432,000
8	$L_0 = 200,000$ $L_1 = 0$							59	1,500,000	85	600,000	632,000[a]

[a]Includes $200,000 for salvage values.

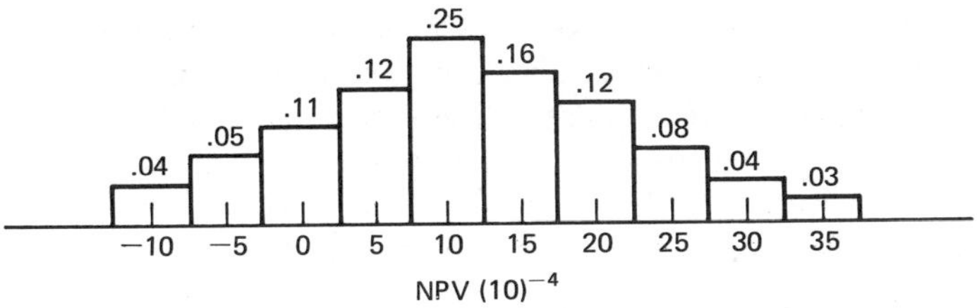

Figure 12-3.
Frequency distribution of net present value.

distribution of the net present value as shown in Figure 12-3; or if desired, a frequency distribution of the internal rate of return and payback period.

With a frequency distribution of net present value, such as shown in Figure 12-3, it is possible to make probability statements regarding the net present value. For example, defining a loss as a net present value of zero or less, the probability of a loss is

$$.04 + .05 + .11 = .20 = 20\%$$

It is also possible to calculate the mean and variance of the net present value.

From a decision-making standpoint, a frequency distribution such as shown in Figure 12-3 is a valuable tool to the person(s) evaluating the project.

Comments and Extensions: The simulation discussed in Example 12-11 is relatively simple. There are many possible extensions. For example the gross income could be simulated by estimating distributions for the selling price per unit and volume of yearly sales. The annual costs could be broken down into such components as material, labor, and overhead costs. Working capital considerations could also be included. In actuality, there are very few limitations in the simulation of cash flows except, perhaps, computer time. For, of course, Monte Carlo simulation is definitely computer oriented.

Note in Example 12-11 that the variables are assumed to be independent. This is not a requirement for it is entirely possible to include dependency between variables in Monte Carlo Simulation. The problem when allowing dependency between variables is in the *estimation* and *expression* of the dependency; not in the Monte Carlo procedure itself. Some examples and discussion of including dependencies in the simulation of cash flows are given in references provided at the end of this chapter.

In this example, the assumed distributions are discrete. This, of course, is not a requirement. It is entirely possible to represent variables by continuous distributions.

A FINAL POINT

In this chapter it has been stated several times that probabilities, expected losses, and means and variances of the net present value do not, by themselves, provide a decision regarding the acceptability of a project. They simply provide a decision maker with additional quantitative information on which to evaluate a project. In the final analysis, the acceptability of a project will be decided by the decision maker's attitudes toward risk and the evaluation of intangible factors. There are, theoretically at least, methods for quantifying attitudes towards risk;

for example, the concept of cardinal utility theory. A project will be accepted or rejected on the basis of the decision-maker's utility function. Consequently if this function can be expressed explicitly, the acceptability of a project can be readily determined. The concept of cardinal utility theory is beyond the scope of this text. However, it is strongly recommended that, for persons interested in decision making under risk, a study of this concept be undertaken. For, it is the basis on which many advanced economic analysis problems are discussed.

References

1. Canada, J. R., *Intermediate Economic Analysis for Management and Engineering,* Prentice-Hall, Inc., Englewood Cliffs, N.J., 1971.
2. Hillier, F. S., "The Derivation of Probabilistic Information for the Evaluation of Risky Investments," *Management Science,* April 1963.
3. Jaedicke, R. K., and A. A. Robichek, "Cost-Volume-Profit Analysis Under Conditions of Uncertainty," *The Accounting Review, XXXIX,* October 1964, pp. 917–926.
4. Marshall, A. W. and Olkin, I. "A One-Side Inequality of the Chebyshev Type," *Annals of Mathematical Statistics,* 1960.
5. Schlaifer, R., *Probability and Statistics for Business Decisions,* McGraw-Hill Book Co., New York, 1959.

Simulation References

6. Bonini, C. P., "Simulation of Information and Decision Systems in the Firm," Prentice-Hall, Inc., Englewood Cliffs, N.J., 1963.
7. Hillier, F. S., and G. J. Lieberman, "Introduction to Operations Research," Chap. 14, Holden-Day, Inc., San Francisco, 1968.
8. Mize, J. H., and J. F. Cox, "Essentials of Simulation," Prentice-Hall, Inc., Englewood Cliffs, N.J., 1968.
9. Naylor, T. F., J. L. Balintfy, D. S. Burdick, and C. Kong, "Computer Simulation Techniques," John Wiley & Sons, New York, 1966.
10. Schmidt, J. W., and R. E. Taylor, "Simulation and Analysis of Industrial Systems," Richard D. Irwin, Inc., Homewood, Ill., 1970.
11. Tocher, K. D., "The Art of Simulation," The English University Press, Ltd., London, 1963.

Problems

12-1. For the cash flow means and standard deviations shown below, determine the following assuming the minimum attractive rate of return is 12% and the net present value is normally distributed.

(a) The mean and standard deviation of the net present value.
(b) The probability of a loss.
(c) The expected loss.
(d) The expected loss given that a loss occurs.
(e) The expected gain.

End of Year	Cash Flow	
	Mean	Standard Deviation
0	−$300,000	$30,000
1	70,000	7,000
2	80,000	10,000
3	80,000	10,000
4	80,000	10,000
5	80,000	10,000
6	50,000	12,000
7	50,000	12,000
8	50,000	12,000

12-2. Using the data and assumptions in Problem 12-1, determine the following:
(a) The probability that the rate of return is 12% or less.
(b) The probability that the rate of return is 15% or less.

12-3. If the mean and standard deviation of the net present value are, respectively, $50,000 and $30,000, and there is no knowledge about the distribution of net present value, determine the following:
(a) The probability that the net present value will fall outside the plus and minus three sigma limits. What are these limits?
(b) The probability that there will be a loss.

12-4. Consider the definition of cash flow, X_t, shown below and determine the following:
(a) The mean and variance of the cash flow if the tax rate is a constant and all other variables are independent and random.
(b) The same as part (b) only S, L, and M are dependent.
(c) What additional term would have to be added in part (b) if I and B are dependent?

$$X_t = (VS - VC - VM - H - I) - (VS - VC - VM - H - I - D)\,T - K + L - P + B$$

where

X_t = cash flow in year t
V = volume of sales
S = selling price per unit
C = labor costs per unit
M = material costs per unit
H = overhead

I = interest
D = depreciation
T = tax rate
K = capital investment
L = salvage value
P = principal payment
B = borrowed money received

12-5. For the situation depicted below and the following conditions, derive expressions for the mean and variance of the net present value.

(a) n and T are constants.
(b) The discount rate is a constant.
(c) G_t, C_t, D_t, K_0, K_1, L_0, L_1 are random variables.
(d) K_0 and L_0 are dependent.
(e) K_1 and L_1 are dependent
(f) The gross incomes, G_t, are dependent for the first X years.
(g) The costs, C_t, are dependent for the first X years.
(h) The depreciations, D_t, are dependent for the first X years.
(i) All other variables are independent.

End-of-Year Variables					
0	K_0				
1	G_1,	C_1,	D_1	T,	K
2	G_2	C_2	D_2	T	
.	.	.	.	.	
.	.	.	.	.	
.	.	.	.	.	
t	G_t,	C_t,	D_t,	T	
.	.	.	.	.	
.	.	.	.	.	
.	.	.	.	.	
n	G_n,	C_n,	D_n,	T	

12-6. For a particular capital investment the data shown below has been estimated. Based on this data, determine the mean and standard deviation of the net present value assuming all variables are independent.

	Mean	Standard Deviation
Capital investment in year zero	\$300,000	\$40,000
Life of project	10 years	0
Depreciation (years 1–10)	\$40,000	0
Minimum acceptable return	15%	0
Annual gross income (years 1–10)	\$170,000	\$30,000
Annual costs (years 1–10)	\$50,000	\$10,000
Tax rate	52%	0

12-7. Using the model presented in Example 12-7 and the following data, determine the mean and standard deviation of the net present value.

	Mean	Standard Deviation
Capital investment in year zero	\$300,000	\$40,000
Life of project	10 years	0
Annual gross income (years 1–10)	\$170,000	\$30,000
Annual costs (years 1–10)	\$50,000	\$10,000
Cost of debt capital	10%	0
Cost of equity capital	15%	0
Tax rate	52%	0
Debt Ratio	30%	0
Salvage value (% of capital investment)	5%	0

12-8. Past data indicates that mean and standard deviation, respectively, of yearly sales are \$120,000 and \$24,000. Assuming that the distributions of monthly sales are normal, the same, and independent, determining the probability of monthly sales being less than \$5000.

12-9. Using the data given below for a particular capital investment, determine the following based on the cash flow definition:

$$X_t = (G - C) - (G - C - D)\,T - K_t$$

(a) A decision tree showing all cash flow elements (gross income, costs, etc.).
(b) The net present value for all possible outcomes and their respective probabilities.
(c) The probability of a loss for this investment.
(d) What major assumption is implied in parts (a) and (b)?

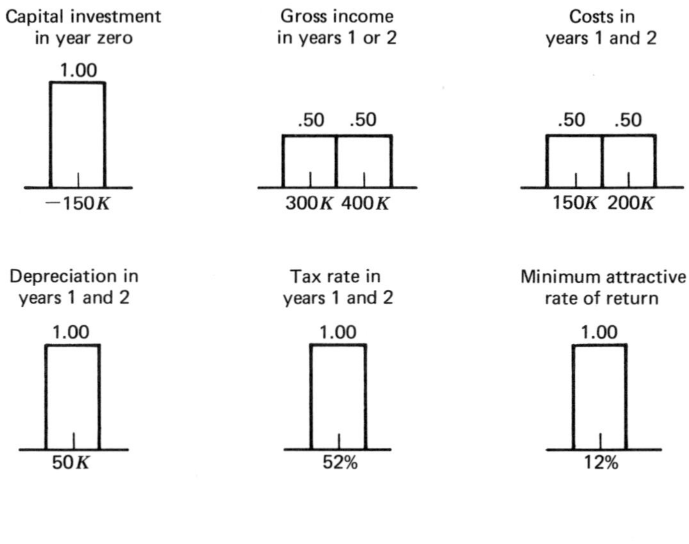

12-10. Listed below are the necessary data for a *C-V-P* model. Assuming the variables are independent and the profit is normally distributed, determine the probability of at least breaking even based on a before-tax analysis.

Variable	Mean	Standard Deviation
Volume of sales per year	60,000	8,000
Selling price per units	$80	$7
Variable cost per unit	$35	$4
Fixed costs per year	$1,900,000	$200,000

12-11. Repeat Problem 12-10 on an after-tax basis assuming a tax rate of 52% assuming the tax and book depreciation are the same.

12-12. Assume that in the *C-V-P* model considered in this chapter the selling price (S) and variable costs (C) are dependent and that all other variables are independent. Derive expressions for the mean and variance of the profit based on these assumptions.

12-13. Assume in Problem 12-11 that the selling price and variable costs are dependent and the correlation coefficient that measures this dependency is .5. Determine the probability of at least breaking even with this assumption.

12-14. Repeat Example 12-11 using the same conditions and the following random numbers:
life = 19
capital investment in years zero = 85
salvage value = 10
capital investment in year one = 16
salvage value = 32

End of Year	Random Numbers[a]	
	Gross Income	Costs
1	45	26
2	65	27
3	06	75
4	59	87
5	33	37
6	17	45
7	13	19
8	44	16

[a]There may be more than needed.

12-15. The Monte Carlo simulation of a particular capital investment has given the frequency distribution shown below.

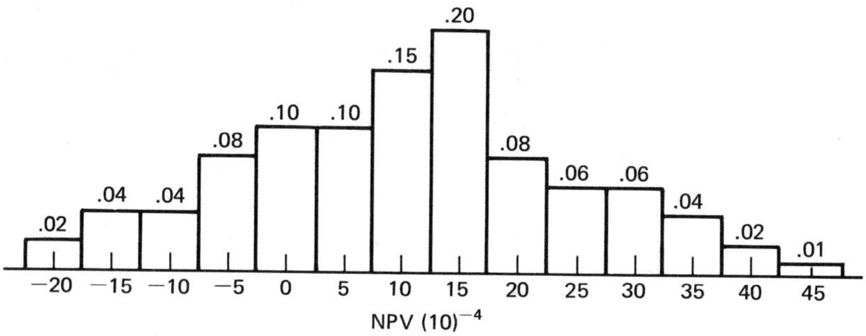

Determine the following:

(a) The probability of a loss. Define a loss as the NPV < 0 since a discrete distribution is involved.

(b) The expected loss.

(c) The expected loss given that a loss occurs.

(d) The mean and variance of the net present value.

APPENDIX A
INTEREST FACTORS

This appendix contains three sections. Section 1 gives values for discrete interest factors. Section 2 gives values for continuous interest factors. Section 3 gives values for funds-flow conversion factors.

SECTION 1 DISCRETE INTEREST FACTORS

Mathematical Expressions

Single-payment compound-amount factor $= FPi,n = (1 + i)^n$

Single-payment present-worth factor $= PFi,n = \dfrac{1}{(1 + i)^n}$

Equal-payment-series compound-amount factor $= FAi,n = \dfrac{(1 + i)^n - 1}{i}$

Equal-payment-series sinking-fund factor $= AFi,n = \dfrac{i}{(1 + i)^n - 1}$

Equal-payment-series present-worth factor $= PAi,n = \dfrac{(1 + i)^n - 1}{(i)(1 + i)^n}$

Equal-payment-series capital recovery factor $= APi,n = \dfrac{(i)(1 + i)^n}{(1 + i)^n - 1}$

Gradient factor $= AGi,n = \left[\dfrac{1}{i} - \dfrac{n}{i}(AFi,n)\right]$

Table A-1

DISCRETE COMPOUNDING: $i = 1/4\%$

	Single Payment		Uniform Series				Gradient Series
	Compound Amount Factor	Present Worth Factor	Compound Amount Factor	Sinking Fund Factor	Present Worth Factor	Capital Recovery Factor	Uniform Series Factor
n	To Find *F* Given *P* *FPi,n*	To Find *P* Given *E* *PFi,n*	To Find *F* Given *A* *FAi,n*	To Find *A* Given *F* *AFi,n*	To Find *P* Given *A* *PAi,n*	To Find *A* Given *P* *APi,n*	To Find *A* Given *G* *AGi,n*
1	1.0025	0.9975	1.0000	1.0000	0.9975	1.0025	0.0000
2	1.0050	0.9950	2.0025	0.4994	1.9925	0.5019	0.4994
3	1.0075	0.9925	3.0075	0.3325	2.9851	0.3350	0.9983
4	1.0100	0.9901	4.0150	0.2491	3.9751	0.2516	1.4969
5	1.0125	0.9876	5.0251	0.1990	4.9627	0.2015	1.9950
6	1.0151	0.9851	6.0376	0.1656	5.9478	0.1681	2.4927
7	1.0176	0.9827	7.0527	0.1418	6.9305	0.1443	2.9900
8	1.0202	0.9802	8.0704	0.1239	7.9107	0.1264	3.4869
9	1.0227	0.9778	9.0905	0.1100	8.8885	0.1125	3.9834
10	1.0253	0.9753	10.1133	0.0989	9.8639	0.1014	4.4794
11	1.0278	0.9729	11.1385	0.0898	10.8368	0.0923	4.9750
12	1.0304	0.9705	12.1664	0.0822	11.8073	0.0847	5.4702
13	1.0330	0.9681	13.1968	0.0758	12.7753	0.0783	5.9650
14	1.0356	0.9656	14.2298	0.0705	13.7410	0.0728	6.4594
15	1.0382	0.9532	15.2654	0.0655	14.7042	0.0680	6.9534
16	1.0408	0.9608	16.3035	0.0613	15.6650	0.0638	7.4469
17	1.0434	0.9584	17.3443	0.0577	16.6235	0.0602	7.9401
18	1.0460	0.9561	18.3876	0.0544	17.5795	0.0569	8.4328
19	1.0486	0.9537	19.4336	0.0515	18.5332	0.0540	8.9251
20	1.0512	0.9513	20.4822	0.0488	19.4845	0.0513	9.4170
21	1.0638	0.9489	21.5334	0.0464	20.4334	0.0489	9.9085
22	1.0565	0.9466	22.5872	0.0443	21.3800	0.0468	10.3995
23	1.0591	0.9442	23.6437	0.0423	22.3241	0.0448	10.8901
24	1.0618	0.9418	24.7028	0.0405	23.2660	0.0430	11.3804
25	1.0644	0.9395	25.7646	0.0388	24.2055	0.0413	11.8702
26	1.0671	0.9371	26.3290	0.0373	25.1426	0.0398	12.3596
27	1.0697	0.9348	27.8961	0.0358	26.0774	0.0383	12.8485
28	1.0724	0.9325	28.9658	0.0345	27.0099	0.0370	13.3371
29	1.0751	0.9301	30.0382	0.0333	27.9400	0.0358	13.8252
30	1.0779	0.9278	31.1133	0.0321	28.8679	0.0346	14.3130
31	1.0805	0.9255	32.1911	0.0311	29.7934	0.0336	14.8003
32	1.0832	0.9232	33.2715	0.0301	30.7166	0.0326	15.2872
33	1.0859	0.9209	34.3547	0.0291	31.6375	0.0316	15.7736
34	1.0386	0.9186	35.4406	0.0282	32.5561	0.0307	16.2597

Table A-1 *(continued)*

DISCRETE COMPOUNDING: $i = 1/4\%$

n	Single Payment		Uniform Series				Gradient Series
	Compound Amount Factor	Present Worth Factor	Compound Amount Factor	Sinking Fund Factor	Present Worth Factor	Capital Recovery Factor	Uniform Series Factor
	To Find F Given P FPi,n	To Find P Given E PFi,n	To Find F Given A FAi,n	To Find A Given F AFi,n	To Find P Given A PAi,n	To Find A Given P APi,n	To Find A Given G AGi,n
35	1.0913	0.9163	36.5292	0.0274	33.4724	0.0299	16.7454
40	1.1050	0.9050	42.0132	0.0238	38.0199	0.0263	19.1673
45	1.1189	0.8937	47.5661	0.0210	42.5109	0.0235	21.5789
50	1.1330	0.8826	53.1887	0.0188	46.9462	0.0213	23.9802
55	1.1472	0.8717	58.8819	0.0170	51.3264	0.0195	26.3710
60	1.1616	0.8609	64.6467	0.0155	55.6524	0.0180	28.7514
65	1.1762	0.8502	70.4839	0.0142	59.9246	0.0167	31.1215
70	1.1910	0.8396	76.3944	0.0131	64.1439	0.0156	33.4812
75	1.2059	0.8292	82.3792	0.0121	68.3108	0.0146	35.8305
80	1.2211	0.8189	88.4392	0.0113	72.4260	0.0138	38.1694
85	1.2364	0.8088	94.5753	0.0106	76.4901	0.0131	40.4980
90	1.2620	0.7987	100.7885	0.0099	80.5038	0.0124	42.8162
95	1.2677	0.7888	107.0796	0.0093	84.4677	0.0118	45.1241
100	1.2836	0.7790	113.4499	0.0088	88.3825	0.0113	47.4216

Reprinted, by permission, from J. A. White et al., *Principles of Engineering Economic Analysis,* pp. 396–447.

Table A-2

DISCRETE COMPOUNDING: $i = \frac{1}{3}\%$

	Single Payment		Uniform Series				Gradient Series
	Compound Amount Factor	Present Worth Factor	Compound Amount Factor	Sinking Fund Factor	Present Worth Factor	Capital Recovery Factor	Uniform Series Factor
n	To Find *F* Given *P* *FPi,n*	To Find *P* Given *F* *PFi,n*	To Find *F* Given *A* *FAi,n*	To Find *A* Given *F* *AFi,n*	To Find *P* Given *A* *PAi,n*	To Find *A* Given *P* *APi,n*	To Find *A* Given *G* *AGi,n*
1	1.0033	0.9967	1.0000	1.0000	0.9967	1.0033	0.0000
2	1.0067	0.9934	2.0033	0.4992	1.9900	0.5025	0.4992
3	1.0100	0.9901	3.0100	0.3322	2.9801	0.3356	0.9978
4	1.0134	0.9868	4.0200	0.2488	3.9669	0.2521	1.4958
5	1.0168	0.9835	5.0334	0.1987	4.9504	0.2020	1.9933
6	1.0202	0.9802	6.0502	0.1653	5.9306	0.1686	2.4903
7	1.0236	0.9770	7.0704	0.1414	6.9076	0.1448	2.9867
8	1.0270	0.9737	8.0940	0.1235	7.8813	0.1269	3.4825
9	1.0304	0.9705	9.1209	0.1096	8.8518	0.1130	3.9778
10	1.0338	0.9673	10.1513	0.0985	9.8191	0.1018	4.4725
11	1.0373	0.9641	11.1852	0.0894	10.7831	0.0927	4.9667
12	1.0407	0.9609	12.2225	0.0818	11.7440	0.0851	5.4603
13	1.0442	0.9577	13.2632	0.0754	12.7017	0.0787	5.9534
14	1.0477	0.9545	14.3074	0.0699	13.6561	0.0732	6.4459
15	1.0512	0.9513	15.3551	0.0651	14.6074	0.0685	6.9379
16	1.0547	0.9481	16.4063	0.0610	15.5556	0.0643	7.4293
17	1.0582	0.9450	17.4610	0.0573	16.5006	0.0606	7.9201
18	1.0617	0.9419	18.5192	0.0540	17.4424	0.0573	8.4104
19	1.0653	0.9387	19.5809	0.0511	18.3812	0.0544	8.9002
20	1.0688	0.9355	20.6462	0.0484	19.3168	0.0518	9.3894
21	1.0724	0.9325	21.7150	0.0461	20.2493	0.0494	9.8780
22	1.0760	0.9294	22.7874	0.0439	21.1787	0.0472	10.3661
23	1.0795	0.9263	23.8633	0.0419	22.1050	0.0452	10.8536
24	1.0831	0.9232	24.9429	0.0401	23.0283	0.0434	11.3406
25	1.0863	0.9202	26.0260	0.0384	23.9484	0.0418	11.8270
26	1.0904	0.9171	27.1128	0.0369	24.8655	0.0402	12.3128
27	1.0940	0.9141	28.2032	0.0355	25.7796	0.0388	12.7981
28	1.0977	0.9110	29.2972	0.0341	26.6906	0.0375	13.2829
29	1.1013	0.9080	30.3948	0.0329	27.5986	0.0362	13.7671
30	1.1050	0.9050	31.4961	0.0317	28.5036	0.0351	14.2507
31	1.1087	0.9020	32.6011	0.0307	29.4056	0.0340	14.7338
32	1.1124	0.8990	33.7098	0.0297	30.3046	0.0330	15.2164
33	1.1161	0.8960	34.8222	0.0287	31.2006	0.0321	15.6983
34	1.1198	0.3930	35.9382	0.0278	32.0936	0.0312	16.1798

Table A-2 *(continued)*

DISCRETE COMPOUNDING: $i = 1/3\%$

n	Single Payment		Uniform Series				Gradient Series
	Compound Amount Factor	Present Worth Factor	Compound Amount Factor	Sinking Fund Factor	Present Worth Factor	Capital Recovery Factor	Uniform Series Factor
	To Find *F* Given *P* *FPi,n*	To Find *P* Given *F* *PFi,n*	To Find *F* Given *A* *FAi,n*	To Find *A* Given *F* *AFi,n*	To Find *P* Given *A* *PAi,n*	To Find *A* Given *P* *APi,n*	To Find *A* Given *G* *AGi,n*
35	1.1235	0.8901	37.0580	0.0270	32.9837	0.0303	16.6606
40	1.1424	0.8754	42.7132	0.0234	37.3898	0.0267	19.0567
45	1.1615	0.8609	48.4633	0.0206	41.7232	0.0240	21.4389
50	1.1810	0.8457	54.3099	0.0184	45.9851	0.0217	23.8073
55	1.2008	0.8327	60.2546	0.0166	50.1767	0.0199	26.1619
60	1.2210	0.8190	66.2990	0.0151	54.2991	0.0184	28.5026
65	1.2415	0.8055	72.4448	0.0138	58.3535	0.0171	30.8295
70	1.2623	0.7922	78.6937	0.0127	62.3409	0.0160	33.1427
75	1.2835	0.7791	85.0475	0.0118	66.2626	0.0151	35.4420
80	1.3050	0.7663	91.5079	0.0109	70.1196	0.0143	37.7275
85	1.3269	0.7536	98.0767	0.0102	73.9129	0.0135	39.9993
90	1.3492	0.7412	104.7557	0.0095	77.6436	0.0129	42.2574
95	1.3718	0.7290	111.5467	0.0090	81.3128	0.0123	44.5017
100	1.3948	0.7169	118.4517	0.0084	84.9214	0.0118	46.7322

Table A-3
DISCRETE COMPOUNDING: $i = 5/12\%$

	Single Payment		Uniform Series				Gradient Series
	Compound Amount Factor	Present Worth Factor	Compound Amount Factor	Sinking Fund Factor	Present Worth Factor	Capital Recovery Factor	Uniform Series Factor
n	To Find F Given P FPi,n	To Find P Given F PFi,n	To Find F Given A FAi,n	To Find A Given F AFi,n	To Find P Given A PAi,n	To Find A Given P APi,n	To Find A Given G AGi,n
1	1.0042	0.9959	1.0000	1.0000	0.9959	1.0042	0.0000
2	1.0084	0.9917	2.0042	0.4990	1.9876	0.5031	0.4990
3	1.0126	0.9876	3.0125	0.3319	2.9752	0.3361	0.9972
4	1.0168	0.9835	4.0251	0.2484	3.9587	0.2526	1.4948
5	1.0210	0.9794	5.0418	0.1983	4.9381	0.2025	1.9917
6	1.0253	0.9754	6.0628	0.1649	5.9135	0.1691	2.4879
7	1.0295	0.9713	7.0881	0.1411	6.8848	0.1452	2.9834
8	1.0338	0.9673	8.1176	0.1232	7.8521	0.1274	3.4782
9	1.0381	0.9633	9.1515	0.1093	8.8158	0.1134	3.9723
10	1.0425	0.9593	10.1896	0.0981	9.7746	0.1023	4.4657
11	1.0468	0.9553	11.2321	0.0890	10.7299	0.0932	4.9584
12	1.0512	0.9513	12.1789	0.0814	11.6812	0.0856	5.4505
13	1.0555	0.9474	13.3300	0.0750	12.6286	0.0792	5.9418
14	1.0599	0.9434	14.3856	0.0695	13.5721	0.0737	6.4324
15	1.0644	0.9395	15.4455	0.0647	14.5116	0.0689	6.9224
16	1.0688	0.9356	16.5099	0.0606	15.4472	0.0647	7.4116
17	1.0732	0.9318	17.5786	0.0569	16.3790	0.0611	7.9002
18	1.0777	0.9279	18.6519	0.0536	17.3069	0.0578	8.3881
19	1.0822	0.9240	19.7296	0.0507	18.2309	0.0549	8.8753
20	1.0867	0.9202	20.8118	0.0480	19.1511	0.0522	9.3618
21	1.0912	0.9164	21.8985	0.0457	20.0675	0.0498	9.8476
22	1.0958	0.9126	22.9898	0.0435	20.9801	0.0477	10.3327
23	1.1004	0.9088	24.0856	0.0415	21.8889	0.0457	10.8171
24	1.1049	0.9050	25.1859	0.0397	22.7939	0.0439	11.3008
25	1.1095	0.9013	26.2909	0.0380	23.6952	0.0422	11.7838
26	1.1142	0.8975	27.4004	0.0365	24.5927	0.0407	12.2662
27	1.1188	0.8938	28.5146	0.0351	25.4865	0.0392	12.7478
28	1.1235	0.8901	29.6354	0.0337	26.3766	0.0379	13.2288
29	1.1282	0.8854	30.7569	0.0325	27.2630	0.0367	13.7090
30	1.1329	0.8827	31.8850	0.0314	28.1457	0.0355	14.1886
31	1.1376	0.8791	33.0179	0.0303	29.0248	0.0345	14.6675
32	1.1423	0.9734	34.1554	0.0293	29.9002	0.0334	15.1456
33	1.1471	0.8718	35.2978	0.0283	30.7720	0.0325	15.6231
34	1.1519	0.8682	36.4448	0.0274	31.6402	0.0316	16.0999

Table A-3 *(continued)*

DISCRETE COMPOUNDING: $i = 5/12\%$

n	Single Payment		Uniform Series				Gradient Series
	Compound Amount Factor	Present Worth Factor	Compound Amount Factor	Sinking Fund Factor	Present Worth Factor	Capital Recovery Factor	Uniform Series Factor
	To Find *F* Given *P* *FPi,n*	To Find *P* Given *F* *PFi,n*	To Find *F* Given *A* *FAi,n*	To Find *A* Given *F* *AFi,n*	To Find *P* Given *A* *PAi,n*	To Find *A* Given *P* *APi,n*	To Find *A* Given *G* *AGi,n*
35	1.1567	0.8645	37.5967	0.0266	32.5047	0.0308	16.5760
40	1.1810	0.8468	43.4283	3.0230	36.7740	0.0272	18.9462
45	1.2058	0.8294	49.3825	0.0203	40.9555	0.0244	21.2991
50	1.2311	0.8123	55.4618	0.0180	45.0509	0.0222	23.6347
55	1.2570	0.7956	61.6687	0.0162	49.0621	0.0204	25.9531
60	1.2834	0.7792	68.0061	0.0147	52.9907	0.0189	28.2542
65	1.3103	0.7632	74.4766	0.0134	56.8385	0.0176	30.5382
70	1.3378	0.7475	31.0830	0.0123	60.6071	0.0165	32.8049
75	1.3660	0.7321	87.8282	0.0114	64.2982	0.0156	35.0544
80	1.3946	0.7170	94.7151	0.0106	67.9134	0.0147	37.2868
85	1.4239	0.7023	101.7467	0.0098	71.4541	0.0140	39.5021
90	1.4539	0.5878	108.9259	0.0092	74.9220	0.0133	41.7002
95	1.4844	0.5737	116.2561	0.0086	78.3186	0.0128	43.8813
100	1.5156	0.5598	123.7402	0.0081	81.6452	0.0122	46.0453

Table A-4

DISCRETE COMPOUNDING: $i = \frac{1}{2}\%$

	Single Payment		Uniform Series				Gradient Series
	Compound Amount Factor	Present Worth Factor	Compound Amount Factor	Sinking Fund Factor	Present Worth Factor	Capital Recovery Factor	Uniform Series Factor
n	To Find F Given P FPi,n	To Find P Given F PFi,n	To Find F Given A FAi,n	To Find A Given F AFi,n	To Find P Given A PAi,n	To Find A Given P APi,n	To Find A Given G AGi,n
1	1.0050	0.9950	1.0000	1.0000	0.9950	1.0050	0.0000
2	1.0100	0.9901	2.0050	0.4988	1.9851	0.5038	0.4988
3	1.0151	0.9851	3.0150	0.3317	2.9702	0.3367	0.9967
4	1.0202	0.9802	4.0301	0.2481	3.9505	0.2531	1.4938
5	1.0253	0.9754	5.0503	0.1980	4.9259	0.2030	1.9900
6	1.0304	0.9705	6.0755	0.1646	5.8954	0.1696	2.4855
7	1.0355	0.9657	7.1059	0.1407	6.8621	0.1457	2.9801
8	1.0407	0.9609	8.1414	0.1228	7.8230	0.1278	3.4738
9	1.0459	0.9561	9.1821	0.1089	8.7791	0.1139	3.9668
10	1.0511	0.9513	10.2280	0.0978	9.7304	0.1028	4.4589
11	1.0564	0.9465	11.2792	0.0887	10.6770	0.0937	4.9501
12	1.0617	0.9419	12.3356	0.0811	11.6189	0.0861	5.4406
13	1.0670	0.9372	13.3972	0.0746	12.5562	0.0796	5.9302
14	1.0723	0.9325	14.4642	0.0691	13.4887	0.0741	6.4190
15	1.0777	0.9279	15.5365	0.0644	14.4166	0.0694	6.9069
16	1.0881	0.9233	16.6142	0.0602	15.3399	0.0652	7.3940
17	1.0885	0.9187	17.6973	0.0565	16.2586	0.0615	7.8803
18	1.0939	0.9141	18.7858	0.0532	17.1728	0.0582	8.3658
19	1.0994	0.9096	19.8797	0.0503	18.0824	0.0553	8.8504
20	1.1049	0.9051	20.9791	0.0477	18.9874	0.0527	9.3342
21	1.1104	0.9006	22.9840	0.0453	19.8880	0.0503	9.8172
22	1.1160	0.8961	23.1944	0.0431	20.7841	0.0481	10.2993
23	1.1216	0.8915	24.3104	0.0411	21.6757	0.0461	10.7806
24	1.1272	0.8872	25.4320	0.0393	22.5629	0.0443	11.2611
25	1.1328	0.8828	26.5591	0.0377	23.4456	0.0427	11.7407
26	1.1385	0.8734	27.6919	0.0361	24.3243	0.0411	12.2195
27	1.1442	0.8740	28.8304	0.0347	25.1980	0.0397	12.6975
28	1.1499	0.8697	29.9745	0.0334	26.0677	0.0384	13.1747
29	1.1556	0.8653	31.1244	0.0321	26.9330	0.0371	13.6510
30	1.1614	0.8610	32.2800	0.0310	27.7941	0.0360	14.1265
31	1.1672	0.8567	33.4414	0.0299	28.6508	0.0349	14.6012
32	1.1730	0.8525	34.6086	0.0289	29.5033	0.0339	15.0750
33	1.1789	0.3482	35.7817	0.0279	30.3515	0.0329	15.5480
34	1.1848	0.8440	36.9606	0.0271	31.1955	0.0321	16.0202

Table A-4 *(continued)*

DISCRETE COMPOUNDING: $i = ½\%$

	Single Payment		Uniform Series				Gradient Series
	Compound Amount Factor	Present Worth Factor	Compound Amount Factor	Sinking Fund Factor	Present Worth Factor	Capital Recovery Factor	Uniform Series Factor
n	To Find F Given P FPi,n	To Find P Given F PFi,n	To Find F Given A FAi,n	To Find A Given F AFi,n	To Find P Given A PAi,n	To Find A Given P APi,n	To Find A Given G AGi,n
35	1.1907	0.8398	38.1454	0.0262	32.0354	0.0312	16.4915
40	1.2208	0.8191	44.1588	0.0226	36.1722	0.0276	18.8359
45	1.2516	0.7990	50.3242	0.0199	40.2072	0.0249	21.2595
50	1.2882	0.7793	56.6452	0.0177	44.1428	0.0227	23.4624
55	1.3158	0.7501	63.1258	0.0158	47.9814	0.0208	25.7447
60	1.3489	0.7414	69.7700	0.0143	51.7256	0.0193	28.0064
65	1.3829	0.7231	76.5821	0.0131	55.3775	0.0181	30.2475
70	1.4178	0.7053	83.5661	0.0120	58.9394	0.0170	32.4680
75	1.4536	0.6879	90.7265	0.0110	62.4136	0.0160	34.6679
80	1.4903	0.6710	98.0677	0.0102	65.8023	0.0152	36.8474
85	1.5280	0.6545	105.5942	0.0095	69.1075	0.0145	39.0065
90	1.5666	0.6383	113.3109	0.0088	72.3313	0.0138	41.1451
95	1.6061	0.6226	121.2224	0.0082	75.4757	0.0132	43.2633
100	1.6467	0.6073	129.3336	0.0077	78.5426	0.0127	45.3613

Table A-5

DISCRETE COMPOUNDING: $i = 7/12\%$

	Single Payment		Uniform Series				Gradient Series
	Compound Amount Factor	Present Worth Factor	Compound Amount Factor	Sinking Fund Factor	Present Worth Factor	Capital Recovery Factor	Uniform Series Factor
n	To Find *F* Given *P* *FPi,n*	To Find *P* Given *F* *PFi,n*	To Find *F* Given *A* *FAi,n*	To Find *A* Given *F* *AFi,n*	To Find *P* Given *A* *PAi,n*	To Find *A* Given *P* *APi,n*	To Find *A* Given *G* *AGi,n*
1	1.0058	0.9942	1.0000	1.0000	0.9942	1.0058	0.0000
2	1.0117	0.9884	2.0058	0.4985	1.9826	0.5044	0.4985
3	1.0176	0.9827	3.0175	0.3314	2.9653	0.3372	0.9961
4	1.0235	0.9770	4.0351	0.2478	3.9423	0.2537	1.4927
5	1.0295	0.9713	5.0587	0.1977	4.9137	0.2035	1.9884
6	1.0355	0.9657	6.0882	0.1643	5.8794	0.1701	2.4830
7	1.0416	0.9601	7.1237	0.1404	6.8395	0.1462	2.9767
8	1.0475	0.9545	8.1653	0.1225	7.7940	0.1283	3.4695
9	1.0537	0.9490	9.2129	0.1085	8.7430	0.1144	3.9612
10	1.0599	0.9435	10.2666	0.0974	9.6865	0.1032	4.4520
11	1.0651	0.9380	11.3265	0.0883	10.6245	0.0941	4.9418
12	1.0723	0.9326	12.3926	0.0807	11.5571	0.0865	5.4307
13	1.0785	0.9272	13.4649	0.0743	12.4843	0.0801	5.9186
14	1.0848	0.9218	14.5434	0.0688	13.4061	0.0746	6.4055
15	1.0912	0.9165	15.6283	0.0640	14.3225	0.0698	6.8914
16	1.0975	0.9111	16.7194	0.0598	15.2337	0.0656	7.3764
17	1.1039	0.9059	17.8170	0.0561	16.1395	0.0620	7.8604
18	1.1104	0.9005	18.9209	0.0529	17.0401	0.0587	8.3435
19	1.1168	0.8054	20.0313	0.0499	17.9355	0.0558	8.8255
20	1.1234	0.8902	21.1481	0.0473	18.8257	0.0531	9.3066
21	1.1299	0.8850	22.2715	0.0449	19.7107	0.0507	9.7868
22	1.1365	0.8799	23.4014	0.0427	20.5906	0.0486	10.2660
23	1.1431	0.8743	24.5379	0.0408	21.4654	0.0466	10.7442
24	1.1498	0.8697	25.6810	0.0389	22.3351	0.0448	11.2214
25	1.1565	0.8647	26.8308	0.0373	23.1998	0.0431	11.6977
26	1.1653	0.8597	27.9874	0.0357	24.0594	0.0416	12.1730
27	1.1700	0.8547	29.1506	0.0343	24.9141	0.0401	12.6473
28	1.1769	0.8497	30.3207	0.0330	25.7638	0.0388	13.1206
29	1.1837	0.8448	31.4975	0.0317	26.6086	0.0376	13.5930
30	1.1906	0.8399	32.6813	0.0306	27.4485	0.0364	14.0645
31	1.1976	0.8350	33.8719	0.0295	28.2835	0.0354	14.5349
32	1.2046	0.8302	35.0695	0.0285	29.1137	0.0343	15.0044
33	1.2116	0.8254	36.2741	0.0276	29.9390	0.0334	15.4730
34	1.2187	0.8206	37.4857	0.0267	30.7595	0.0325	15.9405

Table A-5 *(continued)*

DISCRETE COMPOUNDING: $i = 7/12\%$

	Single Payment		Uniform Series				Gradient Series
	Compound Amount Factor	Present Worth Factor	Compound Amount Factor	Sinking Fund Factor	Present Worth Factor	Capital Recovery Factor	Uniform Series Factor
n	To Find F Given P FPi,n	To Find P Given F PFi,n	To Find F Given A FAi,n	To Find A Given F AFi,n	To Find P Given A PAi,n	To Find A Given P APi,n	To Find A Given G AGi,n
35	1.2258	0.8158	38.7043	0.0258	31.5754	0.0317	16.4071
40	1.2619	0.7924	44.9051	0.0223	35.5840	0.0281	18.7257
45	1.2992	0.7697	51.2889	0.0195	39.4777	0.0253	21.0201
50	1.3375	0.7477	57.8611	0.0173	43.2599	0.0231	23.2904
55	1.3770	0.7252	64.6271	0.0155	46.9336	0.0213	25.5368
60	1.4176	0.7054	71.5929	0.0140	50.5020	0.0198	27.7591
65	1.4595	0.6852	78.7642	0.0127	53.9681	0.0185	29.9575
70	1.5025	0.6655	86.1471	0.0116	57.3349	0.0174	32.1320
75	1.5469	0.5465	93.7479	0.0107	60.6052	0.0165	34.2827
80	1.5925	0.6279	101.5730	0.0098	63.7817	0.0157	36.4095
85	1.6395	0.6099	109.6289	0.0091	66.8672	0.0150	38.5127
90	1.6879	0.5925	117.9226	0.0085	69.8643	0.0143	40.5922
95	1.7377	0.5755	126.4611	0.0079	72.7754	0.0137	42.6482
100	1.7890	0.5590	135.2515	0.0074	75.6031	0.0132	44.6906

Table A-6

DISCRETE COMPOUNDING: $i = 2/3\%$

	Single Payment		Uniform Series				Gradient Series
	Compound Amount Factor	Present Worth Factor	Compound Amount Factor	Sinking Fund Factor	Present Worth Factor	Capital Recovery Factor	Uniform Series Factor
n	To Find F Given P FPi,n	To Find P Given F PFi,n	To Find F Given A FAi,n	To Find A Given F AFi,n	To Find P Given A PAi,n	To Find A Given P APi,n	To Find A Given G AGi,n
1	1.0067	0.9934	1.0000	1.0000	0.9934	1.0067	0.0000
2	1.0134	0.9868	2.0067	0.4983	1.9802	0.5050	0.4983
3	1.0201	0.9803	3.0270	0.3311	2.9604	0.3378	0.9956
4	1.0269	0.9738	4.0402	0.2475	3.9342	0.2542	1.4917
5	1.0338	0.9673	5.0671	0.1974	4.9015	0.2040	1.9867
6	1.0407	0.9609	6.1009	0.1639	5.8625	0.1706	2.4806
7	1.0476	0.9545	7.1416	0.1400	6.8170	0.1467	2.9734
8	1.0546	0.9482	8.1892	0.1221	7.7652	0.1288	3.4651
9	1.0616	0.9420	9.2438	0.1082	8.7072	0.1148	3.9557
10	1.0687	0.9367	10.3054	0.0970	9.6429	0.1037	4.4452
11	1.0758	0.9295	11.3741	0.0879	10.5724	0.0945	4.9336
12	1.0830	0.9234	12.4499	0.0808	11.4958	0.0870	5.4208
13	1.0902	0.9172	13.5329	0.0739	12.4130	0.0806	5.9070
14	1.0975	0.9112	14.6231	0.0684	13.3242	0.0751	6.3920
15	1.1048	0.9051	15.7206	0.7636	14.2293	0.0703	6.8760
16	1.1122	0.8991	16.8254	0.0594	15.1285	0.0661	7.3588
17	1.1196	0.8932	17.9376	0.0557	16.0217	0.0624	7.8406
18	1.1270	0.8873	19.0572	0.0525	16.9089	0.0591	8.3212
19	1.1346	0.8814	20.1842	0.0495	17.7903	0.0562	8.8007
20	1.1421	0.8755	21.3188	0.0469	18.6659	0.0536	9.2791
21	1.1497	0.8698	22.4609	0.0445	19.5357	0.0512	9.7564
22	1.1574	0.8640	23.6107	0.0424	20.3997	0.0490	10.2327
23	1.1651	0.8583	24.7681	0.0404	21.2579	0.0470	10.7078
24	1.1729	0.8528	25.9332	0.0386	22.1105	0.0452	11.1818
25	1.1807	0.8470	27.1061	0.0369	22.9575	0.0436	11.6546
26	1.1886	0.8413	28.2868	0.0354	23.7988	0.0420	12.1264
27	1.1965	0.8358	29.4754	0.0339	24.6346	0.0406	12.5971
28	1.2045	0.8302	30.6719	0.0326	25.4648	0.0393	13.0667
29	1.2125	0.8247	31.8763	0.0314	26.2896	0.0380	13.5352
30	1.2200	0.8193	33.0889	0.0302	27.1089	0.0369	14.0025
31	1.2287	0.8138	34.3094	0.0291	27.9227	0.0358	14.4688
32	1.2369	0.8095	35.5382	0.0281	28.7312	0.0348	14.9340
33	1.2452	0.8031	36.7751	0.0272	29.5343	0.0339	15.3980
34	1.2535	0.7978	38.0203	0.0263	30.3320	0.0330	15.8610

Table A-6 *(continued)*

DISCRETE COMPOUNDING: $i = \frac{2}{3}\%$

	Single Payment		Uniform Series				Gradient Series
	Compound Amount Factor	Present Worth Factor	Compound Amount Factor	Sinking Fund Factor	Present Worth Factor	Capital Recovery Factor	Uniform Series Factor
n	To Find *F* Given *P* *FPi,n*	To Find *P* Given *F* *PFi,n*	To Find *F* Given *A* *FAi,n*	To Find *A* Given *F* *AFi,n*	To Find *P* Given *A* *PAi,n*	To Find *A* Given *P* *APi,n*	To Find *A* Given *G* *AGi,n*
35	1.2618	0.7925	39.2737	0.0255	31.1246	0.0321	16.3229
40	1.3045	0.7666	45.6675	0.0219	35.0090	0.0286	18.6157
45	1.3485	0.7416	52.2773	0.0191	38.7666	0.0258	20.8810
50	1.3941	0.7173	59.1104	0.0169	42.4013	0.0236	23.1188
55	1.4412	0.6939	66.1743	0.0151	45.9173	0.0218	25.3293
60	1.4898	0.6712	73.4769	0.0136	49.3184	0.0203	27.5125
65	1.5402	0.6493	81.0261	0.0123	52.6084	0.0190	29.6684
70	1.5922	0.6291	88.8303	0.0113	55.7909	0.0179	31.7971
75	1.6460	0.6075	96.8982	0.0103	58.8693	0.0170	33.8987
80	1.7016	0.5877	105.2385	0.0095	61.8472	0.0162	35.9734
85	1.7591	0.5685	113.8607	0.0088	64.7277	0.0154	38.0211
90	1.8185	0.5499	122.7741	0.0081	67.5142	0.0148	40.0420
95	1.8799	0.5319	131.9886	0.0076	70.2096	0.0142	42.0362
100	1.9434	0.5146	141.5144	0.0071	72.8169	0.0137	44.0038

Table A-7

DISCRETE COMPOUNDING: $i = 3/4\%$

	Single Payment		Uniform Series				Gradient Series
	Compound Amount Factor	Present Worth Factor	Compound Amount Factor	Sinking Fund Factor	Present Worth Factor	Capital Recovery Factor	Uniform Series Factor
n	To Find F Given P FPi,n	To Find P Given F PFi,n	To Find F Given A FAi,n	To Find A Given F AFi,n	To Find P Given A PAi,n	To Find A Given P APi,n	To Find A Given G AGi,n
1	1.0075	0.9926	1.0000	1.0000	0.9926	1.0075	0.0000
2	1.0151	0.9852	2.0075	0.4981	1.9777	0.5056	0.4981
3	1.0227	0.9778	3.0226	0.3308	2.9556	0.3383	0.9950
4	1.0303	0.9706	4.0452	0.2472	3.9261	0.2547	1.4907
5	1.0381	0.9633	5.0756	0.1970	4.8894	0.2045	1.9851
6	1.0459	0.9562	6.1136	0.1636	5.8456	0.1711	2.4782
7	1.0537	0.9490	7.1595	0.1397	6.7946	0.1472	2.9701
8	1.0616	0.9420	8.2132	0.1218	7.7366	0.1293	3.4608
9	1.0696	0.9350	9.2748	0.1078	8.6716	0.1153	3.9502
10	1.0776	0.9280	10.3443	9.5996	9.5996	0.1042	4.4384
11	1.0857	0.9211	11.4219	0.0876	10.5207	0.0951	4.9253
12	1.0938	0.9142	12.5076	0.0800	11.4349	0.0875	5.4110
13	1.1020	0.9074	13.6014	0.0735	12.3423	0.0810	5.8954
14	1.1103	0.9007	14.7034	0.0680	13.2430	0.0755	6.3786
15	1.1186	0.8940	15.8137	0.0632	14.1370	0.0707	6.8606
16	1.1270	0.8873	16.9323	0.0591	15.0243	0.0666	7.3413
17	1.1354	0.8807	18.0593	0.0554	15.9050	0.0629	7.8207
18	1.1440	0.8742	19.1947	0.0521	16.7792	0.0596	8.2989
19	1.1525	0.8675	20.3387	0.0492	17.6468	0.0567	8.7759
20	1.1612	0.8612	21.4912	0.0465	18.5080	0.0540	9.2516
21	1.1699	0.8548	22.6524	0.0441	19.3628	0.0516	9.7261
22	1.1787	0.8484	23.8223	0.0420	20.2112	0.0495	10.1994
23	1.1875	0.8421	25.0010	0.0400	21.0533	0.0475	10.6714
24	1.1964	0.8358	26.1885	0.0382	21.8891	0.0457	11.1422
25	1.2054	0.8296	27.3849	0.0365	22.7188	0.0440	11.6117
26	1.2144	0.8234	28.5903	0.0350	23.5422	0.0425	12.0800
27	1.2235	0.8173	29.8047	0.0336	24.3595	0.0411	12.5470
28	1.2327	0.8112	31.0282	0.0322	25.1707	0.0397	13.0128
29	1.2420	0.8052	32.2609	0.0310	25.9759	0.0385	13.4774
30	1.2513	0.7992	33.6029	0.0298	26.7751	0.0373	13.9407
31	1.2607	0.7932	34.7542	0.0200	27.5683	0.0363	14.4028
32	1.2701	0.7973	36.0148	0.0278	28.3557	0.0353	14.8636
33	1.2795	0.7815	37.2849	0.0268	29.1371	0.0343	15.3232
34	1.2892	0.7757	38.5646	0.0259	29.9128	0.0334	15.7816

Table A-7 *(continued)*

DISCRETE COMPOUNDING: i = ¾%

n	Single Payment: Compound Amount Factor, To Find F Given P, FPi,n	Single Payment: Present Worth Factor, To Find P Given F, PFi,n	Uniform Series: Compound Amount Factor, To Find F Given A, FAi,n	Uniform Series: Sinking Fund Factor, To Find A Given F, AFi,n	Uniform Series: Present Worth Factor, To Find P Given A, PAi,n	Uniform Series: Capital Recovery Factor, To Find A Given P, APi,n	Gradient Series: Uniform Series Factor, To Find A Given G, AGi,n
35	1.2989	0.7599	39.8538	0.0251	30.6827	0.0326	16.2387
40	1.3483	0.7415	46.4465	0.0215	34.4469	0.0290	18.5058
45	1.3997	0.7145	53.2901	0.0188	38.0732	0.0263	20.7421
50	1.4530	0.6883	60.3943	0.0166	41.5665	0.0241	22.9476
55	1.5083	0.6630	67.7688	0.0148	44.9316	0.0223	25.1223
60	1.5657	0.6387	75.4241	0.0133	48.1734	0.0208	27.2665
65	1.6253	0.6153	83.3708	0.0120	51.2963	0.0195	29.3801
70	1.6872	0.5927	91.6201	0.0109	54.3046	0.0184	31.4634
75	1.7514	0.5710	100.1833	0.0100	57.2027	0.0175	33.5163
80	1.8180	0.5500	109.0725	0.0092	59.9944	0.0167	35.5391
85	1.8873	0.5299	118.3001	0.0085	62.6838	0.0160	37.5318
90	1.9591	0.5104	127.8789	0.0078	65.2746	0.0153	39.4946
95	2.0337	0.4917	137.8224	0.0073	67.7704	0.0148	41.4277
100	2.1111	0.4737	148.1444	0.0068	70.1746	0.0143	43.3311

Table A-8

DISCRETE COMPOUNDING: $i = 1\%$

	Single Payment		Uniform Series				Gradient Series
	Compound Amount Factor	Present Worth Factor	Compound Amount Factor	Sinking Fund Factor	Present Worth Factor	Capital Recovery Factor	Uniform Series Factor
n	To Find F Given P FPi,n	To Find P Given F PFi,n	To Find F Given A FAi,n	To Find A Given F AFi,n	To Find P Givne A PAi,n	To Find A Given P APi,n	To Find A Given G AGi,n
1	1.0100	0.9901	1.0000	1.0000	0.9901	1.0100	0.0000
2	1.0201	0.9803	2.0100	0.4975	1.9704	0.5075	0.4975
3	1.0303	0.9706	3.0301	0.3300	2.9410	0.3400	0.9934
4	1.0406	0.9610	4.0604	0.2463	3.9020	0.2563	1.4876
5	1.0510	0.9515	5.1010	0.1960	4.8534	0.2060	1.9801
6	1.0615	0.9420	6.1520	0.1625	5.7955	0.1725	2.4710
7	1.0721	0.9327	7.2135	0.1386	6.7282	0.1486	2.9602
8	1.0829	0.9235	8.2857	0.1207	7.6517	0.1307	3.4478
9	1.0937	0.9143	9.3685	0.1067	8.5660	0.1167	3.9337
10	1.1046	0.9053	10.4622	0.0956	9.4713	0.1056	4.4179
11	1.1157	0.8963	11.5668	0.0865	10.3676	0.0965	4.9005
12	1.1268	0.8874	12.6825	0.0788	11.2551	0.0888	5.3815
13	1.1381	0.8787	13.8093	0.0724	12.1337	0.0824	5.8607
14	1.1495	0.8700	14.9474	0.0669	13.0037	0.0769	6.3384
15	1.1610	0.8613	16.0969	0.0621	13.8651	0.0721	6.8143
16	1.1726	0.8528	17.2579	0.0579	14.7179	0.0679	7.2886
17	1.1843	0.8444	18.4304	0.0543	15.5623	0.0643	7.7613
18	1.1961	0.8360	19.6147	0.0510	16.3983	0.0610	8.2323
19	1.2081	0.8277	20.8109	0.0481	17.2260	0.0581	8.7017
20	1.2202	0.8195	22.0190	0.0454	18.0455	0.0554	9.1694
21	1.2324	0.8114	23.2392	0.0430	18.8570	0.0530	9.6354
22	1.2447	0.8034	24.4716	0.0409	19.6604	0.0509	10.0998
23	1.2572	0.7954	25.7163	0.0389	20.4558	0.0489	10.5626
24	1.2697	0.7876	26.9735	0.0371	21.2434	0.0471	11.0237
25	1.2824	0.7798	28.2432	0.0354	22.0232	0.0454	11.4831
26	1.2953	0.7720	29.5256	0.0339	22.7952	0.0439	11.9409
27	1.3082	0.7644	30.8209	0.0324	23.5596	0.0424	12.3971
28	1.3213	0.7568	32.1291	0.0311	24.3164	0.0411	12.8516
29	1.3345	0.7493	33.4504	0.0299	25.0658	0.0399	13.3044
30	1.3478	0.7419	34.7849	0.0287	25.8077	0.0387	13.7557
31	1.3613	0.7346	36.1327	0.0277	26.5423	0.0377	14.2052
32	1.3749	0.7273	37.4941	0.0267	27.2696	0.0367	14.6532
33	1.3887	0.7201	38.8690	0.0257	27.9897	0.0357	15.0995
34	1.4026	0.7130	40.2577	0.0248	28.7027	0.0348	15.5441

Table A-8 *(continued)*

DISCRETE COMPOUNDING: $i = 1\%$

n	Single Payment		Uniform Series				Gradient Series
	Compound Amount Factor	Present Worth Factor	Compound Amount Factor	Sinking Fund Factor	Present Worth Factor	Capital Recovery Factor	Uniform Series Factor
	To Find *F* Given *P* *FPi,n*	To Find *P* Given *F* *PFi,n*	To Find *F* Given *A* *FAi,n*	To Find *A* Given *F* *AFi,n*	To Find *P* Givne *A* *PAi,n*	To Find *A* Given *P* *APi,n*	To Find *A* Given *G* *AGi,n*
35	1.4166	0.7059	41.6603	0.0240	29.4086	0.0340	15.9871
40	1.4889	0.6717	48.8864	0.0205	32.8347	0.0305	18.1776
45	1.5648	0.6391	56.4811	0.0177	36.0945	0.0277	20.3273
50	1.6446	0.6080	64.4632	0.0155	39.1961	0.0255	22.4363
55	1.7285	0.5785	72.8525	0.0137	42.1472	0.0237	24.5049
60	1.8167	0.5504	81.6697	0.0122	44.9550	0.0222	26.5333
65	1.9094	0.5237	90.9366	0.0110	47.6266	0.0210	28.5217
70	2.0068	0.4983	100.6763	0.0099	50.1685	0.0199	30.4703
75	2.1091	0.4741	110.9128	0.0090	52.5871	0.0190	32.3793
80	2.2167	0.4511	121.6715	0.0082	54.8882	0.0182	34.2492
85	2.3298	0.4292	132.9789	0.0075	57.0777	0.0175	36.0801
90	2.4486	0.4084	144.8632	0.0069	59.1609	0.0169	37.8724
95	2.5735	0.3886	157.3537	0.0064	61.1430	0.0164	39.6265
100	2.7048	0.3697	170.4813	0.0059	63.0289	0.0159	41.3426

Table A-9

DISCRETE COMPOUNDING: $i = 3/2\%$

	Single Payment		Uniform Series				Gradient Series
n	Compound Amount Factor	Present Worth Factor	Compound Amount Factor	Sinking Fund Factor	Present Worth Factor	Capital Recovery Factor	Uniform Series Factor
	To Find F Given P FPi,n	To Find P Given F PFi,n	To Find F Given A FAi,n	To Find A Given F AFi,n	To Find P Given A PAi,n	To Find A Given P APi,n	To Find A Given G AGi,n
1	1.0150	0.9852	1.0000	1.0000	0.9852	1.0150	0.0000
2	1.0302	0.9707	2.0150	0.4963	1.9559	0.5113	0.4963
3	1.0457	0.9563	3.0452	0.3284	2.9122	0.3434	0.9901
4	1.0614	0.9422	4.0909	0.2444	3.8544	0.2594	1.4814
5	1.0773	0.9233	5.1523	0.1941	4.7826	0.2091	1.9702
6	1.0934	0.9145	6.2296	0.1605	5.6972	0.1755	2.4566
7	1.1098	0.9010	7.3230	0.1366	6.5982	0.1516	2.9405
8	1.1265	0.8877	8.4328	0.1186	7.4859	0.1336	3.4219
9	1.1434	0.8746	9.5593	0.1046	8.3605	0.1196	3.9008
10	1.1605	0.8617	10.7027	0.0934	9.2222	0.1084	4.3772
11	1.1779	0.8489	11.8633	0.0843	10.0711	0.0993	4.8512
12	1.1956	0.8364	13.0412	0.0767	10.9075	0.0917	5.3227
13	1.2136	0.8240	14.2368	0.0702	11.7315	0.0852	5.7917
14	1.2318	0.8118	15.4504	0.0647	12.5434	0.0797	6.2582
15	1.2502	0.7999	16.6821	0.0599	13.3432	0.0749	6.7223
16	1.2690	0.7880	17.9324	0.0558	14.1313	0.0708	7.1839
17	1.2880	0.7754	19.2014	0.0521	14.9076	0.0671	7.6431
18	1.3076	0.7649	20.4894	0.0488	15.6726	0.0638	8.0997
19	1.3270	0.7536	21.7967	0.0459	16.4262	0.0609	8.5589
20	1.3469	0.7425	23.1237	0.0432	17.1686	0.0582	9.0057
21	1.3671	0.7315	24.4705	0.0409	17.9001	0.0559	9.4550
22	1.3876	0.7207	25.8376	0.0387	18.6208	0.0537	9.9018
23	1.4084	0.7100	27.2251	0.0367	19.3309	0.0517	10.3462
24	1.4295	0.6995	28.6335	0.0349	20.0304	0.0499	10.7881
25	1.4509	0.6892	30.0630	0.0333	20.7196	0.0483	11.2276
26	1.4727	0.6790	31.5140	0.0317	21.3986	0.0467	11.6646
27	1.4948	0.6690	32.9867	0.0303	22.0676	0.0453	12.0992
28	1.5172	0.6591	34.4815	0.0290	22.7267	0.0440	12.5313
29	1.5400	0.6494	35.9987	0.0278	23.3761	0.0428	12.9610
30	1.5631	0.6398	37.5387	0.0266	24.0158	0.0416	13.3883
31	1.5865	0.6303	39.1018	0.0256	24.6461	0.0406	13.8131
32	1.6108	0.6210	40.6883	0.0246	25.2671	0.0396	14.2355
33	1.6345	0.6118	42.2986	0.0236	25.8790	0.0386	14.6555
34	1.6590	0.6028	43.9331	0.0228	26.4817	0.0378	15.0731

Table A-9 *(continued)*

DISCRETE COMPOUNDING: $i = 3\frac{1}{2}\%$

n	Single Payment		Uniform Series				Gradient Series
	Compound Amount Factor	Present Worth Factor	Compound Amount Factor	Sinking Fund Factor	Present Worth Factor	Capital Recovery Factor	Uniform Series Factor
	To Find F Given P FPi,n	To Find P Given F PFi,n	To Find F Given A FAi,n	To Find A Given F AFi,n	To Find P Given A PAi,n	To Find A Given P APi,n	To Find A Given G AGi,n
35	1.6839	0.5939	45.5921	0.0219	27.0756	0.0369	15.4882
40	1.8140	0.5513	54.2679	0.0184	29.9158	0.0334	17.5277
45	1.9542	0.5117	63.6142	0.0157	32.5523	0.0307	19.5074
50	2.1052	0.4750	73.6828	0.0136	34.9997	0.0286	21.4277
55	2.2679	0.4409	84.5296	0.0118	37.2715	0.0268	23.2894
60	2.4432	0.4093	96.2146	0.0104	39.3803	0.0254	25.0930
65	2.6320	0.3799	108.8027	0.0092	41.3378	0.0242	26.8393
70	2.6355	0.3527	122.3637	0.0082	43.1549	0.0232	28.5290
75	3.0546	0.3274	136.9727	0.0073	44.8416	0.0223	30.1631
80	3.2907	0.3039	152.7108	0.0065	46.4073	0.0215	31.7423
85	3.5450	0.2821	169.6651	0.0059	47.8607	0.0209	33.2676
90	3.3189	0.2619	187.9298	0.0053	49.2099	0.0203	34.7399
95	4.1141	0.2431	207.6061	0.0048	50.4622	0.0198	36.1602
100	4.4320	0.2256	228.8029	0.0044	51.6247	0.0194	37.5295

Reprinted, by permission, from J. A. White et al., *Principles of Engineering Economic Analysis,* pp. 396–447.

Table A-10

DISCRETE COMPOUNDING: $i = 2\%$

n	Single Payment		Uniform Series				Gradient Series
	Compound Amount Factor	Present Worth Factor	Compound Amount Factor	Sinking Fund Factor	Present Worth Factor	Capital Recovery Factor	Uniform Series Factor
	To Find *F* Given *P* *FPi,n*	To Find *P* Given *F* *PFi,n*	To Find *F* Given *A* *FAi,n*	To Find *A* Given *F* *AFi,n*	To Find *P* Given *A* *PAi,n*	To Find *A* Given *P* *APi,n*	To Find *A* Given *G* *AGi,n*
1	1.0200	0.9804	1.0000	1.0000	0.9804	1.0200	0.0000
2	1.0404	0.9612	2.0200	0.4950	1.9416	0.5150	0.4950
3	1.0612	0.9423	3.0604	0.3268	2.8839	0.3468	0.9868
4	1.0824	0.9238	4.1216	0.2426	3.8077	0.2626	1.4752
5	1.1041	0.9057	5.2040	0.1922	4.7135	0.2122	1.9604
6	1.1262	0.8880	6.3081	0.1585	5.6014	0.1785	2.4423
7	1.1487	0.8706	7.4343	0.1345	6.4720	0.1545	2.9208
8	1.1717	0.8535	8.5830	0.1165	7.3255	0.1365	3.3961
9	1.1951	0.8368	9.7546	0.1025	8.1622	0.1225	3.8681
10	1.2190	0.8203	10.9497	0.0913	8.9826	0.1113	4.3367
11	1.2434	0.8043	12.1687	0.0822	9.7868	0.1022	4.8021
12	1.2682	0.7885	13.4121	0.0746	10.5753	0.0946	5.2642
13	1.2936	0.7730	14.6803	0.0681	11.3484	0.0881	5.7231
14	1.3195	0.7579	15.9739	0.0626	12.1062	0.0826	6.1786
15	1.3459	0.7430	17.2934	0.0578	12.8493	0.0778	6.6309
16	1.3728	0.7284	18.6393	0.0537	13.5777	0.0737	7.0799
17	1.4002	0.7142	20.0121	0.0500	14.2919	0.0700	7.5256
18	1.4282	0.7002	21.4123	0.0467	14.9920	0.0667	7.9681
19	1.4568	0.6864	22.8406	0.0438	15.6785	0.0638	8.4073
20	1.4859	0.6730	24.2974	0.0412	16.3514	0.0612	8.8433
21	1.5157	0.6598	25.7833	0.0388	17.0112	0.0588	9.2760
22	1.5460	0.6468	27.2990	0.0366	17.6580	0.0566	9.7055
23	1.5769	0.6342	28.8450	0.0347	18.2922	0.0547	10.1317
24	1.6084	0.6217	30.4219	0.0329	18.9139	0.0529	10.5547
25	1.6406	0.6095	32.0303	0.0312	19.5235	0.0512	10.9745
26	1.6734	0.5976	33.6709	0.0297	20.1210	0.0497	11.3910
27	1.7069	0.5859	35.3443	0.0283	20.7069	0.0483	11.8043
28	1.7410	0.5744	37.0512	0.0270	21.2813	0.0470	12.2145
29	1.7758	0.5631	38.7922	0.0258	21.8444	0.0458	12.6214
30	1.8114	0.5521	40.5681	0.0246	22.3965	0.0446	13.0251
31	1.8476	0.5412	42.3794	0.0236	22.9377	0.0436	13.4257
32	1.8845	0.5306	44.2270	0.0226	23.4683	0.0426	13.8230
33	1.9222	0.5202	46.1116	0.0217	23.9886	0.0417	14.2172
34	1.9607	0.5100	48.0338	0.0208	24.4986	0.0408	14.6083

Table A-10 *(continued)*

DISCRETE COMPOUNDING: $i = 2\%$

	Single Payment		Uniform Series				Gradient Series
	Compound Amount Factor	Present Worth Factor	Compound Amount Factor	Sinking Fund Factor	Present Worth Factor	Capital Recovery Factor	Uniform Series Factor
n	To Find F Given P FPi,n	To Find P Given F PFi,n	To Find F Given A FAi,n	To Find A Given F AFi,n	To Find P Given A PAi,n	To Find A Given P APi,n	To Find A Given G AGi,n
35	1.9999	0.5000	49.9945	0.0200	24.9986	0.0400	14.9961
40	2.2080	0.4529	60.4020	0.0166	27.3555	0.0366	16.8885
45	2.4379	0.4102	71.8927	0.0139	29.4902	0.0339	18.7034
50	2.6916	0.3715	84.5754	0.0118	31.4236	0.0318	20.4420
55	2.9717	0.3365	98.5865	0.0101	33.1748	0.0301	22.1057
60	3.2810	0.3048	114.0515	0.0088	34.7609	0.0288	23.6961
65	3.6225	0.2761	131.1261	0.0076	36.1975	0.0276	25.2147
70	3.9996	0.2500	149.9779	0.0067	37.4986	0.0267	26.6632
75	4.4158	0.2265	170.7917	0.0059	38.6771	0.0259	28.0434
80	4.8754	0.2051	193.7719	0.0052	39.7445	0.0252	29.3572
85	5.3829	0.1858	219.1439	0.0046	40.7113	0.0246	30.6064
90	5.9431	0.1683	247.1566	0.0040	41.5869	0.0240	31.7929
95	6.5617	0.1524	278.0847	0.0036	42.3800	0.0236	32.9189
100	7.2446	0.1380	312.2322	0.0032	43.0984	0.0232	33.9863

Table A-11

DISCRETE COMPOUNDING: $i = 5/2\%$

n	Single Payment		Uniform Series				Gradient Series
	Compound Amount Factor	Present Worth Factor	Compound Amount Factor	Sinking Fund Factor	Present Worth Factor	Capital Recovery Factor	Uniform Series Factor
	To Find *F* Given *P* *FPi,n*	To Find *P* Given *F* *PFi,n*	To Find *F* Given *A* *FAi,n*	To Find *A* Given *F* *AFi,n*	To Find *P* Given *A* *PAi,n*	To Find *A* Given *P* *APi,n*	To Find *A* Given *G* *AGi,n*
1	1.0250	0.9756	1.0000	1.0000	0.9756	1.0250	0.0000
2	1.0506	0.9518	2.0250	0.4938	1.9274	0.5188	0.4938
3	1.0769	0.9286	3.0756	0.3251	2.8560	0.3501	0.9835
4	1.1038	0.9060	4.1525	0.2408	3.7620	0.2658	1.4691
5	1.1314	0.8839	5.2563	0.1902	4.6458	0.2152	1.9506
6	1.1597	0.8623	6.3877	0.1565	5.5081	0.1815	2.4280
7	1.1887	0.8413	7.5474	0.1325	6.3494	0.1575	2.9013
8	1.2184	0.8207	8.7361	0.1145	7.1701	0.1395	3.3704
9	1.2489	0.8007	9.9545	0.1005	7.9709	0.1255	3.8355
10	1.2801	0.7812	11.2034	0.0893	8.7521	0.1143	4.2965
11	1.3121	0.7621	12.4835	0.801	9.5142	0.1051	4.7534
12	1.3449	0.7436	13.7956	0.0725	10.2578	0.0975	5.2062
13	1.3785	0.7254	15.1404	0.0660	10.9832	0.0910	5.6549
14	1.4130	0.7077	16.5190	0.0605	11.6909	0.0855	6.0995
15	1.4483	0.6905	17.9319	0.0558	12.3814	0.0808	6.5401
16	1.4845	0.6736	19.3802	0.0516	13.0550	0.0766	6.9766
17	1.5216	0.6572	20.8647	0.0479	13.7122	0.0729	7.4091
18	1.5597	0.6412	22.3863	0.0447	14.3534	0.0697	7.8375
19	1.5987	0.6255	23.9460	0.0418	14.9789	0.0668	8.2619
20	1.6386	0.6103	25.5447	0.0391	15.5892	0.0641	8.6823
21	1.6796	0.5954	27.1833	0.0368	16.1845	0.0618	9.0986
22	1.7216	0.5809	28.8629	0.0346	16.7654	0.0596	9.5110
23	1.7646	0.5667	30.5844	0.0327	17.3321	0.0577	9.9193
24	1.8087	0.5529	32.3490	0.0309	17.8850	0.0559	10.3237
25	1.8539	0.5394	34.1578	0.0293	18.4244	0.0543	10.7241
26	1.9003	0.5252	36.0117	0.0278	18.9506	0.0528	11.1205
27	1.9478	0.5134	37.9120	0.0264	19.4640	0.0514	11.5130
28	1.9965	0.5009	39.8598	0.0251	19.9649	0.0501	11.9015
29	2.0464	0.4887	41.8563	0.0239	20.4536	0.0489	12.2861
30	2.0976	0.4767	43.9027	0.0228	20.9303	0.0478	12.6668
31	2.1500	0.4651	46.0003	0.0217	21.3954	0.0467	13.0436
32	2.2038	0.4538	48.1503	0.0208	21.8492	0.0458	13.4166
33	2.2589	0.4427	50.3540	0.0199	22.2919	0.0449	13.7856
34	2.3153	0.4319	52.6129	0.0190	22.7238	0.0440	14.1508

Table A-11 *(continued)*

DISCRETE COMPOUNDING: $i = 5\frac{1}{2}\%$

	Single Payment		Uniform Series				Gradient Series
	Compound Amount Factor	Present Worth Factor	Compound Amount Factor	Sinking Fund Factor	Present Worth Factor	Capital Recovery Factor	Uniform Series Factor
n	To Find F Given P FPi,n	To Find P Given F PFi,n	To Find F Given A FAi,n	To Find A Given F AFi,n	To Find P Given A PAi,n	To Find A Given P APi,n	To Find A Given G AGi,n
35	2.3732	0.4214	54.9282	0.0182	23.1452	0.0432	14.5122
40	2.6851	0.3724	67.4026	0.0148	25.1028	0.0398	16.2620
45	3.0379	0.3292	81.5161	0.0123	26.8330	0.0373	17.9185
50	3.4371	0.2909	97.4843	0.0103	28.3623	0.0353	19.4839
55	3.8888	0.2572	115.5509	0.0087	29.7140	0.0337	20.9608
60	4.3998	0.2273	135.9915	0.0074	30.9087	0.0324	22.3518
65	4.9780	0.2009	159.1183	0.0063	31.9645	0.0313	23.6600
70	5.6321	0.1775	185.2841	0.0054	32.8979	0.0304	24.8881
75	6.3722	0.1569	214.8882	0.0047	33.7227	0.0297	26.0393
80	7.2096	0.1387	248.3826	0.0040	34.4518	0.0290	27.1167
85	8.1570	0.1226	286.2783	0.0035	35.0962	0.0285	28.1235
90	9.2289	0.1084	329.1541	0.0030	35.6658	0.0280	29.0629
95	10.4416	0.0958	377.6641	0.0026	36.1692	0.0276	29.9382
100	11.8137	0.0846	432.5486	0.0023	36.6141	0.0273	30.7525

Table A-12

DISCRETE COMPOUNDING: $i = 3\%$

	Single Payment		Uniform Series				Gradient Series
	Compound Amount Factor	Present Worth Factor	Compound Amount Factor	Sinking Fund Factor	Present Worth Factor	Capital Recovery Factor	Uniform Series Factor
n	To Find *F* Given *P* *FPi,n*	To Find *P* Given *F* *PFi,n*	To Find *F* Given *A* *FAi,n*	To Find *A* Given *F* *AFi,n*	To Find *P* Given *A* *PAi,n*	To Find *A* Given *P* *APi,n*	To Find *A* Given *G* *AGi,n*
1	1.0300	0.9709	1.0000	1.0000	0.9709	1.0300	0.0000
2	1.0609	0.9426	2.0300	0.4926	1.9135	0.5226	0.4926
3	1.0927	0.9151	3.0909	0.3235	2.8286	0.3535	0.9803
4	1.1255	0.8885	4.1836	0.2390	3.7171	0.2690	1.4631
5	1.1593	0.8626	5.3091	0.1884	4.5797	0.2184	1.9409
6	1.1941	0.8375	6.4684	0.1546	5.4172	0.1846	2.4138
7	1.2299	0.8131	7.6625	0.1305	6.2303	0.1605	2.8819
8	1.2668	0.7894	8.8923	0.1125	7.0197	0.1425	3.3450
9	1.3048	0.7664	10.1591	0.0984	7.7861	0.1284	3.8032
10	1.3439	0.7441	11.4639	0.0872	8.5302	0.1172	4.2565
11	1.3842	0.7224	12.8078	0.0781	9.2526	0.1081	4.7049
12	1.4258	0.7014	14.1920	0.0705	9.9540	0.1005	5.1485
13	1.4685	0.6810	15.6178	0.0640	10.6350	0.0940	5.5872
14	1.5126	0.6611	17.0863	0.0585	11.2961	0.0885	6.0210
15	1.5580	0.6419	18.5989	0.0538	11.9379	0.0838	6.4500
16	1.6047	0.6232	20.1569	0.0496	12.5611	0.0795	6.8742
17	1.6528	0.6050	21.7616	0.0460	13.1661	0.0760	7.2936
18	1.7024	0.5874	23.4144	0.0427	13.7535	0.0727	7.7081
19	1.7535	0.5703	25.1169	0.0398	14.3238	0.0698	8.1179
20	1.8061	0.5537	26.8704	0.0372	14.8775	0.0672	8.5229
21	1.8603	0.5375	28.6765	0.0349	15.4150	0.0649	8.9231
22	1.9161	0.5219	30.5368	0.0327	15.9369	0.0627	9.3186
23	1.9736	0.5067	32.4529	0.0308	16.4436	0.0608	9.7093
24	2.0328	0.4919	34.4265	0.0290	16.9355	0.0590	10.0954
25	2.0938	0.4776	36.4593	0.0274	17.4131	0.0574	10.4768
26	2.1566	0.4637	38.5530	0.0259	17.8768	0.0559	10.8535
27	2.2213	0.4502	40.7096	0.0246	18.3270	0.0546	11.2255
28	2.2879	0.4371	42.9309	0.0233	18.7641	0.0533	11.5930
29	2.3566	0.4243	45.2188	0.0221	19.1885	0.0521	11.9558
30	2.4273	0.4120	47.5754	0.0210	19.6004	0.0510	12.3141
31	2.5001	0.4000	50.0027	0.0200	20.0004	0.0500	12.6678
32	2.5751	0.3883	52.5028	0.0190	20.3888	0.0490	13.0169
33	2.6523	0.3770	55.0778	0.0182	20.7658	0.0482	12.3616
34	2.7319	0.3660	57.7302	0.0173	21.1318	0.0473	13.7018

Table A-12 *(continued)*

DISCRETE COMPOUNDING: $i = 3\%$

n	Single Payment		Uniform Series				Gradient Series
	Compound Amount Factor	Present Worth Factor	Compound Amount Factor	Sinking Fund Factor	Present Worth Factor	Capital Recovery Factor	Uniform Series Factor
	To Find F Given P FPi,n	To Find P Given F PFi,n	To Find F Given A FAi,n	To Find A Given F AFi,n	To Find P Given A PAi,n	To Find A Given P APi,n	To Find A Given G AGi,n
35	2.8139	0.3554	60.4621	0.0165	21.4872	0.0465	14.0375
40	3.2620	0.3066	75.4013	0.0133	23.1148	0.0433	15.6502
45	3.7816	0.2644	92.7199	0.0108	24.5187	0.0408	17.1556
50	4.3839	0.2281	112.7968	0.0089	25.7298	0.0389	18.5575
55	5.0821	0.1968	136.0716	0.0073	26.7744	0.0373	19.8600
60	5.8916	0.1697	163.0534	0.0061	27.6756	0.0361	21.0674
65	6.8300	0.1464	194.3327	0.0051	28.4529	0.0351	22.1841
70	7.9178	0.1263	230.5940	0.0043	29.1234	00.343	23.2145
75	9.1789	0.1089	272.6306	0.0037	29.7018	0.0337	24.1634
80	10.6409	0.0940	321.3628	0.0031	30.2008	0.0331	25.0353
85	12.3357	0.0811	377.8567	0.0026	30.6312	0.0326	25.8349
90	14.3005	0.0699	443.3486	0.0023	31.0024	0.0323	26.5667
95	16.5782	0.0603	519.2717	0.0019	31.3227	0.0319	27.2351
100	19.2186	0.0520	607.2876	0.0016	31.5989	0.0316	27.8444

Table A-13

DISCRETE COMPOUNDING: $i = 4\%$

n	Single Payment		Uniform Series				Gradient Series
	Compound Amount Factor	Present Worth Factor	Compound Amount Factor	Sinking Fund Factor	Present Worth Factor	Capital Recovery Factor	Uniform Series Factor
	To Find *F* Given *P* *FPi,n*	To Find *P* Given *F* *PFi,n*	To Find *F* Given *A* *FAi,n*	To Find *A* Given *F* *AFi,n*	To Find *P* Given *A* *PAi,n*	To Find *A* Given *P* *APi,n*	To Find *A* Given *G* *AGi,n*
1	1.0400	0.9615	1.0000	1.0000	0.9615	1.0400	0.0000
2	1.0816	0.9245	2.0400	0.4902	1.8861	0.5302	0.4902
3	1.1249	0.8890	3.1216	0.3203	2.7751	0.3603	0.9739
4	1.1699	0.8548	4.2465	0.2355	3.6299	0.2755	1.4510
5	1.2167	0.8219	5.4163	0.1846	4.4518	0.2246	1.9216
6	1.2653	0.7903	6.6330	0.1508	5.2421	0.1908	2.3857
7	1.3159	0.7599	7.8983	0.1266	6.0021	0.1666	2.8433
8	1.3686	0.7307	9.2142	0.1085	6.7327	0.1485	3.2944
9	1.4283	0.7025	10.5828	0.0945	7.4353	0.1345	3.7391
10	1.4802	0.6766	12.0061	0.0833	8.1109	0.1233	4.1773
11	1.5395	0.6496	13.4864	0.0741	8.7605	0.1141	4.6090
12	1.6010	0.6246	15.0258	0.0666	9.3851	0.1066	5.0343
13	1.6651	0.6006	16.6268	0.0601	9.9856	0.1001	5.4533
14	1.7317	0.5775	18.2919	0.0547	10.5631	0.0947	5.8659
15	1.8009	0.5553	20.0236	0.0499	11.1184	0.0899	6.2721
16	1.8730	0.5339	21.8245	0.0458	11.6523	0.0858	6.6720
17	1.9479	0.5164	23.6975	0.0422	12.1657	0.0822	7.0656
18	2.0258	0.4935	25.6454	0.0390	12.6593	0.0790	7.4530
19	2.1068	0.4746	27.6712	0.0361	13.1339	0.0761	7.8342
20	2.1911	0.4564	29.7781	0.0336	13.5903	0.0736	8.2091
21	2.2788	0.4388	31.9692	0.0313	14.0292	0.0713	8.5779
22	2.3699	0.4220	34.2480	0.0292	14.4511	0.0692	8.9407
23	2.4647	0.4057	36.6179	0.0273	14.8568	0.0673	9.2973
24	2.5633	0.3901	39.0826	0.0256	15.2470	0.0656	9.6479
25	2.6658	0.3751	41.6459	0.0240	15.6221	0.0640	9.9925
26	2.7725	0.3607	44.3117	0.0226	15.9828	0.0626	10.3312
27	2.8834	0.3468	47.0842	0.0212	16.3296	0.0612	10.6640
28	2.9987	0.3335	49.9676	0.0200	16.6631	0.0600	10.9909
29	3.1187	0.3207	52.9663	0.0189	16.9837	0.0589	11.3120
30	3.2434	0.3083	56.0849	0.0178	17.2920	0.0578	11.6274
31	3.3731	0.2965	59.3283	0.0169	17.5885	0.0569	11.9371
32	3.5081	0.2851	62.7015	0.0159	17.8736	0.0559	12.2411
33	3.6484	0.2741	66.2095	0.0151	18.1476	0.0551	12.5396
34	3.7943	0.2636	69.8579	0.0143	18.4112	0.0543	12.8324

Table A-13 *(continued)*

DISCRETE COMPOUNDING: $i = 4\%$

n	Single Payment: Compound Amount Factor — To Find F Given P FPi,n	Single Payment: Present Worth Factor — To Find P Given F PFi,n	Uniform Series: Compound Amount Factor — To Find F Given A FAi,n	Uniform Series: Sinking Fund Factor — To Find A Given F AFi,n	Uniform Series: Present Worth Factor — To Find P Given A PAi,n	Uniform Series: Capital Recovery Factor — To Find A Given P APi,n	Gradient Series: Uniform Series Factor — To Find A Given G AGi,n
35	3.9461	0.2534	73.6522	0.0136	18.6646	0.0536	13.1198
40	4.8010	0.2083	95.0255	0.0105	19.7928	0.0505	14.4765
45	5.8412	0.1712	121.0293	0.0083	20.7200	0.0483	15.7047
50	7.1067	0.1407	152.6670	0.0066	21.4822	0.0466	16.8122
55	8.6464	0.1157	191.1591	0.0052	22.1086	0.0452	17.8070
60	10.5196	0.0951	237.9906	0.0042	22.6235	0.0442	18.6972
65	12.7987	0.0781	294.9683	0.0034	23.0467	0.0434	19.4909
70	15.5716	0.0642	364.2903	0.0027	23.3945	0.0427	20.1981
75	18.9453	0.0528	448.6311	0.0022	23.6804	0.0422	20.8206
80	23.0498	0.0434	551.2449	0.0018	23.9154	0.0418	21.3718
85	28.0436	0.0357	676.0899	0.0015	24.1085	0.0415	21.8569
90	34.1193	0.0293	827.9832	0.0012	24.2673	0.0412	22.2826
95	41.5114	0.0241	1012.7844	0.0010	24.3978	0.0410	22.6550
100	50.5049	0.0198	1237.6235	0.0008	24.5050	0.0408	22.9800

Table A-14
DISCRETE COMPOUNDING: $i = 5\%$

	Single Payment		Uniform Series				Gradient Series
	Compound Amount Factor	Present Worth Factor	Compound Amount Factor	Sinking Fund Factor	Present Worth Factor	Capital Recovery Factor	Uniform Series Factor
n	To Find *F* Given *P* *FPi,n*	To Find *P* Given *F* *PFi,n*	To Find *F* Given *A* *FAi,n)*	To Find *A* Given *F* *AFi,n*	To Find *P* Given *A* *PAi,n*	To Find *A* Given *P* *APi,n*	To Find *A* Given *G* *AGi,n*
1	1.0500	0.9524	1.0000	1.0000	0.9524	1.0500	0.0000
2	1.1025	0.9070	2.0500	0.4878	1.8594	0.5378	0.4878
3	1.1576	0.8638	3.1525	0.3172	2.7232	0.3672	0.9675
4	1.2155	0.8227	4.3101	0.2320	3.5460	0.2820	1.4391
5	1.2763	0.7835	5.5256	0.1810	4.3295	0.2310	1.9025
6	1.3401	0.7462	6.8019	0.1470	5.0757	0.1970	2.3579
7	1.4071	0.7107	8.1420	0.1228	5.7864	0.1728	2.8052
8	1.4775	0.6768	9.5491	0.1047	6.4632	0.1547	3.2445
9	1.5513	0.6446	11.0266	0.0907	7.1078	0.1407	3.6758
10	1.6289	0.6139	12.5779	0.0795	7.7217	0.1295	4.0991
11	1.7103	0.5847	14.2068	0.0704	8.3064	0.1204	4.5144
12	1.7959	0.5568	15.9171	0.0628	8.8633	0.1128	4.9219
13	1.8856	0.5303	17.7130	0.0565	9.3936	0.1065	5.3215
14	1.9799	0.5051	19.5986	0.0510	9.8986	0.1010	5.7133
15	2.0789	0.4810	21.5786	0.0463	10.3797	0.0963	6.0973
16	2.1329	0.4581	23.6575	0.0423	10.8378	0.0923	6.4736
17	2.2920	0.4368	25.8404	0.0387	11.2741	0.0887	6.8423
18	2.4066	0.4155	28.1324	0.0355	11.6896	0.0855	7.2034
19	2.5270	0.3957	30.5390	0.0327	12.0853	0.0827	7.5569
20	2.6533	0.3769	33.0660	0.0302	12.4622	0.0802	7.9030
21	2.7860	0.3589	35.7193	0.0280	12.8212	0.0780	8.2416
22	2.9253	0.3418	38.5052	0.0260	13.1630	0.0760	8.5730
23	3.0715	0.3256	41.4305	0.0241	13.4886	0.0741	8.8971
24	3.2251	0.3101	44.5020	0.0225	13.7986	0.0725	9.2140
25	3.3864	0.2953	47.7271	0.0210	14.0939	0.0710	9.5238
26	3.5557	0.2812	51.1135	0.0196	14.3752	0.0696	9.8266
27	3.7335	0.2678	54.6691	0.0183	14.6430	0.0683	10.1224
28	3.9201	0.2551	58.4026	0.0171	14.8981	0.0671	10.4114
29	4.1161	0.2429	62.3227	0.0160	15.1411	0.0660	10.6936
30	4.3219	0.2314	66.4388	0.0151	15.3725	0.0651	10.9691
31	4.5380	0.2204	70.7608	0.0141	15.5938	0.0641	11.2381
32	4.7649	0.2099	75.2988	0.0133	15.8027	0.0633	11.5005
33	5.0032	0.1999	80.0638	0.0125	16.0025	0.0625	11.7566
34	5.2533	0.1904	85.0670	0.0118	16.1929	0.0618	12.0063

Table A-14 *(continued)*

DISCRETE COMPOUNDING: $i = 5\%$

	Single Payment		Uniform Series				Gradient Series
	Compound Amount Factor	Present Worth Factor	Compound Amount Factor	Sinking Fund Factor	Present Worth Factor	Capital Recovery Factor	Uniform Series Factor
n	To Find F Given P FPi,n	To Find P Given F PFi,n	To Find F Given A $FAi,n)$	To Find A Given F AFi,n	To Find P Given A PAi,n	To Find A Given P APi,n	To Find A Given G AGi,n
35	5.5160	0.1813	90.3203	0.0111	16.3742	0.0611	12.2498
40	7.0400	0.1420	120.7997	0.0083	17.1591	0.0583	13.3775
45	8.9850	0.1113	159.7001	0.0063	17.7741	0.0563	14.3644
50	11.4674	0.0872	209.3479	0.0048	18.2559	0.0548	15.2233
55	14.6358	0.0683	242.7124	0.0037	18.6335	0.0537	15.9664
60	18.6792	0.0535	353.5835	0.0028	18.9293	0.0528	16.6062
65	23.8399	0.0419	456.7979	0.0022	19.1611	0.0522	17.1541
70	30.4264	0.0329	588.5283	0.0017	19.3427	0.0517	17.6212
75	38.8327	0.0258	756.6533	0.0013	19.4850	0.0513	18.0176
80	49.5614	0.0202	971.2285	0.0010	19.5965	0.0510	18.3526
85	63.2543	0.0158	1245.0867	0.0008	19.6838	0.0508	18.6346
90	80.7303	0.0124	1594.6067	0.0006	19.7523	0.0506	18.8712
95	103.0346	0.0097	2040.6926	0.0005	19.8059	0.0505	19.0689
100	131.5012	0.0076	2610.0239	0.0004	19.8479	0.0504	19.2337

Table A-15

DISCRETE COMPOUNDING: $i = 6\%$

n	Single Payment		Uniform Series				Gradient Series
	Compound Amount Factor	Present Worth Factor	Compound Amount Factor	Sinking Fund factor	Present Worth Factor	Capital Recovery Factor	Uniform Series Factor
	To Find *F* Given *P* *FPi,n*	To Find *P* Given *F* *PFi,n*	To Find *F* Given *A* *FAi,n*	To Find *A* Given *F* *AFi,n*	To Find *P* Given *A* *PAi,n*	To Find *A* Given *P* *APi,n*	To Find *A* Given *G* *AGi,n*
1	1.0600	0.9434	1.0000	1.0000	0.9434	1.0600	0.0000
2	1.1236	0.8900	2.0600	0.4854	1.8334	0.5454	0.4854
3	1.1910	0.8396	3.1836	0.3141	2.6730	0.3741	0.9612
4	1.2625	0.7921	4.3746	0.2286	3.4651	0.2886	1.4272
5	1.3382	0.7473	5.6371	0.1774	4.2124	0.2374	1.8836
6	1.4185	0.7050	6.9753	0.1434	4.9173	0.2034	2.3304
7	1.5036	0.6651	8.3938	0.1191	5.5824	0.1791	2.7676
8	1.5938	0.5274	9.8975	0.1010	6.2098	0.1610	3.1952
9	1.6895	0.5919	11.4913	0.0870	6.8017	0.1470	3.6133
10	1.7908	0.5584	13.1808	0.0759	7.3601	0.1359	4.0220
11	1.8988	0.5268	14.9716	0.0668	7.8869	0.1268	4.4213
12	2.0122	0.4970	16.8699	0.0593	8.3838	0.1193	4.8113
13	2.1329	0.4688	18.8821	0.0530	8.8527	0.1130	5.1920
14	2.2609	0.4423	21.0151	0.0476	9.2950	0.1075	5.5635
15	2.3966	0.4173	23.2760	0.0430	9.7122	0.1030	5.9260
16	2.5404	0.3936	25.6725	0.0390	10.1059	0.0990	6.2794
17	2.6926	0.3714	28.2129	0.0354	10.4773	0.0954	6.6240
18	2.8543	0.3503	30.9057	0.0324	10.8276	0.0924	6.9597
19	3.0256	0.3305	33.7600	0.0296	11.1581	0.0896	7.2867
20	3.2071	0.3118	36.7856	0.0272	11.4699	0.0872	7.6051
21	3.3996	0.2942	39.9927	0.0250	11.7641	0.0850	7.9151
22	3.6035	0.2775	43.3923	0.0230	12.0416	0.0830	8.2166
23	3.8197	0.2618	46.9958	0.0213	12.3034	0.0813	8.5099
24	4.0489	0.2470	50.8156	0.0197	12.5504	0.0797	8.7951
25	4.2919	0.2330	54.8645	0.0182	12.7834	0.0782	9.0722
26	4.5494	0.2198	59.1564	0.0169	13.0032	0.0769	9.3414
27	4.8223	0.2074	63.7058	0.0157	13.2105	0.0757	9.6029
28	5.1117	0.1956	68.5281	0.0146	13.4062	0.0746	9.8568
29	5.4184	0.1845	73.6398	0.0136	13.5907	0.0736	10.1032
30	5.7435	0.1741	79.0582	0.0126	13.7648	0.0726	10.3422
31	6.0881	0.1643	84.8017	0.0118	13.9291	0.0718	10.5740
32	6.4534	0.1550	90.8898	0.0110	14.0840	0.0710	10.7988
33	6.8406	0.1462	97.3432	0.0103	14.2302	0.0703	11.0166
34	7.2510	0.1379	104.1838	0.0096	14.3681	0.0696	11.2276

Table A-15 *(continued)*
DISCRETE COMPOUNDING: $i = 6\%$

n	Single Payment		Uniform Series				Gradient Series
	Compound Amount Factor	Present Worth Factor	Compound Amount Factor	Sinking Fund factor	Present Worth Factor	Capital Recovery Factor	Uniform Series Factor
	To Find F Given P FPi,n	To Find P Given F PFi,n	To Find F Given A FAi,n	To Find A Given F AFi,n	To Find P Given A PAi,n	To Find A Given P APi,n	To Find A Given G AGi,n
35	7.6861	0.1301	111.4347	0.0090	14.4982	0.0690	11.4319
40	10.2857	0.0972	154.7619	0.0065	15.0463	0.0665	12.3590
45	13.7646	0.0727	212.7435	0.0047	15.4558	0.0647	13.1413
50	18.4202	0.0543	290.0357	0.0034	15.7619	0.0634	13.7964
55	24.6503	0.0406	394.1719	0.0025	15.9905	0.0625	14.3411
60	32.9877	0.0303	533.1279	0.0019	16.1614	0.0619	14.7909
65	44.1450	0.0227	719.0828	0.0014	16.2891	0.0614	15.1601
70	59.0739	0.0169	967.9319	0.0010	16.3845	0.0610	15.4613
75	79.0569	0.0126	1300.9405	0.0008	16.4558	0.0608	15.7058
80	105.7959	0.0095	1746.5984	0.0006	16.5091	0.0606	15.9033
85	141.5788	0.0071	2342.9807	0.0004	16.5489	0.0604	16.0620
90	189.4644	0.0053	3141.0735	0.0003	16.5787	0.0603	16.1891
95	253.5462	0.0039	4209.1016	0.0002	16.6009	0.0602	16.2905
100	339.3018	0.0029	5638.3633	0.0002	16.6175	0.0602	16.3711

Table A-16

DISCRETE COMPOUNDING: $i = 7\%$

	Single Payment		Uniform Series				Gradient Series
	Compound Amount Factor	Present Worth Factor	Compound Amount Factor	Sinking Fund Factor	Present Worth Factor	Capital Recovery Factor	Uniform Series Factor
n	To Find *F* Given *P* *FPi,n*	To Find *P* Given *F* *PFi,n*	To Find *F* Given *A* *FAi,n*	To Find *A* Given *F* *AFi,n*	To Find *P* Given *A* *PAi,n*	To Find *A* Given *P* *APi,n*	To Find *A* Given *G* *AGi,n*
1	1.0700	0.9346	1.0000	1.0000	0.9346	1.0700	0.0000
2	1.1449	0.8734	2.0700	0.4831	1.8080	0.5531	0.4831
3	1.2250	0.8163	3.2149	0.3111	2.6243	0.3811	0.9549
4	1.3108	0.7629	4.4399	0.2252	3.3872	0.2952	1.4155
5	1.4026	0.7130	5.7507	0.1739	4.1002	0.2439	1.8650
6	1.5007	0.6663	7.1533	0.1398	4.7665	0.2098	2.3032
7	1.6058	0.6227	8.6540	0.1156	5.3893	0.1856	2.7304
8	1.7182	0.5820	10.2598	0.0975	5.9713	0.1675	3.1465
9	1.8385	0.5439	11.9780	0.0835	6.5152	0.1535	3.5517
10	1.9672	0.5083	13.8164	0.0724	7.0236	0.1424	3.9461
11	2.1049	0.4751	15.7836	0.0634	7.4987	0.1334	4.3296
12	2.2522	0.4440	17.8885	0.0559	7.9427	0.1259	4.7025
13	2.4098	0.4150	20.1406	0.0497	8.3577	0.1197	5.0648
14	2.5785	0.3878	22.5505	0.0443	8.7455	0.1143	5.4167
15	2.7590	0.3624	25.1290	0.0398	9.1079	0.1098	5.7583
16	2.9522	0.3387	27.8881	0.0359	9.4466	0.1059	6.0897
17	3.1588	0.3165	30.8402	0.0324	9.7632	0.1024	6.4110
18	3.3799	0.2959	33.9990	0.0294	10.0591	0.0994	6.7225
19	3.6165	0.2765	37.3790	0.0268	10.3356	0.0968	7.0242
20	3.8697	0.2584	40.9955	0.0244	10.5940	0.0944	7.3163
21	4.1406	0.2415	44.8652	0.0223	10.8355	0.0923	7.5990
22	4.4304	0.2257	49.0057	0.0204	11.0612	0.0904	7.8725
23	4.7405	0.2109	53.4361	0.0187	11.2722	0.0887	8.1369
24	5.0724	0.1971	58.1767	0.0172	11.4693	0.0872	8.3923
25	5.4274	0.1842	63.2490	0.0158	11.6536	0.0858	8.6391
26	5.8074	0.1722	68.6765	0.0146	11.8258	0.0845	8.8773
27	6.2139	0.1609	74.4838	0.0134	11.9867	0.0834	9.1072
28	6.6488	0.1504	80.6977	0.0124	12.1371	0.0824	9.3289
29	7.1143	0.1406	87.3465	0.0114	12.2777	0.0814	9.5427
30	7.6123	0.1314	94.4608	0.0106	12.4090	0.0806	9.7487
31	8.1451	0.1228	102.0730	0.0098	12.5318	0.0798	9.9471
32	8.7153	0.1147	110.2181	0.0091	12.6466	0.0791	10.1381
33	9.3253	0.1072	118.9334	0.0084	12.7538	0.0784	10.3219
34	9.9781	0.1002	128.2587	0.0078	12.8540	0.0778	10.4987

Table A-16 *(continued)*

DISCRETE COMPOUNDING: $i = 7\%$

	Single Payment		Uniform Series				Gradient Series
	Compound Amount Factor	Present Worth Factor	Compound Amount Factor	Sinking Fund Factor	Present Worth Factor	Capital Recovery Factor	Uniform Series Factor
n	To Find F Given P FPi,n	To Find P Given F PFi,n	To Find F Given A FAi,n	To Find A Given F AFi,n	To Find P Given A PAi,n	To Find A Given P APi,n	To Find A Given G AGi,n
35	10.6766	0.0937	138.2368	0.0072	12.9477	0.0772	10.6687
40	14.9745	0.0668	199.6350	0.0050	13.3317	0.0750	11.4233
45	21.0024	0.0476	285.7490	0.0035	13.6055	0.0735	12.0360
50	29.4570	0.0339	406.5286	0.0025	13.8007	0.0725	12.5287
55	41.3150	0.0242	575.9282	0.0017	13.9399	0.0717	12.9215
60	57.9464	0.0173	813.5200	0.0012	14.0392	0.0712	13.2321
65	81.2728	0.0123	1146.7546	0.0009	14.1099	0.0709	13.4760
70	113.9893	0.0088	1614.1333	0.0006	14.1604	0.0706	13.6662
75	159.8759	0.0063	2269.6558	0.0004	14.1964	0.0704	13.8136
80	224.2342	0.0045	3189.0608	0.0003	14.2220	0.0703	13.9273
85	314.5000	0.0032	4478.5703	0.0002	14.2403	0.0702	14.0146
90	441.1025	0.0023	6287.1797	0.0002	14.2533	0.0702	14.0812
95	618.6692	0.0016	8823.8477	0.0001	14.2626	0.0701	14.1319
100	867.7156	0.0012	12381.6524	0.0001	14.2693	0.0701	14.1703

Table A-17

DISCRETE COMPOUNDING: $i = 8\%$

n	Single Payment		Uniform Series				Gradient Series
	Compound Amount Factor	Present Worth Factor	Compound Amount Factor	Sinking Fund Factor	Present Worth Factor	Capital Recovery Factor	Uniform Series Factor
	To Find F Given P FPi,n	To Find P Given F PFi,n	To Find F Given A FAi,n	To Find A Given F AFi,n	To Find P Given A PAi,n	To Find A Given P APi,n	To Find A Given G AGi,n
1	1.0800	0.9259	1.0000	1.0000	0.9259	1.0800	0.0000
2	1.1664	0.8573	2.0800	0.4808	1.7833	0.5608	0.4808
3	1.2597	0.7938	3.2464	0.3080	2.5771	0.3880	0.9487
4	1.3605	0.7350	4.5061	0.2219	3.3121	0.3019	1.4040
5	1.4693	0.6806	5.8666	0.1705	3.9927	0.2505	1.8465
6	1.5869	0.6302	7.3359	0.1363	4.6299	0.2163	2.2763
7	1.7138	0.5835	8.9228	0.1121	5.2064	0.1921	2.6937
8	1.8509	0.5403	10.6366	0.0940	5.7466	0.1740	3.0985
9	1.9990	0.5002	12.4876	0.0801	6.2469	0.1601	3.4910
10	2.1689	0.4632	14.4866	0.0690	6.7101	0.1490	3.8713
11	2.3316	0.4289	16.6455	0.0601	7.1390	0.1401	4.2395
12	2.5182	0.3971	18.9771	0.0527	7.5361	0.1327	4.5957
13	2.7196	0.3677	21.4953	0.0465	7.9038	0.1265	4.9402
14	2.9372	0.3405	24.2149	0.0413	8.2442	0.1213	5.2731
15	3.1722	0.3152	27.1521	0.0368	8.5595	0.1168	5.5945
16	3.4259	0.2919	30.3243	0.0330	8.8514	0.1130	5.9046
17	3.7000	0.2703	33.7502	0.0296	9.1216	0.1096	6.2037
18	3.9960	0.2502	37.4502	0.0267	9.3719	0.1067	6.4920
19	4.8157	0.2317	41.4463	0.0241	9.6036	0.1041	6.7697
20	4.6610	0.2145	45.7620	0.0219	9.8181	0.1019	7.0369
21	5.0338	0.1987	50.4229	0.0198	10.0168	0.0998	7.2940
22	5.4365	0.1839	55.4567	0.0180	10.2007	0.0980	7.5412
23	5.8715	0.1703	60.8933	0.0164	10.3711	0.0964	7.7786
24	6.3412	0.1577	66.7647	0.0150	10.5288	0.0950	8.0066
25	6.8485	0.1460	73.1059	0.0137	10.6748	0.0937	8.2254
26	7.3964	0.1352	79.9544	0.0125	10.8100	0.0925	8.4352
27	7.9801	0.1252	87.3507	0.0114	10.9352	0.0914	8.6363
28	8.5271	0.1159	95.3388	0.0105	11.0511	0.0905	8.8289
29	9.3173	0.1073	103.9659	0.0096	11.1584	0.0896	9.0133
30	10.0627	0.0994	113.2831	0.0088	11.2578	0.0888	9.1897
31	10.8677	0.0923	123.3458	0.0081	11.3498	0.0881	9.3584
32	11.7371	0.0852	134.2134	0.0075	11.4350	0.0875	9.5197
33	12.6760	0.0789	145.9505	0.0069	11.5139	0.0869	9.6737
34	13.6901	0.0730	158.6266	0.0063	11.5869	0.0863	9.8208

Table A-17 *(continued)*

DISCRETE COMPOUNDING: $i = 8\%$

n	Single Payment: Compound Amount Factor, To Find F Given P, FPi,n	Single Payment: Present Worth Factor, To Find P Given F, PFi,n	Uniform Series: Compound Amount Factor, To Find F Given A, FAi,n	Uniform Series: Sinking Fund Factor, To Find A Given F, AFi,n	Uniform Series: Present Worth Factor, To Find P Given A, PAi,n	Uniform Series: Capital Recovery Factor, To Find A Given P, APi,n	Gradient Series: Uniform Series Factor, To Find A Given G, AGi,n
35	14.7853	0.0675	172.3167	0.0058	11.6546	0.0858	9.9611
40	21.7245	0.0460	259.0564	0.0039	11.9246	0.0839	10.5699
45	31.9204	0.0313	386.5054	0.0026	12.1084	0.0826	11.0447
50	46.9016	0.0213	573.7698	0.0017	12.2335	0.0817	11.4107
55	68.9138	0.0145	848.9224	0.0012	12.3186	0.0812	11.6902
60	101.2570	0.0099	1253.2122	0.0008	12.3766	0.0808	11.9015
65	148.7796	0.0067	1847.2463	0.0005	12.4160	0.0805	12.0602
70	218.6061	0.0046	2720.0767	0.0004	12.4428	0.0804	12.1783
75	321.2039	0.0031	4002.5518	0.0002	12.4611	0.0802	12.2658
80	471.9541	0.0021	5886.9258	0.0002	12.4735	0.0802	12.3301
85	693.4553	0.0014	8655.6953	0.0001	12.4820	0.0801	12.3772
90	1018.9136	0.0010	12723.9219	0.0001	12.4877	0.0801	12.4116
95	1497.1182	0.0037	18701.4805	0.0001	12.4917	0.0801	12.4365
100	2199.7566	0.0005	27484.4649	0.0000	12.4943	0.0800	12.4545

Reprinted, by permission, from J. A. White et al., *Principles of Engineering Economic Analysis,* pp. 396–447.

Table A-18

DISCRETE COMPOUNDING: $i = 9\%$

	Single Payment		Uniform Series				Gradient Series
	Compound Amount Factor	Present Worth Factor	Compound Amount Factor	Sinking Fund Factor	Present Worth Factor	Capital Recovery Factor	Uniform Series Factor
n	To Find F Given P FPi,n	To Find P Given F PFi,n	To Find F Given A FAi,n	To Find A Given F AFi,n	To Find P Given A PAi,n	To Find A Given P APi,n	To Find A Given G AGi,n
1	1.0900	0.9174	1.0000	1.0000	0.9174	1.0900	0.0000
2	1.1881	0.8417	2.0900	0.4785	1.7591	0.5685	0.4785
3	1.2950	0.7722	3.2781	0.3051	2.5313	0.3951	0.9426
4	1.4116	0.7084	4.5731	0.2187	3.2397	0.3087	1.3925
5	1.5386	0.5499	5.9847	0.1671	3.8897	0.2571	1.8282
6	1.6771	0.5963	7.5233	0.1329	4.4859	0.2229	2.2498
7	1.8280	0.5470	9.2004	0.1087	5.0330	0.1987	2.6574
8	1.9925	0.5019	11.0285	0.0907	5.5348	0.1807	3.0512
9	2.1719	0.4604	13.0210	0.0768	5.9952	0.1668	3.4312
10	2.3674	0.4224	15.1929	0.0658	6.4177	0.1558	3.7978
11	2.5804	0.3875	17.5603	0.0569	6.8052	0.1469	4.1510
12	2.8127	0.3555	20.1407	0.0497	7.1607	0.1397	4.4910
13	3.0658	0.3262	22.9534	0.0436	7.4869	0.1336	4.8182
14	3.3417	0.2992	26.0192	0.0384	7.7862	0.1284	5.1326
15	3.6425	0.2745	29.3609	0.0341	8.0607	0.1241	5.4346
16	3.9703	0.2519	33.0034	0.0303	8.3126	0.1203	5.7245
17	4.3276	0.2311	36.9737	0.0270	8.5436	0.1170	6.0024
18	4.7171	0.2120	41.3013	0.0242	8.7556	0.1142	6.2687
19	5.1417	0.1945	46.0184	0.0217	8.9501	0.1117	6.5236
20	5.6044	0.1784	51.1601	0.0195	9.1285	0.1095	6.7674
21	5.1088	0.1637	56.7645	0.0176	9.2922	0.1076	7.0006
22	6.6586	0.1502	62.8733	0.0159	9.4424	0.1059	7.2232
23	7.2579	0.1378	69.5319	0.0144	9.5802	0.1044	7.4357
24	7.9111	0.1264	76.7898	0.0130	9.7066	0.1030	7.6384
25	8.6231	0.1160	84.7009	0.0118	9.8226	0.1018	7.8316
26	9.3992	0.1064	93.3239	0.0107	9.9290	0.1007	8.0156
27	10.2451	0.0976	102.7231	0.0097	10.0266	0.0997	8.1906
28	11.1671	0.0895	112.9681	0.0089	10.1161	0.0989	8.3571
29	12.1722	0.0822	124.1352	0.0081	10.1983	0.0981	8.5154
30	13.2677	0.0754	136.3074	0.0073	10.2737	0.0973	8.6657
31	14.4618	0.0691	149.5751	0.0067	10.3428	0.0967	8.8083
32	15.7633	0.0634	164.0368	0.0061	10.4062	0.0961	8.9436
33	17.1820	0.0582	179.8002	0.0056	10.4644	0.0956	9.0718
34	18.7284	0.0534	196.9822	0.0051	10.5178	0.0951	9.1933

Table A-18 *(continued)*

DISCRETE COMPOUNDING: $i = 9\%$

n	Single Payment: Compound Amount Factor — To Find F Given P FPi,n	Single Payment: Present Worth Factor — To Find P Given F PFi,n	Uniform Series: Compound Amount Factor — To Find F Given A FAi,n	Uniform Series: Sinking Fund Factor — To Find A Given F AFi,n	Uniform Series: Present Worth Factor — To Find P Given A PAi,n	Uniform Series: Capital Recovery Factor — To Find A Given P APi,n	Gradient Series: Uniform Series Factor — To Find A Given G AGi,n
35	20.4140	0.0490	215.7106	0.0046	10.5668	0.0946	9.3083
40	31.4094	0.0318	337.8821	0.0030	10.7574	0.0930	9.7957
45	48.3272	0.0207	525.8582	0.0019	10.8812	0.0919	10.1603
50	74.3574	0.0134	815.0825	0.0012	10.9617	0.0912	10.4295
55	114.4081	0.0087	1260.0903	0.0008	11.0140	0.0908	10.6261
60	176.0310	0.0057	1944.7886	0.0005	11.0480	0.0905	10.7683
65	270.8455	0.0037	2998.2830	0.0003	11.0701	0.0903	10.8702
70	416.7293	0.0024	4619.2149	0.0002	11.0845	0.0902	10.9427
75	641.1897	0.0016	7113.2188	0.0001	11.0938	0.0901	10.9940
80	986.5496	0.0010	10950.5547	0.0001	11.0999	0.0901	11.0299
85	1517.9287	0.0007	16854.7695	0.0001	11.1038	0.0901	11.0551
90	2335.5198	0.0004	25939.1328	0.0000	11.1064	0.0900	11.0726
95	3593.4878	0.0003	39916.5430	0.0000	11.1080	0.0900	11.0847
100	5529.0234	0.0002	61422.5391	0.0000	11.1091	0.0900	11.0930

Table A-19

DISCRETE COMPOUNDING: i = 10%

n	Single Payment		Uniform Series				Gradient Series
	Compound Amount Factor	Present Worth Factor	Compound Amount Factor	Sinking Fund Factor	Present Worth Factor	Capital Recovery Factor	Uniform Series Factor
	To Find *F* Given *P* *FPi,n*	To Find *P* Given *F* *PFi,n*	To Find *F* Given *A* *FAi,n*	To Find *A* Given *F* *AFi,n*	To Find *P* Given *A* *PAi,n*	To Find *A* Given *P* *APi,n*	To Find *A* Given *G* *AGi,n*
1	1.1000	0.9091	1.0000	1.0000	0.9091	1.1000	0.0000
2	1.2100	0.8264	2.1000	0.4762	1.7355	0.5762	0.4762
3	1.3310	0.7513	3.3100	0.3021	2.4869	0.4021	0.9366
4	1.4641	0.6830	4.6410	0.2155	3.1699	0.3155	1.3812
5	1.6105	0.6209	6.1051	0.1638	3.7908	0.2638	1.8101
6	1.7716	0.5645	7.7156	0.1296	4.3553	0.2296	2.2236
7	1.9487	0.5132	9.4872	0.1054	4.8684	0.2054	2.6216
8	2.1436	0.4665	11.4359	0.0874	5.3349	0.1874	3.0045
9	2.3579	0.4241	13.5795	0.0736	5.7590	0.1736	3.3724
10	2.5937	0.3855	15.9374	0.0627	6.1446	0.1627	3.7255
11	2.8531	0.3505	18.5312	0.0540	6.4951	0.1540	4.0641
12	3.1384	0.3186	21.3843	0.0468	6.8137	0.1468	4.3884
13	3.4523	0.2897	24.5227	0.0408	7.1034	0.1408	4.6988
14	3.7975	0.2633	27.9750	0.0357	7.3667	0.1357	4.9955
15	4.1772	0.2394	31.7725	0.0315	7.6061	0.1315	5.2789
16	4.5950	0.2176	35.9497	0.0278	7.8237	0.1278	5.5493
17	5.0545	0.1978	40.5447	0.0247	8.0216	0.1247	5.8071
18	5.5599	0.1799	45.5992	0.0219	8.2014	0.1219	6.0526
19	5.1159	0.1635	51.1591	0.0195	8.3649	0.1195	6.2861
20	6.7275	0.1486	57.2750	0.0175	8.5136	0.1175	6.5081
21	7.4002	0.1351	64.0025	0.0156	8.6487	0.1156	6.7189
22	8.1403	0.1228	71.4027	0.0140	8.7715	0.1140	6.9189
23	8.9543	0.1117	79.5430	0.0126	8.8832	0.1126	7.1085
24	9.8497	0.1015	88.4973	0.0113	8.9847	0.1113	7.2881
25	10.8347	0.0923	98.3470	0.0102	9.0770	0.1102	7.4580
26	11.9132	0.0809	109.1817	0.0092	9.1609	0.1092	7.6186
27	13.1100	0.0763	121.0998	0.0083	9.2372	0.1083	7.7704
28	14.4210	0.0693	134.2098	0.0075	9.3066	0.1075	7.9137
29	15.8631	0.0630	148.6308	0.0067	9.3696	0.1067	8.0489
30	17.4494	0.0573	164.4939	0.0061	9.4269	0.1061	8.1762
31	19.1943	0.0521	181.9432	0.0055	9.4790	0.1055	8.2962
32	21.1138	0.474	201.1376	0.0050	9.5264	0.1050	8.4091
33	23.2251	0.0431	222.2513	0.0045	9.5694	0.1045	8.5152
34	25.5476	0.0391	245.4765	0.0041	9.6085	0.1041	8.6149

Table A-19 *(continued)*

DISCRETE COMPOUNDING: i = 10%

	Single Payment		Uniform Series				Gradient Series
	Compound Amount Factor	Present Worth Factor	Compound Amount Factor	Sinking Fund Factor	Present Worth Factor	Capital Recovery Factor	Uniform Series Factor
n	To Find F Given P FPi,n	To Find P Given F PFi,n	To Find F Given A FAi,n	To Find A Given F AFi,n	To Find P Given A PAi,n	To Find A Given P APi,n	To Find A Given G AGi,n
35	28.1024	0.0356	271.0239	0.0037	9.6442	0.1037	8.7086
40	45.2592	0.0221	442.5921	0.0023	9.7791	0.1023	9.0962
45	72.8904	0.0137	718.9038	0.0014	9.8628	0.1014	9.3740
50	117.3906	0.0085	1163.9070	0.0009	9.9148	0.1009	9.5704
55	189.0588	0.0053	1880.5872	0.0005	9.9471	0.1005	9.7075
60	304.4810	0.0033	3034.8110	0.0003	9.9672	0.1003	9.8023
65	490.3694	0.0020	4893.6953	0.0002	9.9796	0.1002	9.8672
70	789.7449	0.0013	7887.4531	0.0001	9.9873	0.1001	9.9113
75	1271.8921	0.0008	12708.9258	0.0001	9.9921	0.1001	9.9410
80	2048.3936	0.0005	20473.9531	0.0000	9.9951	0.1000	9.9609
85	3208.9583	0.0000	32979.6094	0.0000	9.9970	0.1000	9.9742
90	5313.0039	0.0002	53120.0742	0.0000	9.9981	0.1000	9.9831
95	8556.6484	0.0001	85556.5000	0.0000	9.9988	0.1000	9.9889
100	13780.5625	0.0001	137795.6875	0.0000	9.9993	0.1000	9.9927

Table A-20
DISCRETE COMPOUNDING: i = 12%

n	Single Payment: Compound Amount Factor — To Find F Given P FPi,n	Single Payment: Present Worth Factor — To Find P Given F PFi,n	Uniform Series: Compound Amount Factor — To Find F Given A FAi,n	Uniform Series: Sinking Fund Factor — To Find A Given F AFi,n	Uniform Series: Present Worth Factor — To Find P Given A PAi,n	Uniform Series: Capital Recovery Factor — To Find A Given P APi,n	Gradient Series: Uniform Series Factor — To Find A Given G AGi,n
1	1.1200	0.8929	1.0000	1.0000	0.8929	1.1200	0.0000
2	1.2544	0.7972	2.1200	0.4717	1.6901	0.5917	0.4717
3	1.4049	0.7110	3.3744	0.2963	2.4018	0.4163	0.9246
4	1.5735	0.6355	4.7793	0.2092	3.0373	0.3292	1.3589
5	1.7623	0.5674	6.3528	0.1574	3.6048	0.2774	1.7746
6	1.9738	0.5066	8.1152	0.1232	4.1114	0.2432	2.1720
7	2.2107	0.4523	10.0890	0.0991	4.5638	0.2191	2.5515
8	2.4760	0.4039	12.2997	0.0813	4.9676	0.2013	2.9131
9	2.7731	0.3606	14.7757	0.0677	5.3283	0.1877	3.2574
10	3.1058	0.3220	17.5487	0.0570	5.6502	0.1770	3.5847
11	3.4785	0.2875	20.6546	0.0484	5.9377	0.1684	3.8953
12	3.8960	0.2567	24.1331	0.0414	6.1944	0.1614	4.1897
13	4.3635	0.2292	28.0291	0.0357	6.4236	0.1557	4.4683
14	4.8871	0.2046	32.3926	0.0309	6.6282	0.1509	4.7317
15	5.4736	0.1827	37.2797	0.0268	6.8109	0.1468	4.9803
16	6.1304	0.1631	42.7533	0.0234	6.9740	0.1434	5.2147
17	6.8660	0.1456	48.8836	0.0205	7.1196	0.1405	5.4353
18	7.6900	0.1300	55.7497	0.0179	7.2497	0.1379	5.6427
19	8.6128	0.1161	63.4396	0.0158	7.3658	0.1358	5.8375
20	9.6463	0.1037	72.0524	0.0139	7.4694	0.1339	6.0202
21	10.8038	0.0926	81.6987	0.0122	7.5620	0.1322	6.1913
22	12.1003	0.0826	92.5025	0.0108	7.6446	0.1308	6.3514
23	13.5523	0.0738	104.6028	0.0096	7.7184	0.1296	6.5010
24	15.1786	0.0659	118.1551	0.0085	7.7843	0.1285	6.6406
25	17.0000	0.0538	133.3337	0.0075	7.8431	0.1275	6.7708
26	19.0400	0.0525	150.3338	0.0067	7.8957	0.1267	6.8921
27	21.3249	0.0469	169.3738	0.0059	7.9426	0.1259	7.0049
28	23.8838	0.0419	190.6987	0.0052	7.9844	0.1252	7.1098
29	26.7499	0.0374	214.5825	0.0047	8.0218	0.1247	7.2071
30	29.9599	0.0334	241.3324	0.0041	8.0552	0.1241	7.2974
31	33.5551	0.0298	271.2922	0.0037	8.0850	0.1237	7.3811
32	37.5817	0.0266	304.8472	0.0033	8.1116	0.1233	7.4586
33	42.0915	0.0238	342.4290	0.0029	8.1354	0.1229	7.5302
34	47.1424	0.0212	384.5203	0.0026	8.1566	0.1226	7.5965
35	52.7995	0.0189	431.6629	0.0023	8.1755	0.1223	7.6577
40	93.0508	0.0107	767.0901	0.0013	8.2438	0.1213	7.8988
45	163.9872	0.0051	1358.2275	0.0007	8.2825	0.1207	8.0572
50	289.0012	0.0035	2400.0127	0.0004	8.3045	0.1204	8.1597

Table A-21
DISCRETE COMPOUNDING: $i = 15\%$

n	Single Payment		Uniform Series				Gradient Series
	Compound Amount Factor	Present Worth Factor	Compound Amount Factor	Sinking Fund Factor	Present Worth Factor	Capital Recovery Factor	Uniform Series Factor
	To Find F Given P FPi,n	To Find P Given F PFi,n	To Find F Given A FAi,n	To Find A Given F AFi,n	To Find P Given A PAi,n	To Find A Given P APi,n	To Find A Given G AGi,n
1	1.1500	0.8696	1.0000	1.0000	0.8696	1.1500	0.0000
2	1.3225	0.7551	2.1500	0.4651	1.6257	0.6151	0.4651
3	1.5209	0.6575	3.4725	0.2880	2.2832	0.4380	0.9071
4	1.7490	0.5718	4.9934	0.2003	2.8550	0.3503	1.3263
5	2.0114	0.4972	6.7424	0.1483	3.3522	0.2983	1.7228
6	2.3131	0.4323	8.7537	0.1142	3.7845	0.2642	2.0972
7	2.6600	0.3759	11.0668	0.0904	4.1604	0.2404	2.4498
8	3.0590	0.3269	13.7268	0.0729	4.4873	0.2229	2.7813
9	3.5179	0.2843	16.7858	0.0596	4.7716	0.2096	3.0922
10	4.0456	0.2472	20.3037	0.0493	5.0188	0.1993	3.3832
11	4.6524	0.2149	24.3493	0.0411	5.2337	0.1911	3.6549
12	5.3502	0.1859	29.0017	0.0345	5.4206	0.1845	3.9082
13	6.1528	0.1625	34.3519	0.0291	5.5831	0.1791	4.1438
14	7.0757	0.1413	40.5047	0.0247	5.7245	0.1747	4.3624
15	8.1371	0.1229	47.5804	0.0210	5.8474	0.1710	4.5650
16	9.3576	0.1069	55.7175	0.0179	5.9542	0.1679	4.7522
17	10.7613	0.0929	65.0751	0.0154	6.0472	0.1654	4.9251
18	12.3754	0.0808	75.8363	0.0132	6.1280	0.1632	5.0843
19	14.2318	0.0703	88.2118	0.0113	6.1982	0.1613	5.2307
20	16.3665	0.0611	102.4436	0.0098	6.2593	0.1598	5.3651
21	18.8215	0.0531	118.8100	0.0084	6.3125	0.1584	5.4883
22	21.6447	0.0462	137.6315	0.0073	6.3587	0.1573	5.6010
23	24.8914	0.0402	159.2763	0.0063	6.3988	0.1563	5.7040
24	28.6252	0.0349	184.1677	0.0054	6.4338	0.1554	5.7979
25	32.9189	0.0304	212.7929	0.0047	6.4642	0.1547	5.8834
26	37.8568	0.0264	245.7118	0.0041	6.4906	0.1541	5.9612
27	43.5353	0.0230	183.5684	0.0035	6.5135	0.1535	6.0319
28	50.0656	0.0200	327.1038	0.0031	6.5335	0.1531	6.0960
29	57.5754	0.0174	377.1694	0.0027	6.5509	0.1527	6.1541
30	66.2117	0.0151	434.7449	0.0023	6.5660	0.1523	6.2066
31	76.1435	0.0131	500.9566	0.0020	6.5791	0.1520	6.2541
32	87.5650	0.0114	577.1001	0.0017	6.5905	0.1517	6.2970
33	100.6998	0.0099	664.6650	0.0015	6.6005	0.1515	6.3357
34	115.8047	0.0086	765.3648	0.0013	6.6091	0.1513	6.3705
35	133.1754	0.0075	881.1694	0.0011	6.6166	0.1511	6.4019
40	267.8633	0.0037	1779.0879	0.0006	6.6418	0.1506	6.5168
45	538.7686	0.0019	3585.1248	0.0003	6.6543	0.1503	6.5830
50	1083.6563	0.0009	7217.7070	0.0001	6.6605	0.1501	6.6205

Table A-22
DISCRETE COMPOUNDING: $i = 20\%$

	Single Payment		Uniform Series				Gradient Series
	Compound Amount Factor	Present Worth Factor	Compound Amount Factor	Sinking Fund Factor	Present Worth Factor	Capital Recovery Factor	Uniform Series Factor
n	To Find *F* Given *P* *FPi,n*	To Find *P* Given *F* *PFi,n*	To Find *F* Given *A* *FAi,n*	To Find *A* Given *F* *AFi,n*	To Find *P* Given *A* *PAi,n*	To Find *A* Given *P* *APi,n*	To Find *A* Given *G* *AGi,n*
1	1.2000	0.8333	1.0000	1.0000	0.8333	1.2000	0.0000
2	1.4400	0.6944	2.2000	0.4545	1.5278	0.6545	0.4545
3	1.7280	0.5787	3.6400	0.2747	2.1065	0.4747	0.8791
4	2.0736	0.4823	5.3680	0.1863	2.5887	0.3863	1.2742
5	2.4883	0.4019	7.4416	0.1344	2.9906	0.3344	1.6405
6	2.9060	0.3349	9.9299	0.1007	3.3255	0.3007	1.9788
7	3.5832	0.2791	12.9159	0.0774	3.6046	0.2774	2.2902
8	4.2998	0.2326	16.4991	0.0606	3.8372	0.2606	2.5756
9	5.1598	0.1938	20.7989	0.0481	4.0310	0.2481	2.8364
10	6.1917	0.1615	25.9587	0.0385	4.1925	0.2385	3.0739
11	7.4301	0.1346	32.1504	0.0311	4.3271	0.2311	3.2893
12	8.9161	0.1122	39.5805	0.0253	4.4392	0.2253	3.4841
13	10.6993	0.0935	48.4966	0.0206	4.5327	0.2206	3.6597
14	12.8392	0.0779	59.1959	0.0169	4.6106	0.2169	3.8175
15	15.4070	0.0649	72.0351	0.0139	4.6755	0.2139	3.9588
16	18.4884	0.0541	87.4421	0.0114	4.7296	0.2114	4.0851
17	22.1861	0.0451	105.9305	0.0094	4.7746	0.2094	4.1976
18	26.6233	0.0375	128.1166	0.0078	4.8122	0.2078	4.2975
19	31.9480	0.0313	154.7399	0.0065	4.8435	0.2065	4.3861
20	38.3376	0.0261	186.6879	0.0054	4.8696	0.2054	4.4643
21	46.0051	0.0217	225.0255	0.0044	4.8913	0.2044	4.5334
22	55.2061	0.0181	271.0305	0.0037	4.9094	0.2037	4.5941
23	66.2474	0.0151	326.2366	0.0031	4.9245	0.2031	4.6475
24	79.4968	0.0126	392.4839	0.0025	4.9371	0.2025	4.6943
25	95.3962	0.0105	471.9807	0.0021	4.9470	0.2021	4.7352
26	114.4754	0.0087	567.3770	0.0018	4.9563	0.2018	4.7709
27	137.3705	0.0073	681.8525	0.0015	4.9636	0.2015	4.8020
28	164.8446	0.0061	819.2229	0.0012	4.9697	0.2012	4.8291
29	197.8135	0.0051	984.0676	0.0010	4.9747	0.2010	4.8527
30	237.3762	0.0042	1181.8811	0.0008	4.9789	0.2008	4.8731
31	284.8513	0.0035	1419.2573	0.0007	4.9824	0.2007	4.8908
32	341.8215	0.0029	1704.1087	0.0006	4.9854	0.2006	4.9061
33	410.1860	0.0024	2045.9295	0.0005	4.9878	0.2005	4.9194
34	492.2232	0.0020	2456.1167	0.0004	4.9898	0.2004	4.9308
35	590.6680	0.0017	2948.3391	0.0003	4.9915	0.2003	4.9406
40	1469.7708	0.0007	7343.8516	0.0001	4.9966	0.2001	4.9728
45	3657.2590	0.0003	18281.3008	0.0001	4.9986	0.2001	4.9877
50	9100.4336	0.0001	45497.1641	0.0000	4.9995	0.2000	4.9945

Table A-23

DISCRETE COMPOUNDING: i = 25%

n	Single Payment		Uniform Series				Gradient Series
	Compound Amount Factor	Present Worth Factor	Compound Amount Factor	Sinking Fund Factor	Present Worth Factor	Capital Recovery Factor	Uniform Series Factor
	To Find F Given P FPi,n	To Find P Given F PFi,n	To Find F Given A FAi,n	To Find A Given F AFi,n	To Find P Given A PAi,n	To Find A Given P APi,n	To Find A Given G AGi,n
1	1.2500	0.8000	1.0000	1.0000	0.8000	1.2500	0.0000
2	1.5825	0.6400	2.2500	0.4444	1.4400	0.6944	0.4444
3	1.9531	0.5120	3.8125	0.2623	1.9520	0.5123	0.8525
4	2.4414	0.4096	5.7656	0.1734	2.3616	0.4234	1.2249
5	3.0518	0.3277	8.2070	0.1218	2.6893	0.3718	1.5631
6	3.8147	0.2621	11.2588	0.0888	2.9514	0.3388	1.8683
7	4.7684	0.2097	15.0735	0.0663	3.1611	0.3163	2.1424
8	5.9605	0.1678	19.8419	0.0504	3.3289	0.3004	2.3872
9	7.4506	0.1342	25.8023	0.0388	3.4631	0.2888	2.6048
10	9.3132	0.1074	33.2529	0.0301	3.5705	0.2801	2.7971
11	11.6415	0.0859	42.5661	0.0235	3.6564	0.2735	2.9663
12	14.5519	0.0687	54.2077	0.0184	3.7251	0.2684	3.1145
13	18.1899	0.0550	68.7596	0.0145	3.7801	0.2645	3.2437
14	22.7374	0.0440	86.9495	0.0115	3.8241	0.2615	3.3559
15	28.4217	0.0352	109.6868	0.0091	3.8593	0.2591	3.4530
16	35.5271	0.0231	138.1085	0.0072	3.8874	0.2572	3.5366
17	44.4089	0.0225	173.6356	0.0058	3.9099	0.2558	3.6084
18	55.5112	0.0180	218.0446	0.0046	3.9279	0.2546	3.6698
19	69.3389	0.0144	273.5557	0.0037	3.9424	0.2537	3.7222
20	86.7362	0.0115	342.9446	0.0029	3.9539	0.2529	3.7667
21	108.4202	0.0092	429.6807	0.0023	3.9631	0.2523	3.8045
22	135.5252	0.0074	538.2008	0.0019	3.9705	0.2519	3.8365
23	169.4065	0.0059	673.6262	0.0015	3.9764	0.2515	3.8634
24	211.7582	0.0047	843.0327	0.0012	3.9811	0.2512	3.8861
25	264.6975	0.0038	1054.7910	0.0009	3.9849	0.2509	3.9052
26	330.6721	0.0030	1319.4888	0.0008	3.9879	0.2508	3.9212
27	413.5901	0.0024	1660.3611	0.0006	3.9903	0.2506	3.9346
28	516.9378	0.0019	2063.9502	0.0005	3.9923	0.2505	3.9457
29	646.2846	0.0015	2580.9390	0.0004	3.9938	0.2504	3.9551
30	807.7035	0.0012	3227.1726	0.0003	3.9950	0.2503	3.9628
31	1009.7417	0.0010	4034.9663	0.0002	3.9960	0.2502	3.9693
32	1262.1773	0.0008	5044.7070	0.0002	3.9968	0.2502	3.9746
33	1577.7217	0.0006	6306.8828	0.0002	3.9975	0.2502	3.9791
34	1972.1519	0.0005	7834.6055	0.0001	3.9980	0.2501	3.9828
35	2455.1887	0.0004	9856.7578	0.0001	3.9984	0.2501	3.9858

Table A-24
DISCRETE COMPOUNDING: $i = 30\%$

n	Single Payment		Uniform Series				Gradient Series
	Compound Amount Factor	Present Worth Factor	Compound Amount Factor	Sinking Fund Factor	Present Worth Factor	Capital Recovery Factor	Uniform Series Factor
	To Find F Given P FPi,n	To Find P Given F PFi,n	To Find F Given A FAi,n	To Find A Given F AFi,n	To Find P Given A PAi,n	To Find A Given P APi,n	To Find A Given G AGi,n
1	1.3000	0.7692	1.0000	1.0000	0.7692	1.3000	0.0000
2	1.6900	0.5917	2.3000	0.4348	1.3609	0.7348	0.4348
3	2.1970	0.4552	3.9900	0.2506	1.8161	0.5506	0.8271
4	2.8561	0.3501	6.1870	0.1616	2.1662	0.4616	1.1783
5	3.7129	0.2693	9.0431	0.1106	2.4356	0.4106	1.4903
6	4.8268	0.2072	12.7560	0.0784	2.6427	0.3784	1.7654
7	6.2749	0.1594	17.5828	0.0569	2.8021	0.3569	2.0063
8	8.1573	0.1226	23.8577	0.0419	2.9247	0.3419	2.2156
9	10.6045	0.0943	32.0150	0.0312	3.0190	0.3312	2.3963
10	13.7858	0.0725	42.6195	0.0235	3.0915	0.3235	2.5512
11	17.9216	0.0558	56.4053	0.0177	3.1473	0.3177	2.6833
12	23.2981	0.0429	74.3269	0.0135	3.1903	0.3135	2.7952
13	30.2875	0.0330	97.6250	0.0102	3.2233	0.3102	2.8895
14	39.3737	0.0254	127.9124	0.0078	3.2487	0.3078	2.9685
15	51.1859	0.0195	167.2862	0.0060	3.2682	0.3060	3.0344
16	66.5416	0.0150	218.4721	0.0046	3.2832	0.3046	3.0892
17	86.5041	0.0116	285.0137	0.0035	3.2948	0.3035	3.1345
18	112.4553	0.0089	371.5176	0.0027	3.3037	0.3027	3.1718
19	146.1919	0.0068	483.9729	0.0021	3.3105	0.3021	3.2025
20	190.0494	0.0053	630.1648	0.0016	3.3158	0.3016	3.2275
21	247.0643	0.0040	820.2144	0.0012	3.3198	0.3012	3.2480
22	321.1836	0.0031	1067.2788	0.0009	3.3230	0.3009	3.2646
23	417.5386	0.0024	1288.4624	0.0007	3.3254	0.3007	3.2781
24	542.8001	0.0018	1806.0000	0.0006	3.3272	0.3006	3.2890
25	705.6401	0.0014	2348.7998	0.0004	3.3286	0.3004	3.2979
26	917.3323	0.0011	3054.4414	0.0003	3.3297	0.3003	3.3050
27	1192.5320	0.0008	3971.7727	0.0003	3.3305	0.3003	3.3107
28	1550.2915	0.0006	5164.3047	0.0002	3.3312	0.3002	3.3153
29	2015.3775	0.0005	6714.5938	0.0001	3.3317	0.3001	3.3189
30	2619.9920	0.0004	8729.9727	0.0001	3.3321	0.3001	3.3219
31	3405.9902	0.0003	11349.9688	0.0001	3.3324	0.3001	3.3242
32	4427.7852	0.0002	14755.9570	0.0001	3.3326	0.3001	3.3261
33	5756.1211	0.0002	19183.7461	0.0001	3.3328	0.3001	3.3276
34	7482.9570	0.0001	24939.8672	0.0000	3.3329	0.3000	3.3288
35	9727.8438	0.0001	32422.8086	0.0000	3.3330	0.3000	3.3297

SECTION 2 CONTINUOUS INTEREST FACTORS

Mathematical Expressions

Single payment compound-amount factor $= FPr,n = e^{rn}$

Single payment present-worth factor $= PFr,n = \dfrac{1}{e^{rn}}$

Equal-payment-series compound amount factor $= FAr,n = \dfrac{e^{rn} - 1}{e^{r} - 1}$

Equal-payment-series sinking-fund factor $= AFr,n = \dfrac{e^{r} - 1}{e^{rn} - 1}$

Equal-payment series present-worth factor $= PAr,n = \dfrac{1 - e^{-rn}}{e^{r} - 1}$

Equal-payment-series capital-recovery factor $= APr,n = \dfrac{e^{r} - 1}{1 - e^{-rn}}$

Gradient factor $= AGr,n = \dfrac{1}{e^{r} - 1} - \dfrac{n}{e^{rn} - 1}$

Table A-25

CONTINUOUS COMPOUNDING: $r = \frac{1}{2}\%$

	Single Payment		Uniform Series				Gradient Series
	Compound Amount Factor	Present Worth Factor	Compound Amount Factor	Sinking Fund Factor	Present Worth Factor	Capital Recovery Factor	Uniform Series Factor
n	To Find F Given P FPr,n	To Find P Given F PFr,n	To Find F Given A FAr,n	To Find A Given F AFr,n	To Find P Given A PAr,n	To Find A Given P APr,n	To Find A Given G AGr,n
1	1.0050	0.9950	1.0000	1.0000	0.9950	1.0050	0.0000
2	1.0100	0.9910	2.0049	0.4988	1.9850	0.5038	0.4988
3	1.0151	0.9851	3.0150	0.3317	2.9702	0.3367	0.9967
4	1.0202	0.9802	4.0301	0.2481	3.9503	0.2531	1.4938
5	1.0253	0.9753	5.0502	0.1980	4.9255	0.2030	1.9900
6	1.0305	0.9704	6.0755	0.1646	5.8960	0.1596	2.4854
7	1.0356	0.9656	7.1060	0.1407	6.8615	0.1457	2.9800
8	1.0408	0.9608	8.1417	0.1228	7.8225	0.1278	3.4738
9	1.0460	0.9560	9.1825	0.1089	8.784	0.1139	3.9667
10	1.0513	0.9512	10.2285	0.0978	9.7297	0.1028	4.4588
11	1.0565	0.9465	11.2799	0.0887	10.6762	0.0937	4.9500
12	1.0618	0.9418	12.3364	0.0811	11.6180	0.0861	5.4404
13	1.0672	0.9371	13.3982	0.0746	12.5550	0.0796	5.9300
14	1.0725	0.9324	14.4654	0.0691	13.4874	0.0741	6.4188
15	1.0779	0.9277	15.5379	0.0644	14.4152	0.0694	6.9067
16	1.0833	0.9231	16.6157	0.0602	15.3382	0.0652	7.3938

Table A-25
CONTINUOUS COMPOUNDING: $r = \frac{1}{2}\%$

n	Single Payment: Compound Amount Factor, To Find F Given P, FPr,n	Single Payment: Present Worth Factor, To Find P Given F, PFr,n	Uniform Series: Compound Amount Factor, To Find F Given A, FAr,n	Uniform Series: Sinking Fund Factor, To Find A Given F, AFr,n	Uniform Series: Present Worth Factor, To Find P Given A, PAr,n	Uniform Series: Capital Recovery Factor, To Find A Given P, APr,n	Gradient Series: Uniform Series Factor, To Find A Given G, AGr,n
17	1.0887	0.9185	17.6990	0.0565	16.2568	0.0615	7.8800
18	1.0942	0.9139	18.7877	0.0532	17.1707	0.0582	8.3654
19	1.0997	0.9094	19.8818	0.0503	18.0800	0.0553	8.8500
20	1.1052	0.9048	20.9815	0.0477	18.9849	0.0527	9.3338
21	1.1107	0.9003	22.0867	0.0453	19.8853	0.0503	9.8167
22	1.1163	0.8958	23.1975	0.0431	20.7811	0.0481	10.2988
23	1.1219	0.8914	24.3137	0.0411	21.6725	0.0461	10.7800
24	1.1275	0.8869	25.4357	0.0393	22.5594	0.0443	11.2605
25	1.1331	0.8825	26.5632	0.0376	23.4419	0.0427	11.7401
26	1.1388	0.8781	27.6963	0.0361	24.3200	0.0411	12.2188
27	1.1445	0.8737	28.8350	0.0347	25.1936	0.0397	12.6968
28	1.1503	0.8694	29.9796	0.0334	26.0631	0.0384	13.1739
29	1.1560	0.8650	31.1299	0.0321	26.9281	0.0371	13.6501
30	1.1618	0.8607	32.2859	0.0310	27.7888	0.0360	14.1256
31	1.1677	0.8554	33.4479	0.0299	28.6453	0.0349	14.6002
32	1.1735	0.8521	34.6155	0.0289	29.4974	0.0339	15.0739
33	1.1794	0.8479	35.7890	0.0279	30.3453	0.0330	15.5469
34	1.1853	0.8437	36.9684	0.0271	31.1889	0.0321	16.0190
35	1.1912	0.8395	38.1537	0.0262	32.0284	0.0312	16.4903
40	1.2214	0.8187	44.1699	0.0226	36.1633	0.0277	18.8342
45	1.2523	0.7985	50.3385	0.0199	40.1961	0.0249	21.1574
50	1.2840	0.7788	56.6632	0.0176	44.1294	0.0227	23.4598
55	1.3165	0.7596	63.1480	0.0158	47.9655	0.0208	25.7416
60	1.3499	0.7408	69.7970	0.0143	51.7069	0.0193	28.0027
65	1.3840	0.7225	76.6143	0.0131	55.3560	0.0181	30.2431
70	1.4191	0.7047	83.6042	0.0120	58.9149	0.0170	32.4629
75	1.4550	0.6873	90.7709	0.0110	62.3859	0.0160	34.6621
80	1.4918	0.6703	98.1193	0.0102	65.7713	0.0152	36.8408
85	1.5296	0.6538	105.6535	0.0095	69.0731	0.0145	38.9990
90	1.5683	0.6376	113.3786	0.0088	72.2934	0.0138	41.1368
95	1.6080	0.5219	121.2990	0.0082	75.4341	0.0133	43.2541
100	1.6487	0.6065	129.4202	0.0077	78.4974	0.0127	45.3510

Table A-26

CONTINUOUS COMPOUNDING: $r = 3/4\%$

n	Single Payment: Compound Amount Factor, To Find F Given P, FPr,n	Single Payment: Present Worth Factor, To Find P Given F, PFr,n	Uniform Series: Compound Amount Factor, To Find F Given A, FAr,n	Uniform Series: Sinking Fund Factor, To Find A Given F, AFr,n	Uniform Series: Present Worth Factor, To Find P Given A, PAr,n	Uniform Series: Capital Recovery Factor, To Find A Given P, APr,n	Gradient Series: Uniform Series Factor, To Find A Given G, AGr,n
1	1.0075	0.9925	1.0000	1.0000	0.9925	1.0075	0.0000
2	1.0151	0.9851	2.0077	0.4981	1.9778	0.5055	0.4981
3	1.0228	0.9770	3.0229	0.3308	2.9557	0.3383	0.9950
4	1.0305	0.9704	4.0457	0.2472	3.9262	0.2547	1.4906
5	1.0382	0.9632	5.0764	0.1970	4.8896	0.2045	1.9850
6	1.0460	0.9560	6.1147	0.1635	5.8456	0.1711	2.4781
7	1.0539	0.9489	7.1608	0.1396	6.7945	0.1472	2.9700
8	1.0618	0.9418	8.2149	0.1217	7.7365	0.1293	3.4606
9	1.0698	0.9347	9.2768	0.1078	8.6713	0.1153	3.9500
10	1.0779	0.9277	10.3468	0.0966	9.5991	0.1042	4.4381
11	1.0860	0.9208	11.4248	0.0875	10.5201	0.0951	4.9250
12	1.0942	0.9139	12.5108	0.0799	11.4341	0.0875	5.4106
13	1.1024	0.9071	13.6052	0.0735	12.3413	0.0810	5.8950
14	1.1107	0.9003	14.7077	0.0680	13.2417	0.0755	6.3781
15	1.1191	0.8936	15.8186	0.0632	14.1354	0.0707	6.8600
16	1.1275	0.8869	16.9378	0.0590	15.0225	0.0666	7.3407
17	1.1360	0.8803	18.0654	0.0554	15.9028	0.0629	7.8200
18	1.1445	0.8737	19.2014	0.0521	16.7766	0.0596	8.2982
19	1.1532	0.8672	20.3461	0.0491	17.6439	0.0567	8.7751
20	1.1618	0.8607	21.4994	0.0465	18.5047	0.0540	9.2507
21	1.1706	0.8543	22.6613	0.0441	19.3591	0.0517	9.7251
22	1.1794	0.8479	23.8321	0.0420	20.2071	0.0495	10.1983
23	1.1883	0.8416	25.0116	0.0400	21.0488	0.0475	10.6702
24	1.1972	0.8353	26.2000	0.0382	21.8841	0.0457	11.1408
25	1.2062	0.8290	27.3973	0.0365	22.7132	0.0440	11.6102
26	1.2153	0.8228	28.6037	0.0350	23.5361	0.0425	12.0784
27	1.2245	0.8167	29.8192	0.0335	24.3530	0.0411	12.5453
28	1.2337	0.8106	31.0438	0.0322	25.1636	0.0397	13.0110
29	1.2430	0.8045	32.2776	0.0310	25.9682	0.0385	13.4754
30	1.2523	0.7985	33.5207	0.0298	26.7668	0.0374	13.9386
31	1.2618	0.7925	34.7732	0.0288	27.5595	0.0363	14.4005
32	1.2712	0.7866	36.0351	0.0278	28.3462	0.0353	14.8612
33	1.2808	0.7808	37.3065	0.0268	29.1270	0.0343	15.3207
34	1.2905	0.7749	38.5873	0.0259	29.9020	0.0334	15.7789

Table A-26

CONTINUOUS COMPOUNDING: $r = \frac{3}{4}\%$

n	Single Payment		Uniform Series				Gradient Series
	Compound Amount Factor	Present Worth Factor	Compound Amount Factor	Sinking Fund Factor	Present Worth Factor	Capital Recovery Factor	Uniform Series Factor
	To Find *F* Given *P* *FPr,n*	To Find *P* Given *F* *PFr,n*	To Find *F* Given *A* *FAr,n*	To Find *A* Given *F* *AFr,n*	To Find *P* Given *A* *PAr,n*	To Find *A* Given *P* *APr,n*	To Find *A* Given *G* *AGr,n*
35	1.3002	0.7691	39.8780	0.0251	30.6712	0.0326	16.2359
40	1.3499	0.7408	46.4783	0.0215	34.4320	0.0290	10.5021
45	1.4014	0.7136	53.3307	0.0188	38.0542	0.0263	20.7374
50	1.4550	0.6873	60.4449	0.0165	41.5432	0.0241	22.9418
55	1.5106	0.6620	67.8311	0.0147	44.9038	0.0223	25.1153
60	1.5683	0.6376	75.4995	0.0132	48.1407	0.0208	27.2582
65	1.6282	0.6142	83.4609	0.0120	51.2584	0.0195	29.3704
70	1.6905	0.5916	91.7266	0.0109	54.2614	0.0184	31.4521
75	1.7551	0.5698	100.3080	0.0100	57.1538	0.0175	33.5034
80	1.8221	0.5488	109.2174	0.0092	59.9398	0.0167	35.5244
85	1.8917	0.5286	118.4672	0.0084	62.6233	0.0160	37.5153
90	1.9640	0.5092	128.0705	0.0078	65.2080	0.0153	39.4762
95	2.0391	0.4904	138.0407	0.0072	67.6975	0.0148	41.4072
100	2.1170	0.4724	148.3919	0.0067	70.0955	0.0143	43.3084

Table A-27

CONTINUOUS COMPOUNDING: $r = 1\%$

	Single Payment		Uniform Series				Gradient Series
	Compound Amount Factor	Present Worth Factor	Compound Amount Factor	Sinking Fund Factor	Present Worth Factor	Capital Recovery Factor	Uniform Series Factor
n	To Find *F* Given *P* *FPr,n*	To Find *P* Given *F* *PFr,n*	To Find *F* Given *A* *FAr,n*	To Find *A* Give *F* *AFr,n*	To Find *P* Given *A* *PAr,n*	To Find *A* Given *P* *APr,n*	To Find *A* Given *G* *AGr,n*
1	1.0100	0.9901	1.0000	1.0000	0.9901	1.0100	0.0000
2	1.0202	0.9802	2.0101	0.4975	1.9703	0.5075	0.4975
3	1.0305	0.0704	3.0303	0.3300	2.9407	0.3401	0.9933
4	1.0408	0.9608	4.0608	0.2463	3.9015	0.2563	1.4875
5	1.0513	0.9512	5.1016	0.1960	4.8528	0.2061	1.9800
6	1.0618	0.9418	6.1530	0.1625	5.7947	0.1726	2.4708
7	1.0725	0.9324	7.2148	0.1386	6.7271	0.1487	2.9600
8	1.0833	0.9231	8.2873	0.1207	7.6502	0.1307	3.4475
9	1.0942	0.9139	9.3707	0.1067	8.5641	0.1168	3.9333
10	1.1052	0.9048	10.4649	0.0956	9.4690	0.1056	4.4175
11	1.1163	0.8958	11.5701	0.0864	10.3649	0.0965	4.9000
12	1.1275	0.8869	12.6865	0.0788	11.2519	0.0889	5.3809
13	1.1388	0.8781	13.8140	0.0724	12.1300	0.0824	5.8600
14	1.1503	0.8694	14.9528	0.0669	12.9994	0.0769	6.3376
15	1.1618	0.8607	16.1031	0.0621	13.8601	0.0721	6.8134
16	1.1735	0.8521	17.2650	0.0579	14.7123	0.0680	7.2876
17	1.1853	0.8437	18.4386	0.0542	15.5560	0.0643	7.7601
18	1.1972	0.8353	19.6239	0.0510	16.3913	0.0610	8.2310
19	1.2092	0.8270	20.8212	0.0480	17.2183	0.0581	8.7002
20	1.2214	0.8187	22.0305	0.0454	18.0370	0.0554	9.1677
21	1.2337	0.8106	23.2519	0.0430	18.8477	0.0531	9.6336
22	1.2461	0.8025	24.4857	0.0408	19.6502	0.0509	10.0978
23	1.2586	0.7945	25.7317	0.0389	20.4447	0.0489	10.5604
24	1.2712	0.7866	26.9904	0.0371	21.2314	0.0471	11.0213
25	1.2840	0.7788	28.2617	0.0354	22.0103	0.0454	11.4805
26	1.2969	0.7711	29.5457	0.0338	22.7813	0.0439	11.9361
27	1.3100	0.7634	30.8428	0.0324	23.5447	0.0425	12.3941
28	1.3231	0.7558	32.1528	0.0311	24.3006	0.0412	12.8484
29	1.3364	0.7483	33.4759	0.0299	25.0488	0.0399	13.3010
30	1.3499	0.7408	34.8124	0.0287	25.7897	0.0388	13.7520
31	1.3634	0.7334	36.1623	0.0277	26.5231	0.0377	14.2013
32	1.3771	0.7261	37.5258	0.0266	27.2493	0.0367	14.6490
33	1.3910	0.7189	38.9029	0.0257	27.9682	0.0358	15.0950
34	1.4049	0.7113	40.2940	0.0248	28.6801	0.0349	15.5394

Table A-27
CONTINUOUS COMPOUNDING: $r = 1\%$

	Single Payment		Uniform Series				Gradient Series
	Compound Amount Factor	Present Worth Factor	Compound Amount Factor	Sinking Fund Factor	Present Worth Factor	Capital Recovery Factor	Uniform Series Factor
n	To Find *F* Given *P* *FPr,n*	To Find *P* Given *F* *PFr,n*	To Find *F* Given *A* *FAr,n*	To Find *A* Give *F* *AFr,n*	To Find *P* Given *A* *PAr,n*	To Find *A* Given *P* *APr,n*	To Find *A* Given *G* *AGr,n*
35	1.4191	0.7047	41.6990	0.0240	29.3848	0.0340	15.9821
40	1.4918	0.6703	48.9386	0.0204	32.8045	0.0305	18.1710
45	1.5683	0.6376	56.5494	0.0177	36.0575	0.0277	20.3190
50	1.6487	0.6065	64.5505	0.0155	39.1519	0.0255	22.4261
55	1.7333	0.5769	72.9617	0.0137	42.0953	0.0238	24.4926
60	1.8221	0.5488	81.8042	0.0122	44.8951	0.0223	26.5187
65	1.9165	0.5220	91.1001	0.0110	47.5585	0.0210	23.5045
70	2.0138	0.4966	100.8727	0.0099	50.0919	0.0200	30.4505
75	2.1170	0.4724	111.1462	0.0090	52.5018	0.0190	32.3567
80	2.2255	0.4493	121.9464	0.0082	54.7941	0.0183	34.2235
85	2.3396	0.4274	133.3005	0.0075	56.9746	0.0176	36.0513
90	2.4596	0.4066	145.2366	0.0069	59.0488	0.0169	37.8402
95	2.5857	0.3867	157.7847	0.0063	61.0219	0.0164	39.5907
100	2.7183	0.3679	170.9762	0.0058	62.8987	0.0159	41.3032

Table A-28

CONTINUOUS COMPOUNDING: $r = 3\frac{1}{2}\%$

	Single Payment		Uniform Series				Gradient Series
	Compound Amount Factor	Present Worth Factor	Compound Amount Factor	Sinking Fund Factor	Present Worth Factor	Capital Recovery Factor	Uniform Series Factor
n	To Find *F* Given *P* *Fr,n*	To Find *P* Given *F* *PFr,n*	To Find *F* Given *A* *FAr,n*	To Find *A* Given *F* *AFr,n*	To Find *P* Given *A* *PAr,n*	To Find *A* Given *P* *APr,n*	To Find *A* Given *G* *AGr,n*
1	1.0151	0.9851	1.0000	1.0000	0.9851	1.0151	0.0000
2	1.0305	0.9704	2.0151	0.4963	1.9555	0.5114	0.4963
3	1.0460	0.9560	3.0456	0.3283	2.9116	0.3435	0.9900
4	1.0618	0.9418	4.0916	0.2444	3.8534	0.2595	1.4813
5	1.0779	0.9277	5.1535	0.1940	4.7811	0.2092	1.9700
6	1.0942	0.9139	6.2313	0.1605	5.6950	0.1756	2.4563
7	1.1107	0.9003	7.3236	0.1365	6.5954	0.1516	2.9400
8	1.1275	0.8869	8.4363	0.1185	7.4823	0.1336	3.4213
9	1.1445	0.8737	9.5638	0.1046	8.3560	0.1197	3.9000
10	1.1618	0.8607	10.7083	0.0934	9.2168	0.1085	4.3763
11	1.1794	0.8479	11.8702	0.0842	10.0647	0.0994	4.8501
12	1.1972	0.8353	13.0496	0.0766	10.8999	0.0917	5.3213
13	1.2153	0.8228	14.2468	0.0702	11.7228	0.0853	5.7901
14	1.2337	0.8106	15.4622	0.0647	12.5334	0.0798	6.2564
15	1.2523	0.7985	16.6958	0.0599	13.3319	0.0750	6.7202
16	1.2712	0.7866	17.9482	0.0557	14.1186	0.0708	7.1816
17	1.2905	0.7749	19.2195	0.0520	14.8935	0.0671	7.6404
18	1.3100	0.7634	20.5099	0.0488	15.6569	0.0639	8.0967
19	1.3298	0.7520	21.8199	0.0458	16.4089	0.0609	8.5506
20	1.3499	0.7408	23.1497	0.0432	17.1497	0.0583	9.0020
21	1.3703	0.7298	24.4995	0.0408	17.8795	0.0559	9.4509
22	1.3910	0.7189	25.8698	0.0387	18.5984	0.0538	9.8973
23	1.4120	0.7082	27.2608	0.0367	19.3067	0.0518	10.3413
24	1.4333	0.6977	28.6728	0.0349	20.0043	0.0500	10.7828
25	1.4550	0.6873	30.1061	0.0332	20.6916	0.0483	11.2218
26	1.4770	0.6771	31.5612	0.0317	21.3687	0.0468	11.6584
27	1.4993	0.6670	33.0382	0.0303	22.0357	0.0454	12.0925
28	1.5220	0.6570	34.5375	0.0290	22.6928	0.0441	12.5241
29	1.5450	0.6473	36.0595	0.0277	23.3400	0.0428	12.9533
30	1.5683	0.5376	37.6045	0.0266	23.9777	0.0417	13.3800
31	1.5920	0.6281	39.1728	0.0255	24.6058	0.0406	13.8043
32	1.6161	0.5188	40.7648	0.0245	25.2246	00396	14.2261
33	1.6405	0.6096	42.3809	0.0236	25.8342	0.0387	14.6455
34	1.6653	0.6005	44.0215	0.0227	26.4347	0.0378	15.0625

Table A-28

CONTINUOUS COMPOUNDING: $r = 3\frac{1}{2}\%$

	Single Payment		Uniform Series				Gradient Series
	Compound Amount Factor	Present Worth Factor	Compound Amount Factor	Sinking Fund Factor	Present Worth Factor	Capital Recovery Factor	Uniform Series Factor
n	To Find F Given P Fr,n	To Find P Given F PFr,n	To Find F Given A FAr,n	To Find A Given F AFr,n	To Find P Given A PAr,n	To Find A Given P APr,n	To Find A Given G AGr,n
35	1.6905	0.5916	45.6868	0.0219	27.0262	0.0370	15.4770
40	1.8221	0.5488	54.3985	0.0184	29.8545	0.0335	17.5131
45	1.9640	0.5092	63.7888	0.0157	32.4785	0.0308	19.4890
50	2.1170	0.4724	73.9104	0.0135	34.9128	0.0286	21.4052
55	2.2819	0.4382	84.8203	0.0118	37.1713	0.0269	23.2622
60	2.4596	0.4065	96.5800	0.0104	39.2665	0.0255	25.0609
65	2.6412	0.3772	109.2555	0.0092	41.2104	0.0243	26.8018
70	2.8576	0.3499	122.9183	0.0081	43.0138	0.0232	28.4859
75	3.0802	0.3247	137.6452	0.0073	44.6869	0.0224	30.1140
80	3.3201	0.3012	153.5190	0.0065	46.2391	0.0216	31.6869
85	3.5787	0.2794	170.6293	0.0059	47.6791	0.0210	33.2056
90	3.8574	0.2592	189.0721	0.0053	49.0151	0.0204	34.6710
95	4.1579	0.2405	208.9513	0.0048	50.2546	0.0199	36.0842
100	4.4817	0.2231	230.3787	0.0043	51.4045	0.0195	37.4462

Table A-29

CONTINUOUS COMPOUNDING: $r = 2\%$

	Single Payment		Uniform Series				Gradient Series
	Compound Amount Factor	Present Worth Factor	Compound Amount Factor	Sinking Fund Factor	Present Worth Factor	Capital Recovery Factor	Uniform Series Factor
n	To Find *F* Given *P* *FPr,n*	To Find *P* Given *F* *PFr,n*	To Find *F* Given *A* *FAr,n*	To Find *A* Given *F* *AFr,n*	To Find *P* Given *A* *PAr,n*	To Find *A* Given *P* *APr,n*	To Find *A* Given *G* *AGr,n*
1	1.0202	0.9802	1.0000	1.0000	0.9802	1.0202	0.0000
2	1.0408	0.9609	2.0203	0.4950	1.9410	0.5152	0.4950
3	1.0618	0.9418	3.0611	0.3267	2.8828	0.3469	0.9867
4	1.0833	0.9231	4.1229	0.2425	3.8060	0.2627	1.4750
5	1.1052	0.9048	5.2063	0.1921	4.7108	0.2123	1.9600
6	1.1275	0.8869	6.3115	0.1584	5.5978	0.1786	2.4417
7	1.1503	0.8694	7.4390	0.1344	6.4672	0.1546	2.9200
8	1.1735	0.8521	8.5893	0.1164	7.3193	0.1366	3.3950
9	1.1972	0.8353	9.7629	0.1024	8.1546	0.1226	3.8667
10	1.2214	0.8187	10.9601	0.0912	8.9734	0.1114	4.3351
11	1.2461	0.8025	12.1816	0.0821	9.7759	0.1023	4.8002
12	1.2712	0.7865	13.4277	0.0745	10.5626	0.0947	5.2619
13	1.2969	0.7711	14.6989	0.0680	11.3336	0.0882	5.7203
14	1.3231	0.7553	15.9959	0.0625	12.0895	0.0827	6.1754
15	1.3499	0.7403	17.3191	0.0577	12.8303	0.0779	6.6272
16	1.3771	0.7251	18.6690	0.0536	13.5565	0.0738	7.0757
17	1.4049	0.7118	20.0462	0.0499	14.2683	0.0701	7.5209
18	1.4333	0.6977	21.4511	0.0466	14.9660	0.0668	7.9628
19	1.4623	0.6839	22.8845	0.0437	15.6498	0.0639	8.4014
20	1.4918	0.6703	24.3468	0.0411	16.3202	0.0613	8.8368
21	1.5220	0.6570	25.8387	0.0387	16.9772	0.0589	9.2688
22	1.5527	0.6440	27.3607	0.0365	17.6213	0.0567	9.6976
23	1.5841	0.5313	28.9135	0.0346	18.2526	0.0548	10.1231
24	1.6161	0.6188	30.4976	0.0328	18.8714	0.0530	10.5453
25	1.6487	0.6065	32.1137	0.0311	19.4780	0.0513	10.9643
26	1.6820	0.5945	33.7625	0.0296	20.0725	0.0498	11.3800
27	1.7160	0.5827	35.4446	0.0282	20.6553	0.0484	11.7925
28	1.7507	0.5712	37.1606	0.0269	21.2265	0.0471	12.2018
29	1.7860	0.5599	38.9113	0.0257	21.7864	0.0459	12.6078
30	1.8221	0.5438	40.6974	0.0246	22.3352	0.0448	13.0106
31	1.8589	0.5379	42.5196	0.0235	22.8732	0.0437	13.4102
32	1.8965	0.5273	44.3786	0.0225	23.4005	0.0427	13.8065
33	1.9348	0.5169	46.2791	0.0216	23.9174	0.0418	14.1997
34	1.9739	0.5066	48.2100	0.0207	24.4240	0.0409	14.5897

Table A-29

CONTINUOUS COMPOUNDING: $r = 2\%$

n	Single Payment: Compound Amount Factor, To Find F Given P, FPr,n	Single Payment: Present Worth Factor, To Find P Given F, PFr,n	Uniform Series: Compound Amount Factor, To Find F Given A, FAr,n	Uniform Series: Sinking Fund Factor, To Find A Given F, AFr,n	Uniform Series: Present Worth Factor, To Find P Given A, PAr,n	Uniform Series: Capital Recovery Factor, To Find A Given P, APr,n	Gradient Series: Uniform Series Factor, To Find A Given G, AGr,n
35	2.0138	0.4966	50.1839	0.0199	24.9206	0.0401	14.9765
40	2.2255	0.4493	60.6581	0.0165	27.2599	0.0367	16.8630
45	2.4596	0.4066	72.2549	0.0138	29.3767	0.0340	18.6714
50	2.7183	0.3679	85.0603	0.0118	31.2919	0.0320	20.4028
55	3.0042	0.3339	99.2125	0.0101	33.0250	0.0303	22.0580
60	3.3201	0.3012	114.8530	0.0087	34.5931	0.0289	23.6409
65	3.6693	0.2725	132.1385	0.0076	36.0120	0.0278	25.1507
70	4.0552	0.2466	151.2420	0.0066	37.2958	0.0268	26.5899
75	4.4817	0.2231	172.3545	0.0058	38.4575	0.0260	27.9604
80	4.9530	0.2019	195.6875	0.0051	39.5085	0.0253	29.2640
85	5.4739	0.1827	221.4745	0.0045	40.4597	0.0247	30.5028
90	6.0496	0.1653	249.9734	0.0040	41.3203	0.0242	31.6766
95	6.8359	0.1496	281.4692	0.0036	42.0990	0.0238	32.7937
100	7.3891	0.1353	316.2781	0.0032	42.8036	0.0234	33.8499

Table A-30

CONTINUOUS COMPOUNDING: $r = 3\%$

	Single Payment		Uniform Series				Gradient Series
	Compound Amount Factor	Present Worth Factor	Compound Amount Factor	Sinking Fund Factor	Present Worth Factor	Capital Recovery Factor	Uniform Series Factor
n	To Find *F* Given *P* *FPr,n*	To Find *P* Given *F* *PFr,n*	To Find *F* Given *A* *FAr,n*	To Find *A* Given *F* *AFr,n*	To Find *P* Given *A* *PAr,n*	To Find *A* Given *P* *APr,n*	To Find *A* Given *G* *AGr,n*
1	1.0305	0.9704	1.0000	1.0000	0.9704	1.0305	0.0000
2	1.0618	0.9418	2.0305	0.4925	1.9123	0.5229	0.4925
3	1.0492	0.9139	3.0923	0.3234	2.8262	0.3538	0.9800
4	1.1275	0.8869	4.1866	0.2389	3.7132	0.2693	1.4625
5	1.1618	0.8607	5.3141	0.1882	4.5739	0.2185	1.9400
6	1.1972	0.8353	6.4760	0.1544	5.4092	0.1849	2.4125
7	1.2337	0.8106	7.6732	0.1303	6.2198	0.1608	2.8801
8	1.2712	0.7866	8.9069	0.1123	7.0064	0.1427	3.3427
9	1.3100	0.7634	10.1782	0.0982	7.7698	0.1287	3.8002
10	1.3499	0.7408	11.4882	0.0870	8.5107	0.1175	4.2529
11	1.3910	0.7189	12.8381	0.0779	9.2295	0.1083	4.7005
12	1.4333	0.6977	14.2291	0.0703	9.9273	0.1007	5.1433
13	1.4770	0.6771	15.6625	0.0638	10.6044	0.0943	5.5811
14	1.5220	0.6570	17.1395	0.0583	11.2615	0.0888	6.0139
15	1.5683	0.6376	18.6615	0.0536	11.8991	0.0840	6.4419
16	1.6161	0.6188	20.2298	0.0494	12.5179	0.0799	6.8649
17	1.6653	0.6005	21.8460	0.0458	13.1184	0.0762	7.2831
18	1.7160	0.5827	23.5113	0.0425	13.7012	0.0730	7.6964
19	1.7683	0.5655	25.2274	0.0396	14.2667	0.0701	8.1048
20	1.8221	0.5488	26.9957	0.0370	14.8156	0.0675	8.5084
21	1.8776	0.5326	28.8179	0.0347	15.3482	0.0652	8.9072
22	1.9348	0.5169	30.6955	0.0326	15.8650	0.0630	9.3012
23	1.9937	0.5016	32.6304	0.0306	16.3666	0.0611	9.6904
24	2.0544	0.4868	34.6241	0.0289	16.8534	0.0593	10.0748
25	2.1170	0.4724	36.6786	0.0273	17.3258	0.0577	10.4545
26	2.1815	0.4584	38.7957	0.0258	17.7842	0.0562	10.8294
27	2.2479	0.4449	40.9772	0.0244	18.2290	0.0549	11.1990
28	2.3164	0.4317	43.2252	0.0231	18.6608	0.0536	11.5652
29	2.3869	0.4190	45.5416	0.0220	19.0797	0.0524	11.9201
30	2.4596	0.4066	47.9286	0.0209	19.4863	0.0513	12.2823
31	2.5345	0.3946	50.3883	0.0198	19.8809	0.0503	12.6339
32	2.6117	0.3829	52.9228	0.0189	20.2638	0.0493	12.9810
33	2.6912	0.3716	55.5346	0.0180	20.6354	0.0485	13.3205
34	2.7732	0.3606	58.2259	0.0172	20.9960	0.0475	13.6614

Table A-30

CONTINUOUS COMPOUNDING: $r = 3\%$

	Single Payment		Uniform Series				Gradient Series
	Compound Amount Factor	Present Worth Factor	Compound Amount Factor	Sinking Fund Factor	Present Worth Factor	Capital Recovery Factor	Uniform Series Factor
n	To Find F Given P FPr,n	To Find P Given F PFr,n	To Find F Given A FAr,n	To Find A Given F AFr,n	To Find P Given A PAr,n	To Find A Given P APr,n	To Find A Given G AGr,n
35	2.8577	0.3499	60.9992	0.0164	21.3459	0.0468	13.9948
40	3.3201	0.3012	76.1851	0.0131	22.9465	0.0436	15.5953
45	3.8574	0.2592	93.8285	0.0107	24.3241	0.0411	17.0874
50	4.4817	0.2231	114.3273	0.0087	25.5099	0.0392	18.4750
55	5.2070	0.1920	138.1435	0.0072	26.5304	0.0377	19.7623
60	6.0496	0.1653	165.8139	0.0060	27.4089	0.0365	20.9538
65	7.0287	0.1423	197.9624	0.0051	28.1649	0.0355	22.0541
70	8.1662	0.1225	135.3136	0.0042	28.8157	0.0347	23.0677
75	9.4877	0.1054	278.7092	0.0036	29.3758	0.0340	23.9996
80	11.0232	0.0907	329.1282	0.0030	29.8579	0.0335	24.8543
85	12.8071	0.0781	387.7066	0.0026	30.2728	0.0330	25.6368
90	14.8797	0.0672	455.7647	0.0022	30.6299	0.0326	26.3516
95	17.2878	0.0578	534.8369	0.0019	30.9373	0.0323	27.0032
100	20.0855	0.0498	626.7063	0.0016	31.2019	0.0320	27.5963

Table A-31
CONTINUOUS COMPOUNDING: r = 4%

	Single Payment		Uniform Series				Gradient Series
	Compound Amount Factor	Present Worth Factor	Compound Amount Factor	Sinking Fund Factor	Present Worth Factor	Capital Recovery Factor	Uniform Series Factor
n	To Find F Given P FPr,n	To Find P Given F PFr,n	To Find F Given A FAr,n	To Find A Given F AFr,n	To Find P Given A PAr,n	To Find A Given P APr,n	To Find A Given G AGr,n
1	1.0408	0.9608	1.0000	1.0000	0.9608	1.0408	0.0000
2	1.0833	0.9231	2.0408	0.4900	1.8839	0.5308	0.4900
3	1.1275	0.8869	3.1241	0.3201	2.7708	0.3609	0.9733
4	1.1735	0.8521	4.2516	0.2352	3.6230	0.2760	1.4500
5	1.2214	0.8187	5.4251	0.1843	4.4417	0.2251	1.9201
6	1.2712	0.7866	6.6465	0.1505	5.2283	0.1913	2.3834
7	1.3231	0.7558	7.9178	0.1263	5.9841	0.1571	2.8402
8	1.3771	0.7261	9.2409	0.1082	6.7103	0.1490	3.2904
9	1.4333	0.5977	10.6180	0.0942	7.4080	0.1350	3.7339
10	1.4918	0.5703	12.0514	0.0830	8.0783	0.1238	4.1709
11	1.5527	0.6440	13.5432	0.0738	8.7223	0.1146	4.6013
12	1.6161	0.6188	15.0959	0.0662	9.3411	0.1071	5.0252
13	1.6820	0.5945	16.7120	0.0598	9.9356	0.1006	5.4425
14	1.7507	0.5712	18.3941	0.0544	10.5069	0.0952	5.8534
15	1.8221	0.5488	20.1447	0.0496	11.0557	0.0905	6.2578
16	1.8965	0.5273	21.9669	0.0455	11.5830	0.0863	6.6558
17	1.9739	0.5066	23.8633	0.0419	12.0896	0.0827	7.0473
18	2.0544	0.4858	25.8372	0.0387	12.5763	0.0795	7.4326
19	2.1383	0.4677	27.8917	0.0359	13.0440	0.0767	7.8114
20	2.2255	0.4493	30.0300	0.0333	13.4933	0.0741	8.1840
21	2.3164	0.4317	32.2555	0.0310	13.9250	0.0718	8.5503
22	2.4109	0.4148	34.5719	0.0289	14.3398	0.0697	8.9104
23	2.5093	0.3985	36.9828	0.0270	14.7384	0.0679	9.2644
24	2.6117	0.3829	39.4921	0.0253	15.1213	0.0661	9.6122
25	2.7183	0.3679	42.1038	0.0238	15.4891	0.0646	9.9539
26	2.8292	0.3535	44.8221	0.0223	15.8426	00631	10.2896
27	2.9447	0.3396	47.6513	0.0210	16.1822	0.0618	10.6193
28	3.0649	0.3263	50.5960	0.0198	16.5085	0.0606	10.9431
29	3.1899	0.3135	53.6609	0.0186	16.8219	0.0594	11.2609
30	3.3201	0.3012	56.8508	0.0176	17.1231	0.0584	11.5730
31	3.4556	0.2894	60.1710	0.0166	17.4125	0.0574	11.8792
32	3.5966	0.2780	63.6266	0.0157	17.6906	0.0565	12.1797
33	3.7434	0.2671	67.2233	0.0149	17.9577	0.0557	12.4746
34	3.8962	0.2567	70.9667	0.0141	18.2144	0.0549	12.7638

Table A-31

CONTINUOUS COMPOUNDING: $r = 4\%$

	Single Payment		Uniform Series				Gradient Series
n	Compound Amount Factor	Present Worth Factor	Compound Amount Factor	Sinking Fund Factor	Present Worth Factor	Capital Recovery Factor	Uniform Series Factor
	To Find F Given P FPr,n	To Find P Given F PFr,n	To Find F Given A FAr,n	To Find A Given F AFr,n	To Find P Given A PAr,n	To Find A Given P APr,n	To Find A Given G AGr,n
35	4.0552	0.2466	74.8629	0.0134	18.4610	0.0542	13.0475
40	4.9530	0.2019	96.8629	0.0103	19.5563	0.0511	14.3845
45	6.0496	0.1653	123.7337	0.0081	20.4530	0.0489	15.5918
50	7.6891	0.1353	156.5538	0.0064	21.1873	0.0472	16.6775
55	9.0250	0.1108	196.6404	0.0051	21.7884	0.0459	17.6498
60	11.0232	0.0907	245.6023	0.0041	22.2805	0.0449	18.5172
65	13.4637	0.0743	305.4041	0.0033	22.6835	0.0441	19.2882
70	16.4446	0.0608	378.4465	0.0026	23.0134	0.0435	19.9710
75	20.0855	0.0498	467.6609	0.0021	23.2835	0.0429	20.5737
80	24.5325	0.0408	576.6275	0.0017	23.5045	0.0425	21.1038
85	29.9641	0.0334	709.7195	0.0014	23.6857	0.0422	21.5687
90	36.5982	0.0273	872.2788	0.0011	23.8339	0.0420	21.9751
95	44.7012	0.0224	1070.8289	0.0009	23.9553	0.0417	22.3295
100	54.5981	0.0183	1313.3386	0.0008	24.0546	0.0416	22.6376

Table A-32

CONTINUOUS COMPOUNDING: $r = 5\%$

	Single Payment		Uniform Series				Gradient Series
	Compound Amount Factor	Present Worth Factor	Compound Amount Factor	Sinking Fund Factor	Present Worth Factor	Capital Recovery Factor	Uniform Series Factor
n	To Find F Given P FPr,n	To Find P Given F PFr,n	To Find F Given A FAr,n	To Find A Given F AFr,n	To Find P Given A PAr,n	To Find A Given P APr,n	To Find A Given G AGr,n
1	1.0513	0.9512	1.000	1.0000	0.9512	1.0513	0.0000
2	1.1052	0.9048	2.0513	0.4875	1.8561	0.5383	0.4875
3	1.1618	0.8607	3.1565	0.3168	2.7168	0.3681	0.9667
4	1.2214	0.8187	4.3183	0.2316	3.5355	0.2828	1.4375
5	1.2840	0.7788	5.5397	0.1805	4.3144	0.2318	1.9001
6	1.3499	0.7408	6.8238	0.1465	5.0552	0.1978	2.3544
7	1.4191	0.7047	8.1737	0.1223	5.7599	0.1736	2.8004
8	1.4918	0.6703	9.5927	0.1042	6.4302	0.1555	3.2382
9	1.5683	0.6376	11.0846	0.0902	7.0678	0.1415	3.6678
10	1.6487	0.6065	12.6529	0.0790	7.6744	0.1303	4.0892
11	1.7333	0.5769	14.3016	0.0699	8.2513	0.1212	4.5025
12	1.8221	0.5488	16.0349	0.0624	8.8002	0.1136	4.9077
13	1.9155	0.5220	17.8571	0.0560	9.3222	0.1073	5.3049
14	2.0138	0.4966	19.7726	0.0506	9.8188	0.1018	5.6941
15	2.1170	0.4724	21.7864	0.0459	10.2912	0.0972	6.0753
16	2.2255	0.4493	23.9034	0.0418	10.7405	0.0931	6.4487
17	2.3396	0.4274	26.1290	0.0383	11.1679	0.0895	6.8143
18	2.4596	0.4065	28.4687	0.0351	11.5745	0.0864	7.1720
19	2.5857	0.3867	30.9283	0.0323	11.9612	0.0836	7.5221
20	2.7183	0.3679	33.5140	0.0298	12.3291	0.0811	7.8646
21	2.8577	0.3499	36.2324	0.0276	12.6791	0.0789	8.1996
22	3.0042	0.3329	39.0900	0.0256	13.0119	0.0769	8.5270
23	3.1582	0.3165	42.0942	0.0238	13.3286	0.0750	8.8471
24	3.3201	0.3012	45.2525	0.0221	13.6298	0.0734	9.1599
25	3.4903	0.2865	48.5726	0.0206	13.9163	0.0719	9.4654
26	3.6693	0.2725	52.0630	0.0192	14.1888	0.0705	9.7638
27	3.8574	0.2592	55.7323	0.0179	14.4481	0.0692	10.0591
28	4.0552	0.2466	59.5898	0.0168	14.6947	0.0681	10.3395
29	4.2631	0.2346	63.6451	0.0157	14.9292	0.0670	10.6170
30	4.4817	0.2231	67.9082	0.0147	15.1524	0.0660	10.6877
31	4.7115	0.2122	72.3900	0.0138	15.3646	0.0651	11.1517
32	4.9530	0.2019	77.1015	0.0130	15.5665	0.0642	11.4091
33	5.2070	0.1920	82.0546	0.0122	15.7586	0.0635	11.6601
34	5.4739	0.1827	87.2616	0.0115	15.9413	0.0627	11.9046

Table A-32

CONTINUOUS COMPOUNDING: $r = 5\%$

	Single Payment		Uniform Series				Gradient Series
n	Compound Amount Factor	Present Worth Factor	Compound Amount Factor	Sinking Fund Factor	Present Worth Factor	Capital Recovery Factor	Uniform Series Factor
	To Find F Given P FPr,n	To Find P Given F PFr,n	To Find F Given A FAr,n	To Find A Given F AFr,n	To Find P Given A PAr,n	To Find A Given P APr,n	To Find A Given G AGr,n
35	5.7546	0.1738	92.7356	0.0108	16.1150	0.0621	12.1429
40	7.3891	0.1353	124.6146	0.0080	16.8648	0.0593	13.2435
45	9.4877	0.1054	165.5401	0.0060	17.4486	0.0573	14.2024
50	12.1825	0.0821	218.1077	0.0046	17.9034	0.0559	15.0329
55	15.6426	0.0639	285.5952	0.0035	18.2575	0.0548	15.7480
60	20.0855	0.0498	372.2512	0.0027	18.5333	0.0540	16.3604
65	25.7903	0.0388	483.5200	0.0021	18.7481	0.0533	16.8822
70	33.1154	0.0302	626.3919	0.0016	18.9154	0.0529	17.3245
75	42.5211	0.0235	809.8430	0.0012	19.0457	0.0525	17.6979
80	54.5981	0.0183	1045.3989	0.0010	19.1472	0.0522	18.0116
85	70.1054	0.0143	1347.8584	0.0007	19.2262	0.0520	18.2742
90	90.0171	0.0111	1736.2239	0.0006	19.2877	0.0518	18.4931
95	115.5842	0.0087	2234.8960	0.0004	19.3356	0.0517	18.6751
100	148.4130	0.0067	2875.2031	0.0003	19.3730	0.0516	18.8258

Table A-33
CONTINUOUS COMPOUNDING: $r = 6\%$

	Single Payment		Uniform Series				Gradient Series
	Compound Amount Factor	Present Worth Factor	Compound Amount Factor	Sinking Fund Factor	Present Worth Factor	Capital Recovery Factor	Uniform Series Factor
n	To Find F Given P FPr,n	To Find P Given F PFr,n	To Find F Given A FAr,n	To Find A Given F AFr,n	To Find P Given A PAr,n	To Find A Given P APr,n	To Find A Given G AGr,n
1	1.0618	0.9418	1.0000	1.0000	0.9418	1.0618	0.0000
2	1.1275	0.8869	2.0618	0.4850	1.8287	0.5468	0.4850
3	1.1972	0.8353	3.1893	0.3135	2.6640	0.3754	0.9600
4	1.2712	0.7866	4.3866	0.2280	3.4506	0.2898	1.4251
5	1.3499	0.7408	5.6578	0.1767	4.1914	0.2386	1.8802
6	1.4333	0.6977	7.0077	0.1427	4.8891	0.2045	2.3254
7	1.5220	0.6570	8.4410	0.1185	5.5461	0.1803	2.7607
8	1.6161	0.6188	9.9630	0.1004	6.1649	0.1622	3.1862
9	1.7160	0.5827	11.5791	0.0864	6.7477	0.1482	3.6020
10	1.8221	0.5488	13.2951	0.0752	7.2965	0.1371	4.0080
11	1.9348	0.5169	15.1172	0.0661	7.8134	0.1280	4.4043
12	2.0544	0.4868	17.0520	0.0586	8.3001	0.1205	4.7911
13	2.1815	0.4584	19.1065	0.0523	8.7585	0.1142	5.1684
14	2.3164	0.4317	21.2879	0.0470	9.1902	0.1088	5.5363
15	2.4596	0.4066	23.6043	0.0424	9.5968	0.1042	5.8949
16	2.6117	0.3829	26.0639	0.0384	9.9797	0.1002	6.2442
17	2.7732	0.3606	28.6756	0.0349	10.3403	0.0967	6.5845
18	2.9447	0.3396	31.4489	0.0318	10.6799	0.0936	6.9156
19	3.1268	0.3198	34.3935	0.0291	10.9997	0.0909	7.2379
20	3.3201	0.3012	37.5203	0.0267	11.3009	0.0885	7.5514
21	3.5254	0.2837	40.8405	0.0245	11.5846	0.0863	7.8562
22	3.7434	0.2671	44.3659	0.0225	11.8517	0.0844	8.1525
23	3.9749	0.2516	48.1093	0.0208	12.1033	0.0826	8.4403
24	4.2207	0.2369	52.0843	0.0192	12.3402	0.0810	8.7199
25	4.4817	0.2231	56.3050	0.0178	12.5633	0.0796	8.9912
26	4.7586	0.2101	60.7867	0.0165	12.7735	0.0783	9.2546
27	5.0531	0.1979	65.5455	0.0153	12.9714	0.0771	9.5101
28	5.3656	0.1864	70.5986	0.0142	13.1578	0.0760	9.7578
29	5.6973	0.1755	75.9642	0.0132	13.3333	0.0750	9.9980
30	6.0496	0.1653	81.6616	0.0122	13.4986	0.0741	10.2307
31	6.4237	0.1557	87.7113	0.0114	13.6542	0.0732	10.4560
32	6.8210	0.1466	94.1350	0.0106	13.8009	0.0725	10.6743
33	7.2427	0.1381	100.9560	0.0099	13.9389	0.0717	10.8835
34	7.6906	0.1300	108.1988	0.0092	14.0690	0.0711	11.0899

Table A-33

CONTINUOUS COMPOUNDING: $r = 6\%$

	Single Payment		Uniform Series				Gradient Series
	Compound Amount Factor	Present Worth Factor	Compound Amount Factor	Sinking Fund Factor	Present Worth Factor	Capital Recovery Factor	Uniform Series Factor
n	To Find *F* Given *P* *FPr,n*	To Find *P* Given *F* *PFr,n*	To Find *F* Given *A* *FAr,n*	To Find *A* Given *F* *AFr,n*	To Find *P* Given *A* *PAr,n*	To Find *A* Given *P* *APr,n*	To Find *A* Given *G* *AGr,n*
35	8.1662	0.1225	115.8894	0.0086	14.1914	0.0705	11.2876
40	11.0232	0.0907	162.0922	0.0062	14.7047	0.0680	12.1809
45	14.8797	0.0672	224.4594	0.0045	15.0849	0.0663	12.9295
50	20.0855	0.0498	308.6460	0.0032	15.3666	0.0651	13.5519
55	27.1126	0.0369	422.2864	0.0024	15.5753	0.0642	14.0654
60	36.5982	0.0273	575.6851	0.0017	15.7299	0.0636	14.4862
65	49.4024	0.0202	782.7515	0.0013	15.8444	0.0631	14.8288
70	66.6863	0.0150	1062.2620	0.0009	15.9292	0.0628	15.1060
75	90.0171	0.0111	1439.5615	0.0007	15.9921	0.0625	15.3291
80	121.5103	0.0082	1948.8623	0.0005	16.0387	0.0623	15.5078
85	164.0218	0.0061	2636.3472	0.0004	16.0732	0.0622	15.6503
90	221.4063	0.0045	3564.3550	0.0003	16.0987	0.0621	15.7633
95	298.8670	0.0033	4817.0274	0.0002	16.1176	0.0620	15.8527
100	403.4282	0.0025	6507.9649	0.0002	16.1317	0.0620	15.9232

Table A-34
CONTINUOUS COMPOUNDING: $r = 7\%$

n	Single Payment		Uniform Series				Gradient Series
	Compound Amount Factor	Present Worth Factor	Compound Amount Factor	Sinking Fund Factor	Present Worth Factor	Capital Recovery Factor	Uniform Series Factor
	To Find F Given P FPr,n	To Find P Given F PFr,n	To Find F Given A FAr,n	To Find A Given F AFr,n	To Find P Given A PAr,n	To Find A Given P APr,n	To Find A Given G AGr,n
1	1.0725	0.9324	1.0000	1.0000	0.9324	1.0725	0.0000
2	1.1503	0.8694	2.0725	0.4825	1.8018	0.5550	0.4825
3	1.2337	0.8106	3.2228	0.3103	2.6123	0.3828	0.9534
4	1.3231	0.7558	4.4565	0.2244	3.3681	0.2969	1.4126
5	1.4191	0.7047	5.7796	0.1730	4.0728	0.2455	1.8603
6	1.5220	0.6570	7.1987	0.1389	4.7299	0.2114	2.2904
7	1.6323	0.6126	8.7206	0.1147	5.3425	0.1872	2.7211
8	1.7507	0.5712	10.3530	0.0966	5.9137	0.1691	3.1344
9	1.8776	0.5326	12.1037	0.0826	6.4463	0.1551	3.5364
10	2.0138	0.4966	13.9813	0.0715	6.9429	0.1440	3.9272
11	2.1598	0.4630	15.9950	0.0625	7.4059	0.1350	4.3069
12	2.3164	0.4317	18.1548	0.0551	7.8376	0.1276	4.6755
13	2.4843	0.4025	20.4712	0.0488	8.2401	0.1214	5.0333
14	2.6645	0.3753	22.9555	0.0436	8.6155	0.1161	5.3804
15	2.8576	0.3499	25.6200	0.0390	8.9654	0.1115	5.7160
16	3.0649	0.3263	28.4776	0.0351	9.2917	0.1076	6.0428
17	3.2871	0.3042	31.5425	0.0317	9.5959	0.1042	6.3585
18	3.5254	0.2837	34.8296	0.0287	9.8796	0.1012	6.6640
19	3.7810	0.2645	38.3550	0.0261	10.1440	0.0986	6.9596
20	4.0552	0.2466	42.1361	0.0237	10.3906	0.0962	7.2453
21	4.3492	0.2299	46.1913	0.0216	10.6206	0.0942	7.5215
22	4.6640	0.2144	50.5406	0.0198	10.8349	0.0923	7.7881
23	5.0028	0.1999	55.2052	0.0181	11.0348	0.0906	8.0456
24	5.3656	0.1864	60.2080	0.0166	11.2212	0.0891	8.2940
25	5.7546	0.1738	65.5736	0.0153	11.3950	0.0878	8.5335
26	6.1719	0.1620	71.3282	0.0140	11.5570	0.0865	8.7643
27	6.6194	0.1511	77.5001	0.0129	11.7081	0.0854	8.9867
28	7.0993	0.1409	84.1195	0.0119	11.8489	0.0844	9.2009
29	7.6141	0.1313	91.2189	0.0110	11.9803	0.0835	9.4070
30	8.1662	0.1225	98.8330	0.0101	12.1027	0.0826	9.6052
31	8.7583	0.1142	106.9991	0.0093	12.2169	0.0819	9.7958
32	9.3933	0.1065	115.7574	0.0086	12.3234	0.0811	9.9790
33	10.0744	0.0993	125.1508	0.0080	12.4226	0.0805	10.1550
34	10.8049	0.0926	135.2253	0.0074	12.5152	0.0799	10.3239

Table A-34

CONTINUOUS COMPOUNDING: $r = 7\%$

	Single Payment		Uniform Series				Gradient Series
	Compound Amount Factor	Present Worth Factor	Compound Amount Factor	Sinking Fund Factor	Present Worth Factor	Capital Recovery Factor	Uniform Series Factor
n	To Find *F* Given *P* *FPr,n*	To Find *P* Given *F* *PFr,n*	To Find *F* Given *A* *FAr,n*	To Find *A* Given *F* *AFr,n*	To Find *P* Given *A* *PAr,n*	To Find *A* Given *P* *APr,n*	To Find *A* Given *G* *AGr,n*
35	11.5883	0.0863	146.0302	0.0068	12.6015	0.0794	10.4860
40	16.4446	0.0608	213.0062	0.0047	12.9529	0.0772	11.2017
45	23.6360	0.0429	308.0496	0.0032	13.2006	0.0758	11.7769
50	33.1154	0.0302	442.9231	0.0023	13.3751	0.0748	12.2347
55	46.9930	0.0213	634.3174	0.0016	13.4981	0.0741	12.5957
60	66.6863	0.0150	905.9192	0.0011	13.5848	0.0736	12.8781
65	94.6323	0.0106	1291.3401	0.0008	13.6459	0.0733	13.0973
70	134.2896	0.0074	1838.2783	0.0005	13.6889	0.0731	13.2664
75	190.5661	0.0052	2614.4207	0.0004	13.7192	0.0729	13.3959
80	270.4258	0.0037	3715.8174	0.0003	13.7406	0.0728	13.4946
85	383.7527	0.0026	5278.7734	0.0002	13.7557	0.0727	13.5695
90	544.5711	0.0018	7496.7149	0.0001	13.7663	0.0726	13.6260
95	772.7832	0.0013	10644.1289	0.0001	13.7738	0.0726	13.6685
100	1096.6318	0.0009	15110.5234	0.0001	13.7790	0.0726	13.7003

Table A-35

CONTINUOUS COMPOUNDING: $r = 8\%$

	Single Payment		Uniform Series				Gradient Series
	Compound Amount Factor	Present Worth Factor	Compound Amount Factor	Sinking Fund Factor	Present Worth Factor	Capital Recovery Factor	Uniform Series Factor
n	To Find *F* Given *P* *FPr,n*	To Find *P* Given *F* *PFr,n*	To Find *F* Given *A* *FAr,n*	To Find *A* Given *F* *AFr,n*	To Find *P* Given *A* *PAr,n*	To Find *A* Given *P* *APr,n*	To Find *A* Given *G* *AGr,n*
1	1.0833	0.9231	1.0000	1.0000	0.9231	1.0833	0.0000
2	1.1735	0.8521	2.0833	0.4800	1.7753	0.5633	0.4800
3	1.2712	0.7866	3.2568	0.3070	2.5619	0.3903	0.9467
4	1.3771	0.7261	4.5281	0.2208	3.2881	0.3041	1.4002
5	1.4918	0.6703	5.9052	0.1693	3.9584	0.2526	1.8404
6	1.6161	0.6188	7.3971	0.1352	4.5772	0.2185	2.2676
7	1.7507	0.5712	9.0131	01109	5.1484	0.1942	2.6817
8	1.8965	0.5273	10.7638	0.0929	5.6757	0.1762	3.0829
9	2.0544	0.4868	12.6603	0.0790	6.1624	0.1623	3.4713
10	2.2255	0.4493	14.7140	0.0680	6.6118	0.1512	3.8470
11	2.4109	0.4148	16.9404	0.0590	7.0266	0.1423	4.2102
12	2.6117	0.3829	19.3513	0.0517	7.4095	0.1350	4.5611
13	2.8292	0.3535	21.9630	0.0455	7.7629	0.1288	4.8998
14	3.0649	0.3263	24.7922	0.0403	8.0892	0.1236	5.2265
15	3.3201	0.3012	27.8571	0.0359	8.3904	0.1192	5.5415
16	3.5966	0.2780	31.1772	0.0321	8.6684	0.1154	5.8449
17	3.8962	0.2567	34.7739	0.0288	8.9251	0.1120	6.1369
18	4.2207	0.2369	38.6702	0.0259	9.1620	0.1091	6.4178
19	4.5722	0.2187	42.8909	0.0233	9.3808	0.1065	6.6879
20	4.9530	0.2019	47.4632	0.0211	9.5827	0.1044	6.9473
21	5.3656	0.1864	52.4162	0.0191	9.7690	0.1024	7.1963
22	5.8124	0.1720	57.7818	0.0173	9.9411	0.1006	7.4352
23	6.2965	0.1588	63.5943	0.0157	10.0999	0.0990	7.6642
24	6.8210	0.1466	69.8909	0.0143	10.2465	0.0976	7.8838
25	7.3891	0.1353	76.7119	0.0130	10.3818	0.0963	8.0937
26	8.0045	0.1249	84.1010	0.0119	10.5068	0.0952	8.2947
27	8.6711	0.1153	92.1056	0.0109	10.6221	0.0941	8.4870
28	9.3933	0.1065	100.7768	0.0099	10.7286	0.0932	8.6707
29	10.1757	0.0983	110.1702	0.0091	10.8268	0.0924	8.8461
30	11.0232	0.0907	120.3459	0.0083	10.9175	0.0916	9.0136
31	11.9413	0.0837	131.3692	0.0076	11.0013	0.0909	9.1734
32	12.9358	0.0773	143.3106	0.0070	11.0786	0.0903	9.3257
33	14.0132	0.0714	156.2465	0.0064	11.1500	0.0897	9.4708
34	15.1803	0.0659	170.2598	0.0059	11.2158	0.0892	9.6090

Table A-35

CONTINUOUS COMPOUNDING: $r = 8\%$

	Single Payment		Uniform Series				Gradient Series
	Compound Amount Factor	Present Worth Factor	Compound Amount Factor	Sinking Fund Factor	Present Worth Factor	Capital Recovery Factor	Uniform Series Factor
n	To Find *F* Given *P* *FPr,n*	To Find *P* Given *F* *PFr,n*	To Find *F* Given *A* *FAr,n*	To Find *A* Given *F* *AFr,n*	To Find *P* Given *A* *PAr,n*	To Find *A* Given *P* *APr,n*	To Find *A* Given *G* *AGr,n*
35	16.4446	0.0608	185.4402	0.0054	11.2766	0.0887	9.7405
40	24.5325	0.0408	282.5491	0.0035	11.5174	0.0868	10.3069
45	36.5982	0.0273	427.4192	0.0023	11.6787	0.0856	10.7426
50	54.5981	0.0183	643.5400	0.0016	11.7869	0.0848	11.0738
55	81.4508	0.0123	965.9544	0.0010	11.8594	0.0843	11.3230
60	121.5102	0.0082	1446.9397	0.0007	11.9080	0.0840	11.5088
65	181.2720	0.0055	2164.4846	0.0005	11.9405	0.0837	11.6461
70	270.4258	0.0037	3234.9375	0.0003	11.9624	0.0836	11.7469
75	403.4277	0.0025	4831.8594	0.0002	11.9770	0.0835	11.8203
80	601.8438	0.0017	7214.1953	0.0001	11.9868	0.0834	11.8735
85	897.8455	0.0011	10768.2227	0.0001	11.9934	0.0834	11.9119
90	1339.4280	0.0007	16070.2070	0.0001	11.9978	0.0833	11.9394
95	1998.1919	0.0005	23979.8399	0.0000	12.0008	0.0833	11.9591
100	2980.9619	0.0003	35779.6094	0.0000	12.0028	0.0833	11.9731

Table A-36
CONTINUOUS COMPOUNDING: $r = 9\%$

	Single Payment		Uniform Series				Gradient Series
	Compound Amount Factor	Present Worth Factor	Compound Amount Factor	Sinking Fund Factor	Present Worth Factor	Capital Recovery Factor	Uniform Series Factor
n	To Find F Given P FPr,n	To Find P Given F PFr,n	To Find F Given A FAr,n	To Find A Given F AFr,n	To Find P Given A PAr,n	To Find A Given P APr,n	To Find A Given G AGr,n
1	1.0942	0.9139	1.0000	1.0000	0.9139	1.0942	0.0000
2	1.1972	0.8353	2.0942	0.4775	1.7492	0.5717	0.4775
3	1.3100	0.7634	3.2914	0.3038	2.5126	0.3980	0.9401
4	1.4333	0.6977	4.6014	0.2173	3.2103	0.3115	1.3878
5	1.5683	0.6376	6.0347	0.1657	3.8479	0.2599	1.8206
6	1.7160	0.5827	7.6031	0.1315	4.4307	0.2257	2.2388
7	1.8776	0.5326	9.3191	0.1073	4.9633	0.2015	2.6424
8	2.0544	0.4868	11.1967	0.0893	5.4500	0.1835	3.0316
9	2.2479	0.4449	13.2512	0.0755	5.8949	0.1696	3.4065
10	2.4596	0.4066	15.4991	0.0645	6.3015	0.1587	3.7674
11	2.6912	0.3716	17.9587	0.0557	6.6730	0.1499	4.1145
12	2.9447	0.3396	20.6500	0.0484	7.0126	0.1426	4.4479
13	3.2220	0.3104	23.5947	0.0424	7.3230	0.1366	4.7680
14	3.5254	0.2837	26.8167	0.0373	7.6067	0.1315	5.0750
15	3.8574	0.2592	30.3421	0.0330	7.8659	0.1271	5.3691
16	4.2207	0.2369	34.1996	0.0292	8.1028	0.1234	5.6507
17	4.6182	0.2165	38.4203	0.0260	8.3194	0.1202	5.9201
18	5.0531	0.1979	43.0385	0.0232	8.5173	0.1174	6.1776
19	5.5290	0.1809	48.0917	0.0208	8.6981	0.1150	6.4234
20	6.0496	0.1653	53.6207	0.0186	8.8634	0.1128	6.6579
21	6.5194	0.1511	59.6704	0.0168	9.0145	0.1109	6.8815
22	7.2427	0.1381	66.2898	0.0151	9.1526	0.1093	7.0945
23	7.9248	0.1262	73.5326	0.0136	9.2788	0.1078	7.2972
24	8.6711	0.1153	81.4575	0.0123	9.3941	0.1064	7.4900
25	9.4877	0.1054	90.1287	0.0111	9.4995	0.1053	7.6732
26	10.3812	0.0963	99.6165	0.0100	9.5958	0.1042	7.8471
27	11.3589	0.0880	109.9978	0.0091	9.6839	0.1033	8.0122
28	12.4286	0.0805	121.3567	0.0082	9.7643	0.1024	8.1686
29	13.5990	0.0735	133.7854	0.0075	9.8379	0.1016	8.3168
30	14.8797	0.0672	147.3846	0.0068	9.9051	0.1010	8.4572
31	16.2810	0.0614	162.2643	0.0062	9.9665	0.1003	8.5899
32	17.8143	0.0561	178.5456	0.0056	10.0226	0.0998	8.7155
33	19.4919	0.0513	196.3600	0.0051	10.0739	0.0993	8.8340
34	21.3275	0.0469	215.8520	0.0046	10.1208	0.0988	8.9460

Table A-36

CONTINUOUS COMPOUNDING: $r = 9\%$

n	Single Payment		Uniform Series				Gradient Series
	Compound Amount Factor	Present Worth Factor	Compound Amount Factor	Sinking Fund Factor	Present Worth Factor	Capital Recovery Factor	Uniform Series Factor
	To Find *F* Given *P* *FPr,n*	To Find *P* Given *F* *PFr,n*	To Find *F* Given *A* *FAr,n*	To Find *A* Given *F* *AFr,n*	To Find *P* Given *A* *PAr,n*	To Find *A* Given *P* *APr,n*	To Find *A* Given *G* *AGr,n*
35	23.3360	0.0429	237.1797	0.0042	10.1637	0.0984	9.0516
40	36.5982	0.0273	378.0064	0.0026	10.3286	0.0968	9.4950
45	57.3974	0.0174	598.8667	0.0017	10.4337	0.0958	9.8207
50	90.0170	0.0111	945.2449	0.0011	10.5007	0.0952	10.0569
55	141.1747	0.0071	1488.4739	0.0007	10.5435	0.0948	10.2262
60	221.4060	0.0045	2340.4255	0.0004	10.5707	0.0946	10.3464
65	347.2334	0.0029	3676.5518	0.0003	10.5881	0.0944	10.4309
70	544.5706	0.0018	5772.0117	0.0002	10.5992	0.0943	10.4898
75	854.0567	0.0012	9058.3555	0.0001	10.6063	0.0943	10.5307
80	1339.4273	0.0007	14212.3633	0.0001	10.6108	0.0942	10.5588
85	2100.6399	0.0005	22295.4570	0.0000	10.6137	0.0942	10.5781
90	3294.4590	0.0003	34972.2617	0.0000	10.6155	0.0942	10.5913
95	5166.7344	0.0002	54853.4492	0.0000	10.6167	0.0942	10.6002
100	8103.0547	0.0001	86033.3125	0.0000	10.6174	0.0942	10.6063

Table A-37

CONTINUOUS COMPOUNDING: $r = 10\%$

	Single Payment		Uniform Series				Gradient Series
	Compound Amount Factor	Present Worth Factor	Compound Amount Factor	Sinking Fund Factor	Present Worth Factor	Capital Recovery Factor	Uniform Series Factor
n	To Find *F* Given *P* *FPr,n*	To Find *P* Given *F* *PFr,n*	To Find *F* Given *A* *FAr,n*	To Find *A* Given *F* *AFr,n*	To Find *P* Given *A* *PAr,n*	To Find *A* Given *P* *APr,n*	To Find *A* Given *G* *AGr,n*
1	1.1052	0.9048	1.0000	1.0000	0.9048	1.1052	0.0000
2	1.2214	0.8187	2.1052	0.4750	1.7236	0.5802	0.4750
3	1.3499	0.7408	3.3266	0.3006	2.4644	0.4058	0.9334
4	1.4918	0.6703	4.6765	0.2138	3.1347	0.3190	1.3754
5	1.6487	0.6065	6.1683	0.1621	3.7413	0.2673	1.8009
6	1.8221	0.5488	7.8170	0.1279	4.2901	0.2331	2.2101
7	2.0138	0.4966	9.6392	0.1037	4.7867	0.2089	2.6033
8	2.2255	0.4493	11.6529	0.0858	5.2360	0.1910	2.9806
9	2.4596	0.4066	13.8785	0.0721	5.6426	0.1772	3.3423
10	2.7183	0.3679	16.3381	0.0612	6.0104	0.1664	3.6886
11	3.0042	0.3329	19.0564	0.0525	6.3433	0.1576	4.0198
12	3.3201	0.3012	22.0606	0.0453	6.6445	0.1505	4.3362
13	3.6693	0.2725	25.3807	0.0394	6.9171	0.1446	4.6381
14	4.0552	0.2466	29.0500	0.0344	7.1637	0.1396	4.9260
15	4.4817	0.2231	33.1052	0.0302	7.3868	0.1354	5.2001
16	4.9530	0.2019	37.5869	0.0266	7.5887	0.1318	5.4608
17	5.4739	0.1827	42.5400	0.0235	7.7714	0.1287	5.7086
18	6.0496	0.1653	48.0140	0.0208	7.9367	0.1260	5.9437
19	6.6859	0.1496	54.0637	0.0185	8.0862	0.1237	6.1667
20	7.3890	0.1353	60.7496	0.0165	8.2216	0.1216	6.3780
21	8.1662	0.1225	68.1387	0.0147	8.3440	0.1198	6.5779
22	9.0250	0.1108	76.3049	0.0131	8.4548	0.1183	6.7669
23	9.9742	0.1003	85.3300	0.0117	8.5551	0.1169	6.9454
24	11.0232	0.0907	95.3042	0.0105	8.6458	0.1157	7.1139
25	12.1825	0.0821	106.3274	0.0094	8.7279	0.1146	7.2727
26	13.4637	0.0743	118.5099	0.0084	8.8022	0.1136	7.4223
27	14.8797	0.0672	131.9737	0.0076	8.8694	0.1127	7.5630
28	16.4446	0.0608	146.8535	0.0068	8.9302	0.1120	7.6954
29	18.1741	0.0550	163.2982	0.0061	8.9852	0.1113	7.8197
30	20.0855	0.0498	181.4725	0.0055	9.0350	0.1107	7.9365
31	22.1979	0.0450	201.5581	0.0050	9.0800	0.1101	8.0459
32	24.5325	0.0408	223.7561	0.0045	9.1208	0.1096	8.1485
33	27.1126	0.0369	248.2887	0.0040	9.1577	0.1092	8.2446
34	29.9641	0.0334	275.4011	0.0036	9.1911	0.1088	8.3345

Table A-37

CONTINUOUS COMPOUNDING: r = 10%

n	Single Payment		Uniform Series				Gradient Series
	Compound Amount Factor	Present Worth Factor	Compound Amount Factor	Sinking Fund Factor	Present Worth Factor	Capital Recovery Factor	Uniform Series Factor
	To Find F Given P FPr,n	To Find P Given F PFr,n	To Find F Given A FAr,n	To Find A Given F AFr,n	To Find P Given A PAr,n	To Find A Given P APr,n	To Find A Given G AGr,n
35	33.1154	0.0302	305.3655	0.0033	9.2213	0.1084	8.4185
40	54.5981	0.0183	509.6309	0.0020	9.3342	0.1071	8.7620
45	90.0170	0.0111	846.4080	0.0012	9.4028	0.1064	9.0028
50	148.4128	0.0067	1401.6589	0.0007	9.4443	0.1059	9.1691
55	244.6914	0.0041	2317.1118	0.0004	9.4695	0.1056	9.2826
60	403.4275	0.0025	3826.4400	0.0003	9.4848	0.1054	9.3592
65	665.1397	0.0015	6314.8984	0.0002	9.4941	0.1053	9.4105
70	1096.6299	0.0009	10417.6758	0.0001	9.4997	0.1053	9.4444
75	1808.0366	0.0006	17182.0117	0.0001	9.5031	0.1052	9.4668
80	2980.9470	0.0003	28334.5117	0.0000	9.5052	0.1052	9.4815
85	4914.7461	0.0002	46721.8399	0.0000	9.5065	0.1052	9.4910
90	8103.0508	0.0001	77037.4375	0.0000	9.5072	0.1052	9.4972
95	13359.6719	0.0001	127019.4375	0.0000	9.5077	0.1052	9.5012
100	22026.3750	0.0000	209425.8750	0.0000	9.5080	0.1052	9.5038

Table A-38

CONTINUOUS COMPOUNDING: r = 12%

	Single Payment		Uniform Series				Gradient Series
	Compound Amount Factor	Present Worth Factor	Compound Amount Factor	Sinking Fund Factor	Present Worth Factor	Capital Recovery Factor	Uniform Series Factor
n	To Find F Given P FPr,n	To Find P Given F PFr,n	To Find F Given A FAr,n	To Find A Given F AFr,n	To Find P Given A PAr,n	To Find A Given P APr,n	To Find A Given G AGr,n
1	1.1275	.0.8869	1.0000	1.0000	0.8869	1.1275	0.0000
2	1.2712	0.7866	2.1275	0.4700	1.6735	0.5975	0.4700
3	1.4333	0.6977	3.3987	0.2942	2.3712	0.4217	0.9202
4	1.6161	0.6188	4.8321	0.2070	2.9900	0.3344	1.3506
5	1.8221	0.5488	6.4481	0.1551	3.5388	0.2826	1.7615
6	2.0544	0.4868	8.2703	0.1209	4.0256	0.2484	2.1531
7	2.3164	0.4317	10.3247	0.0969	4.4573	0.2244	2.5257
8	2.6117	0.3829	12.6411	0.0791	4.8402	0.2066	2.8796
9	2.9447	0.3396	15.2528	0.0656	5.1798	0.1931	3.2153
10	3.3201	0.3012	18.1974	0.0550	5.4810	0.1824	3.5332
11	3.7434	0.2671	21.5176	0.0465	5.7481	0.1740	3.8337
12	4.2207	0.2369	25.2610	0.0396	5.9850	0.1671	4.1174
13	4.7588	0.2101	29.4817	0.0339	6.1952	0.1614	4.3848
14	5.3656	0.1864	34.2405	0.0292	6.3815	0.1567	4.6364
15	6.0496	0.1653	39.6060	0.0252	6.5468	0.1527	4.8728
16	6.8210	0.1466	45.6557	0.0219	6.6934	0.1494	5.0946
17	7.6906	0.1300	52.4767	0.0191	6.8235	0.1466	5.3025
18	8.6711	0.1153	60.1673	0.0166	6.9388	0.1441	5.4969
19	9.7767	0.1023	68.8384	0.0145	7.0411	0.1420	5.6785
20	11.0232	0.0907	78.6151	0.0127	7.1318	0.1402	5.8480
21	12.4286	0.0805	89.6382	0.0112	7.2123	0.1387	6.0058
22	14.0132	0.0714	102.0668	0.0098	7.2836	0.1373	6.1527
23	15.7998	0.0633	116.0800	0.0086	7.3469	0.1361	6.2893
24	17.8142	0.0561	131.8797	0.0076	7.4031	0.1351	6.4160
25	20.0855	0.0498	149.6940	0.0067	7.4528	0.1342	6.5334
26	22.6463	0.0442	169.7795	0.0059	7.4970	0.1334	6.6422
27	25.5337	0.0392	192.4259	0.0052	7.5362	0.1327	6.7428
28	28.7891	0.0347	217.9596	0.0046	7.5709	0.1321	6.8357
29	32.4597	0.0308	246.7487	0.0041	7.6017	0.1315	6.9215
30	36.5982	0.0273	279.2080	0.0036	7.6290	0.1311	7.0006
31	41.2643	0.0242	315.8064	0.0032	7.6533	0.1307	7.0734
32	46.5254	0.0215	357.0706	0.0028	7.6748	0.1303	7.1404
33	52.4572	0.0191	403.5960	0.0025	7.6938	0.1300	7.2020
34	59.1453	0.0169	456.0532	0.0022	7.7107	0.1297	7.2586
35	66.6862	0.0150	515.1987	0.0019	7.7257	0.1294	7.3105
40	121.5101	0.0082	945.2014	0.0011	7.7788	0.1286	7.5114
45	221.4058	0.0045	1728.7168	0.0006	7.8079	0.1281	7.6392
50	403.4273	0.0025	3156.3743	0.0003	7.8239	0.1278	7.7191

Table A-39

CONTINUOUS COMPOUNDING: $r = 15\%$

	Single Payment		Uniform Series				Gradient Series
	Compound Amount Factor	Present Worth Factor	Compound Amount Factor	Sinking Fund Factor	Present Worth Factor	Capital Recovery Factor	Uniform Series Factor
n	To Find F Given P FPr,n	To Find P Given F PFr,n	To Find F Given A FAr,n	To Find A Given F AFr,n	To Find P Given A PAr,n	To Find A Given P APr,n	To Find A Given G AGr,n
1	1.1618	0.8607	1.0000	1.0000	0.8607	1.1618	0.0000
2	1.3499	0.7408	2.1618	0.4626	1.6015	0.6244	0.4626
3	1.5683	0.6376	3.5117	0.2848	2.2392	0.4466	0.9004
4	1.8221	0.5488	5.0800	0.1968	2.7880	0.3587	1.3137
5	2.1170	0.4724	6.9021	0.1449	3.2603	0.3067	1.7029
6	2.4596	0.4066	9.0191	0.1109	3.6669	0.2727	2.0685
7	2.8576	0.3499	11.4788	0.0871	4.0169	0.2490	2.4110
8	3.3201	0.3012	14.3364	0.0698	4.3180	0.2316	2.7311
9	3.8574	0.2592	17.6565	0.0566	4.5773	0.2185	3.0295
10	4.4817	0.2231	21.5140	0.0465	4.8004	0.2083	3.3070
11	5.2070	0.1920	25.9957	0.0385	4.9925	0.2003	3.5645
12	6.0496	0.1653	31.2027	0.0320	5.1578	0.1939	3.6028
13	7.0287	0.1423	37.2523	0.0268	5.3000	0.1887	4.0228
14	8.1662	0.1225	44.2810	0.0226	5.4225	0.1844	4.2255
15	9.4877	0.1054	52.4472	0.0191	5.5279	0.1809	4.4119
16	11.0232	0.0907	61.9350	0.0161	5.6186	0.1780	4.5829
17	12.8071	0.0781	72.9582	0.0137	5.6967	0.1755	4.7394
18	14.8797	0.0672	85.7653	0.0117	5.7639	0.1735	4.8823
19	17.2878	0.0578	100.6450	0.0099	5.8217	0.1718	5.0126
20	20.0855	0.0498	117.9328	0.0085	5.8715	0.1703	5.1312
21	23.3360	0.0429	138.0184	0.0072	5.9144	0.1691	5.2390
22	27.1126	0.0369	161.3545	0.0062	5.9513	0.1680	5.3367
23	31.5004	0.0317	188.4672	0.0053	5.9830	0.1671	5.4251
24	36.5982	0.0273	219.9677	0.0045	6.0103	0.1664	5.5050
25	42.5210	0.0235	256.5657	0.0039	6.0339	0.1657	5.5771
26	49.4024	0.0202	299.0869	0.0033	6.0541	0.1652	5.6420
27	57.3974	0.0174	348.4893	0.0029	6.0715	0.1647	5.7004
28	66.6863	0.0150	405.8870	0.0025	6.0865	0.1643	5.7529
29	77.4784	0.0129	472.5735	0.0021	6.0994	0.1639	5.8000
30	90.0170	0.0111	550.0520	0.0018	6.1105	0.1637	5.8421
31	104.5849	0.0096	640.0693	0.0016	6.1201	0.1634	5.8799
32	121.5102	0.0082	744.6546	0.0013	6.1283	0.1632	5.9136
33	141.1748	0.0071	866.1650	0.0012	6.1354	0.1630	5.9437
34	164.0217	0.0061	1007.3403	0.0010	6.1415	0.1628	5.9706
35	190.5660	0.0052	1171.3626	0.0009	6.1468	0.1627	5.9045
40	403.4280	0.0025	2486.6751	0.0004	6.1639	0.1622	6.0798
45	854.0574	0.0012	5271.1914	0.0002	6.1719	0.1620	6.1264
50	1808.0383	0.0006	11166.0195	0.0001	6.1758	0.1619	6.1515

Table A-40
CONTINUOUS COMPOUNDING: $r = 20\%$

	Single Payment		Uniform Series				Gradient Series
	Compound Amount Factor	Present Worth Factor	Compound Amount Factor	Sinking Fund Factor	Present Worth Factor	Capital Recovery Factor	Uniform Series Factor
n	To Find F Given P FPr,n	To Find P Given F PFr,n	To Find F Given A FAr,n	To Find A Given F AFr,n	To Find P Given A PAr,n	To Find A Given P APr,n	To Find A Given G AGr,n
1	1.2214	0.8187	1.0000	1.0000	0.8187	1.2214	0.0000
2	1.4918	0.6703	2.2214	0.4502	1.4891	0.6716	0.4502
3	1.8221	0.5488	3.7132	0.2693	2.0379	0.4907	0.8675
4	2.2255	0.4493	5.5354	0.1807	2.4872	0.4021	1.2528
5	2.7183	0.3679	7.7609	0.1289	2.8551	0.3503	1.6068
6	3.3201	0.3012	10.4792	0.0954	3.1563	0.3168	1.9306
7	4.0552	0.2466	13.7993	0.0725	3.4029	0.2939	2.2255
8	4.9530	0.2019	17.8545	0.0560	3.6048	0.2774	2.4929
9	6.0496	0.1653	22.8076	0.0438	3.7701	0.2652	2.7344
10	7.3891	0.1353	28.8572	0.0347	3.9054	0.2561	2.9515
11	9.0250	0.1108	36.2463	0.0276	4.0162	0.2490	3.1459
12	11.0232	0.0907	45.2713	0.0221	4.1069	0.2435	3.3194
13	13.4637	0.0743	56.2945	0.0178	4.1812	0.2392	3.4736
14	16.4446	0.0608	69.7583	0.0143	4.2420	0.2357	3.6102
15	20.0855	0.0498	86.2030	0.0116	4.2918	0.2330	3.7307
16	24.5325	0.0408	106.2885	0.0094	4.3326	0.2308	3.8367
17	29.9641	0.0334	130.8211	0.0076	4.3659	0.2290	3.9297
18	36.5982	0.0273	160.7853	0.0062	4.3933	0.2276	4.0110
19	44.7012	0.0224	197.3835	0.0051	4.4156	0.2265	4.0819
20	54.5981	0.0183	242.0849	0.0041	4.4339	0.2255	4.1435
21	66.6863	0.0150	296.6829	0.0034	4.4489	0.2248	4.1970
22	81.4508	0.0123	363.3694	0.0028	4.4612	0.2242	4.2432
23	99.4843	0.0101	444.8203	0.0022	4.4713	0.2237	4.2831
24	121.5103	0.0082	544.3049	0.0018	4.4795	0.2232	4.3175
25	148.4130	0.0067	665.8154	0.0015	4.4862	0.2229	4.3471
26	181.2721	0.0055	814.2290	0.0012	4.4918	0.2226	4.3724
27	221.4062	0.0045	995.5017	0.0010	4.4963	0.2224	4.3942
28	270.4260	0.0037	1216.9085	0.0008	4.5000	0.2222	4.4127
29	330.2991	0.0030	1487.3352	0.0007	4.5030	0.2221	4.4286
30	403.4282	0.0025	1817.6350	0.0006	4.5055	0.2220	4.4421
31	492.7483	0.0020	2221.0640	0.0005	4.5075	0.2219	4.4536
32	601.8442	0.0017	2713.8142	0.0004	4.5092	0.2218	4.4634
33	735.0945	0.0014	3315.6607	0.0003	4.5105	0.2217	4.4717
34	897.8464	0.0011	4050.7566	0.0002	4.5116	0.2216	4.4767
35	1096.6321	0.0009	4948.6016	0.0002	4.5125	0.2216	4.4847
40	2980.9551	0.0003	13459.4649	0.0001	4.5152	0.2215	4.5032
45	8103.0703	0.0001	36594.3516	0.0000	4.5161	0.2214	4.5111
50	22026.4414	0.0000	99481.5000	0.0000	4.5165	0.2214	4.5144

Reprinted, by permission, from J. A. White et al., *Principles of Engineering Economic Analysis*, pp. 396–447.

Table A-41

CONTINUOUS COMPOUNDING: $r = 25\%$

	Single Payment		Uniform Series				Gradient Series
	Compound Amount Factor	Present Worth Factor	Compound Amount Factor	Sinking Fund Factor	Present Worth Factor	Capital Recovery Factor	Uniform Series Factor
n	To Find *F* Given *P* *FPr,n*	To Find *P* Given *F* *PFr,n*	To Find *F* Given *A* *FAr,n*	To Find *A* Given *F* *AFr,n*	To Find *P* Given *A* *PAr,n*	To Find *A* Given *P* *APr,n*	To Find *A* Given *G* *AGr,n*
1	1.2840	0.7788	1.0000	1.0000	0.7788	1.2840	0.0000
2	1.6487	0.6065	2.2840	0.4378	1.3853	0.7218	0.4378
3	2.1170	0.4724	3.9327	0.2543	1.8577	0.5383	0.8350
4	2.7183	0.3679	6.0498	0.1653	2.2256	0.4493	1.1929
5	3.4903	0.2865	8.7680	0.1141	2.5121	0.3981	1.5131
6	4.4817	0.2231	12.2584	0.0816	2.7352	0.3656	1.7975
7	5.7546	0.1738	16.7401	0.0597	2.9090	0.3438	2.0486
8	7.3891	0.1353	22.4947	0.0445	3.0443	0.3285	2.2687
9	9.4877	0.1054	29.8837	0.0335	3.1497	0.3175	2.4605
10	12.1825	0.0821	39.3715	0.0254	3.2318	0.3094	2.6266
11	15.6426	0.0639	51.5540	0.0194	3.2957	0.3034	2.7696
12	20.0855	0.0498	67.1966	0.0149	3.3455	0.2989	2.8921
13	25.7903	0.0388	87.2821	0.0115	3.3843	0.2955	2.9964
14	33.1154	0.0302	113.0725	0.0088	3.4145	0.2929	3.0849
15	42.5211	0.0235	146.1879	0.0068	3.4380	0.2909	3.1595
16	54.5981	0.0183	188.7090	0.0053	3.4563	0.2893	3.2223
17	70.1054	0.0143	243.3072	0.0041	3.4706	0.2881	3.2748
18	90.0171	0.0111	313.4124	0.0032	3.4817	0.2872	3.3186
19	115.5842	0.087	403.4295	0.0025	3.4904	0.2865	3.3550
20	148.4131	0.0067	519.0139	0.0019	3.4971	0.2860	3.3851
21	190.5662	0.0052	667.4273	0.0015	3.5023	0.2855	3.4100
22	244.6918	0.0041	857.9934	0.0012	3.5064	0.2852	3.4305
23	314.1902	0.0032	1102.6851	0.0009	3.5096	0.2849	3.4474
24	403.4285	0.0025	1416.8765	0.0007	3.5121	0.2847	3.4612
25	518.0125	0.0019	1820.3054	0.0005	3.5140	0.2846	3.4725
26	665.1411	0.0015	2338.3167	0.0004	3.5155	0.2845	3.4817
27	854.0581	0.0012	3003.4590	0.0003	3.5167	0.2844	3.4892
28	1096.6326	0.0009	3857.5183	0.0003	3.5176	0.2843	3.4953
29	1408.1043	0.0007	4954.1484	0.0002	3.5183	0.2842	3.5002
30	1808.0415	0.0006	6362.2539	0.0002	3.5189	0.2842	3.5042
31	2321.5711	0.0004	8170.2969	0.0001	3.5193	0.2841	3.5075
32	2980.9568	0.0003	10491.8711	0.0001	3.5196	0.2841	3.5101
33	3827.6238	0.0003	13472.8281	0.0001	3.5199	0.2841	3.5122
34	4914.7617	0.0002	17300.4492	0.0001	3.5201	0.2841	3.5139
35	6310.6797	0.0002	22215.2188	0.0000	3.5203	0.2841	3.5153

Table A-42
CONTINUOUS COMPOUNDING: $r = 30\%$

n	Single Payment: Compound Amount Factor, To Find F Given P, FPr,n	Single Payment: Present Worth Factor, To Find P Given F, PFr,n	Uniform Series: Compound Amount Factor, To Find F Given A, FAr,n	Uniform Series: Sinking Fund Factor, To Find A Given F, AFr,n	Uniform Series: Present Worth Factor, To Find P Given A, PAr,n	Uniform Series: Capital Recovery Factor, To Find A Given P, APr,n	Gradient Series: Uniform Series Factor, To Find A Given G, AGr,n
1	1.3499	0.7408	1.0000	1.0000	0.7408	1.3499	0.0000
2	1.8221	0.5488	2.3499	0.4256	1.2896	0.7754	0.4256
3	2.4596	0.4066	4.1720	0.2397	1.6962	0.5896	0.8029
4	3.3201	0.3012	6.6316	0.1508	1.9974	0.5007	1.1342
5	4.4817	0.2231	9.9517	0.1005	2.2205	0.4503	1.4222
6	6.0496	0.1653	14.4334	0.0693	2.3858	0.4191	1.6701
7	8.1662	0.1225	20.4830	0.0488	2.5083	0.3987	1.8815
8	11.0232	0.0907	28.6492	0.0349	2.5990	0.3848	2.0601
9	14.8797	0.0672	39.6724	0.0252	2.6662	0.3751	2.2099
10	20.0855	0.0498	54.5522	0.0183	2.7169	0.3682	2.3343
11	27.1126	0.0369	74.6377	0.0134	2.7529	0.3633	2.4370
12	36.5982	0.0273	101.7504	0.0098	2.7802	0.3597	2.5212
13	49.4024	0.0202	138.3486	0.0072	2.8004	0.3571	2.5897
14	66.6863	0.0150	187.7510	0.0053	2.8154	0.3552	2.6452
15	90.0170	0.0111	254.4374	0.0039	2.8265	0.3538	2.6898
16	121.5102	0.0082	344.4544	0.0029	2.8348	0.3528	2.7255
17	164.0217	0.0061	465.9646	0.0021	2.8409	0.3520	2.7540
18	221.4061	0.0045	629.9866	0.0016	2.8454	0.3514	2.7766
19	298.8667	0.0033	851.3928	0.0012	2.8487	0.3510	2.7945
20	403.4280	0.0025	1150.2603	0.0009	2.8512	0.3507	2.8086
21	544.5708	0.0018	1553.6890	0.0006	2.8531	0.3505	2.8197
22	735.0940	0.0014	2098.2608	0.0005	2.8544	0.3503	2.8283
23	992.2730	0.0010	2833.3550	0.0004	2.8554	0.3502	2.8351
24	1339.4285	0.0007	3825.6804	0.0003	2.8562	0.3501	2.8404
25	1808.0383	0.0006	5165.0547	0.0002	2.8567	0.3501	2.8445
26	2440.5967	0.0004	6973.0977	0.0001	2.8571	0.3500	2.8476
27	3294.4622	0.0003	9413.6992	0.0001	2.8574	0.3500	2.8501
28	4447.0547	0.0002	12708.1641	0.0001	2.8577	0.3499	2.8520
29	6002.8945	0.0002	17165.2227	0.0001	2.8578	0.3499	2.8535
30	8103.0625	0.0001	23158.1328	0.0000	2.8579	0.3499	2.8546
31	10937.9922	0.0001	31261.2070	0.0000	2.8580	0.3499	2.8555
32	14764.7500	0.0001	42199.2188	0.0000	2.8581	0.3499	2.8561
33	19930.3281	0.0001	56963.9922	0.0000	2.8582	0.3499	2.8566
34	26903.1133	0.0000	76894.2500	0.0000	2.8582	0.3499	2.8570
35	36315.4141	0.0000	103797.4375	0.0000	2.8582	0.3499	2.8573

SECTION 3 FUNDS FLOW CONVERSION FACTOR

Mathematical Expression

Funds flow conversion factor $= A\bar{A}r = \dfrac{e^r - 1}{r}$

Table A-43
FUNDS FLOW CONVERSION FACTOR

r	$A\bar{A}r$
1	1.005020
2	1.010065
3	1.015150
4	1.020270
5	1.025422
6	1.030608
7	1.035831
8	1.041088
9	1.046381
10	1.051709
11	1.057073
12	1.062474
13	1.067910
14	1.073384
15	1.078894
16	1.084443
17	1.090028
18	1.095652
19	1.101313
20	1.107014
21	1.112752
22	1.118530
23	1.124347
24	1.130204
25	1.136101
26	1.142038
27	1.148016
28	1.154035
29	1.160094
30	1.166196
31	1.172339
32	1.178524
33	1.184751
34	1.191022
35	1.197335

APPENDIX B
INTERNAL RATE OF RETURN

This appendix provides a computer program for determining the internal rate of return. It prints the error message ''no single feasible solution exists'' in the case of multiple rates of return. The input data consists of the number of years (N) and the yearly cash flows. The periods used in this program are beginning of the year periods. Some examples are provided.

```
$JOB
C
C
C
C        INTERNAL RATE OF RETURN
         DIMENSION PV(50),CF(50)
         READ(5,10) N,(CF(I),I=1,N)
      10 FORMAT(I5,/,(F15.2))
         WRITE(6,20) (I,CF(I),I=1,N)
      20 FORMAT(1H1,4X,'PERIOD',6X,'      CASH FLOW ',//,(/,4X,I6,6X,F15.2))
         WRITE(6,30)
      30 FORMAT(////,1X,'ITERATION',6X,'  RATE OF RETURN',//)
         ERROR=CF(1)*.00001
         ITER=1
         RATE=1.
         IFLAG=1
         PV(1)=CF(1)
         IF(PV(1)) 1,2,2
       1 DO 40 J=2,N
         PV(J)=PV(J-1)*(1.0+RATE)+CF(J)
         IF(PV(J)+ERROR) 40,40,50
      40 CONTINUE
         WRITE(6,60) ITER,RATE
      60 FORMAT(/,1X,I9,6X,F16.5)
         ITER=ITER+1
         IF(PV(N)-ERROR) 35,55,55
      35 IFLAG=0
         M=N
      11 ARATE=PV(1)
         MK=M-1
         ML=MK-1
         IF(ML) 45,15,45
      45 DO 17 I=2,MK
      17 ARATE=ARATE*(1.+RATE)+PV(I)
      15 RATE=RATE-PV(M)/ARATE
         GO TO 1
      50 IF(IFLAG) 55,19,46
      46 M=J
         GO TO 11
       2 J=1
      19 WRITE(6,21)
      21 FORMAT(///,2X,'NO SINGLE FEASIBLE SOLUTION EXISTS.')
      55 WRITE(6,18)
      18 FORMAT(1H1)
         STOP
         END

$ENTRY
```

PERIOD	CASH FLOW
1	-5000.00
2	885.00
3	885.00
4	885.00
5	885.00
6	885.00
7	885.00
8	885.00
9	885.00
10	885.00
11	885.00

ITERATION	RATE OF RETURN
1	1.00000
2	0.80820
3	0.63790
4	0.48830
5	0.35975
6	0.25444
7	0.17725
8	0.13396
9	0.12104
10	0.12003
11	0.12002

```
PERIOD              CASH FLOW

     1               -1000.00

     2                6000.00

     3              -11000.00

     4                6000.00


ITERATION          RATE OF RETURN


     1                  5.00000

NO SINGLE FEASIBLE SOLUTION EXISTS.
```

APPENDIX C
MINIMUM ANNUAL REVENUE REQUIREMENTS: FLOW-THROUGH METHOD

This appendix gives a computer program for generating the minimum annual revenue requirements using the flow-through method. Three depreciation models are possible with this program: (1) straight-line depreciation, (2) sum of the years-digits, and (3) double declining balance depreciation with a switch to straight-line depreciation. An example of the output from this program is also presented.

```
      K=NX
      GO TO 117
 1562 N=NX
      K=NY
      DO 200 I=1,N
      IF(I.LE.K) GO TO 220
      GO TO 210
  220 DEP(I)=2.*(NY-I+1.)/NY/(NY+1.)*(INC-SALV)
      GO TO 200
  210 DEP(I)=0.
  200 CONTINUE
      GO TO 230
  117 DO 201 I=1,N
      DEP(I)=2.*(NY-I+1.)/NY/(NY+1.)*(INC-SALV)
  201 CONTINUE
  230 RETURN
      END

$ENTRY
```

YEAR	BOOK DEP	S.L DEP	TAX DEP	TAX CREDIT
1	14000.00	14000.00	25454.54	15000.00
2	14000.00	14000.00	22909.07	0.00
3	14000.00	14000.00	20363.63	0.00
4	14000.00	14000.00	17818.17	0.00
5	14000.00	14000.00	15272.72	0.00
6	14000.00	14000.00	12727.27	0.00
7	14000.00	14000.00	10181.82	0.00
8	14000.00	14000.00	7636.36	0.00
9	14000.00	14000.00	5090.91	0.00
10	14000.00	14000.00	2545.45	0.00

YEAR	CAP INV	CAP INT	RET ON DEBT	RET ON EQTY
	150000.00			
1	136000.00	0.00	5400.00	17100.00
2	122000.00	0.00	4896.00	15504.00
3	108000.00	0.00	4392.00	13908.00
4	94000.00	0.00	3888.00	12312.00
5	80000.00	0.00	3384.00	10716.00
6	66000.00	0.00	2880.00	9120.00
7	52000.00	0.00	2376.00	7524.00
8	38000.00	0.00	1872.00	5928.00
9	24000.00	0.00	1368.00	4332.00
10	10000.00	0.00	864.00	2736.00

YEAR	TAX	OPRT.COST	ANNUAL COST OR MIN ANN REV
1	-25134.08	40000.00	51365.91
2	7144.50	40000.00	81544.44
3	8173.06	40000.00	80473.00
4	9201.64	40000.00	79401.63
5	10230.21	40000.00	78330.19
6	11258.79	40000.00	77258.75
7	12287.36	40000.00	76187.31
8	13315.93	40000.00	75115.88
9	14344.51	40000.00	74044.50
10	15373.08	40000.00	72973.06

MIN. ANNUAL REV. REQMT BY FLOW THROUGH METHOD

INPUT DATA;

INITIAL INVESTMENT COST= 150000.00   SALVAGE VALUE= 10000.00
NO. OF USEFUL LIFE FOR BOOK PURPOSE= 10.
NO. OF USEFUL LIFE FOR TAX PURPOSE= 10.
RATE OF RETURN ON EQUITY=0.190   RATE OF RETURN ON DEBT=0.090
DEBT RATIO=0.400   TAX RATE=0.520
CAPITAL INTEREST RATE=0.000   INVESTMENT TAX CREDIT RATE=0.100
TAX SALVAGE VALUE= 10000.00

BOOK DEP. METHOD ; STRAIGHT LINE DEP.

TAX DEP. METHOD ; SUM OF THE YEARS DIGITS DEP.

COST DATA;

40000.0	40000.0	40000.0	40000.0	40000.0
40000.0	40000.0	40000.0	40000.0	40000.0

APPENDIX D
MINIMUM ANNUAL REVENUE REQUIREMENTS: NORMALIZING METHOD

This appendix gives a computer program for generating minimum annual revenue requirements using the normalizing method. This program provides for the same depreciation models mentioned in Appendix C. An example of the output from this program is also presented.

```
$JOB
C       MINIMUM ANNUAL REVENUE REQUIREMENT BY NORMALIZING METHOD
        REAL INITC,INVTCR,NB,NT,INC
        DIMENSION DEPB(50),DEPS(50),DEPT(50),DTX(50),TXC(50),AMTXC(50),
       1CAPINV(50),FD(50),FE(50),TAX(50),COST(50),RN(50),CAPINT(50)
       2,SDEPT(50)
C       INPUT DATA ENTERED AS FOLLOW;
C       "INITC" REPRESENTS INITIAL INVESTMENT COST.
        READ(5,5000) INITC
C       "SALV" REPRESENTS THE SALVAGE VALUE AT THE END OF THE USEFUL LIFE.
        READ(5,5000) SALV
C       "TXSALV" REPRESENTS THE ESTIMATED SALVAGE VALUE FOR TAX PURPOSE.
        READ(5,5000) TXSALV
C       "NB" REPRESENTS THE USEFUL LIFE FOR BOOK PURPOSE.
        READ(5,5000) NB
C       "NT" REPRESENTS THE USEFUL LIFE FOR TAX PURPOSE.
        READ(5,5000) NT
C       "RE" REPRESENTS THE RATE OF RETURN ON EQUITY.
        READ(5,5000) RE
C       "RD" REPRESENTS THE RATE OF RETURN ON DEBT.
        READ(5,5000) RD
C       "DR" REPRESENTS THE RATIO OF DEBT PORTION.
        READ(5,5000) DR
C       "T" REPRESENTS THE TAX RATE.
        READ(5,5000) T
C       "CAPIRA" REPRESENTS THE CAPITAL INTEREST RATE.
        READ(5,5000) CAPIRA
C       "INVTCR" REPRESENTS THE INVESTMENT TAX CREDIT RATE.
        READ(5,5000) INVTCR
C       "BDEP" REPRESENTS THE BOOK DEPRECIATION METHOD.
        READ(5,5000) BDEP
C       "TDEP" REPRESENTS THE TAX DEPRECIATION METHOD.
        READ(5,5000) TDEP
        IF(NB.GE.NT) GO TO 31
        N=NT
        K=NB
        GO TO 32
     31 N=NB
        K=NT
     32 READ(5,5000) (COST(I),I=1,N)
        PINT=INITC*CAPIRA
        CAP=INITC+PINT
        INT=PINT
        CINT=INT
        DELTA=PINT-CINT
        IF(DELTA-0.9) 2,1,1
      1 CAP=CAP+1.0
      2 CALL SLDEP(DEPS,INITC,SALV,NB,NT)
C       CHOOSE APPROPRIATE DEPRECIATION METHOD FOR BOOK AND TAX PURPOSES
C       FROM THE SUBROUTINES.
C       SUBROUTINE "SLDEP" REPRESENTS THE STRAIGHT LINE DEPRECIATION
C       METHOD.( BDEP OR TDEP AS 1.)
C       SUBROUTINE "SYDEP" REPRESENTS THE SUM-OF-THE -YEARS-DIGITS METHOD.
C       ( BDEP OR TDEP AS 2.)
C       SUBROUTINE "DDBDEP" REPRESENTS THE DOUBLE DECLINING BALANCE
C       DEPRECIATION METHOD.( BDEP OR TDEP AS 3.)
C       "DEPB" REPRESENTS BOOK DEPRECIATION AMOUNT FOR A PARTICULAR YEAR
C       "DEPS" REPRESENTS THE STRAIGHT LINE DEPRECIATION AMOUNT FOR A
C       PARTICULAR YEAR.
C       "DEPT" REPRESENTS TAX DEPRECIATION AMOUNT FOR A PARTICULAR
C       YEAR.
```

```
C       "CAP" REPRESENTS CAPITALIZED INVESTMENT COST FOR NORMALIZING
C       METHOD
C       THIS "CAP" ALSO REPRESENTS INITIAL INVESTMENT COST FOR
C       FLOW THROUGH METHOD.
        IBDEP=BDEP
        ITDEP=TDEP
        GO TO (111,222,333),IBDEP
   111  CALL SLDEP(DEPB,CAP,SALV,NB,NT)
        GO TO 1234
   222  CALL SYDDEP(DEPB,CAP,SALV,NB,NT)
        GO TO 1234
   333  CALL DDBDEP(DEPB,CAP,SALV,NB,NT)
  1234  GO TO (1111,2222,3333),ITDEP
  1111  CALL SLDEP(DEPT,INITC,TXSALV,NB,NT)
        GO TO 4321
  2222  CALL SYDDEP(DEPT,INITC,TXSALV,NB,NT)
        GO TO 4321
  3333  CALL DDBDEP(DEPT,INITC,TXSALV,NB,NT)
  4321  TXC(1)=INITC*INVTCR
        DO 200 I=2,N
        TXC(I)=0.
   200  CAPINT(I)=0.
        CAPINT(1)=CAP-INITC
        DO 500 I=1,N
        AMTXC(I)=TXC(1)/NB
        IF(I.EQ.1) GO TO 300
        XCAP=CAPINV(I-1)
        GO TO 400
   300  XCAP=CAP
   400  CONTINUE
        DTX(I)=(DEPT(I)-DEPS(I))*T
        CAPINV(I)=XCAP-DEPB(I)-TXC(I)-DTX(I)+AMTXC(I)
        FD(I)=XCAP*DR*RD
        FE(I)=XCAP*(1.-DR)*RE
        TAX(I)=(T/(1.-T))*(FE(I)+DEPB(I)+DTX(I)+TXC(I)-AMTXC(I)-DEPT(I)
       1-CAPINT(I))-TXC(I)/(1.-T)
        RN(I)=DEPB(I)+DTX(I)+TXC(I)-AMTXC(I)+FD(I)+FE(I)+TAX(I)+COST(I)
   500  CONTINUE
        IF(NT.GT.NB) GO TO 663
        GO TO 670
   663  SDEPT(1)=DEPT(1)
        DO 655 I=2,K
        SDEPT(I)=DEPT(I)+SDEPT(I-1)
        TXBV=INITC-SDEPT(I)
   655  CONTINUE
        TAX(K)=(T/(1.-T))*(FE(K)+DEPB(K)+DTX(K)+TXC(K)-AMTXC(K)-DEPT(K)
       1-CAPINT(K))-TXC(K)/(1.-T)+(SALV-TXBV)*(T/(1.-T))
        RN(K)=DEPB(K)+DTX(K)+TXC(K)-AMTXC(K)+FD(K)+FE(K)+TAX(K)+COST(K)
        M=NB+1
        DO 665 I=M,N
        CAPINV(I)=0.0
        FD(I)=0.0
        FE(I)=0.0
        TAX(I)=0.0
        RN(I)=0.0
   665  CONTINUE
   670  WRITE(6,6000)
        DO 700 I=1,N
        WRITE(6,6100) I,DEPB(I),DEPS(I),DEPT(I),DTX(I)
   700  CONTINUE
        WRITE(6,6200) CAP
```

```
      DO 800 I=1,N
      WRITE(6,6100) I,TXC(I),AMTXC(I),CAPINV(I),CAPINT(I)
  800 CONTINUE
 5000 FORMAT(F10.0)
 6100 FORMAT(I5,8F12.2)
      WRITE(6,6201)
      DO 801 I=1,N
      WRITE(6,6100) I,FD(I),FE(I),TAX(I),COST(I),RN(I)
  801 CONTINUE
 6000 FORMAT(1H1,' YEAR   BOOK DEP      S.L.DEP    TAX DEP DEFERRED TAX')
 6201 FORMAT(53X,'ANNUAL COST OR',/,1X,'YEAR RET ON DEBT  RET ON EQTY
     1  TAX      OPRT.COST MIN ANN REV')
 6200 FORMAT(/,1X,'YEAR  TAX CREDIT  AMT TAX CR    CAP INV     CAP INT',
     1/,29X,F12.2)
      WRITE(6,3900)
 3900 FORMAT(/,10X,'MIN. ANNUAL REV. REQMT BY NORMALIZING METHOD')
      WRITE(6,4000) INITC,SALV,NB,NT,RE,RD,DR,T,CAPIRA,INVTCR,TXSALV
 4000 FORMAT(///,1X,'INPUT DATA;',//,1X,'INITIAL INVESTMENT COST=',F10.2
     1,5X,'SALVAGE VALUE=',F10.2,/,1X,'NO. OF USEFUL LIFE FOR BOOK PURPO
     1SE=',F5.0,/,1X,'NO. OF USEFUL LIFE FOR TAX PURPOSE=',F5.0,/,1X,
     2'RATE OF RETURN ON EQUITY=',F5.3,3X,'RATE OF RETURN ON DEBT=',
     3F5.3,/,1X,'DEBT RATIO=',F5.3,5X,'TAX RATE=',F5.3,5X,/,1X,
     4'CAPITAL INTEREST RATE=',F5.3,5X,'INVESTMENT TAX CREDIT RATE='
     5,F5.3,5X,/,1X,'TAX SALVAGE VALUE=',F10.2)
      GO TO (555,666,777),IBDEP
  555 WRITE(6,8000)
 8000 FORMAT(/,1X,'BOOK DEP. METHOD ; STRAIGHT LINE DEP.')
      GO TO 8301
  666 WRITE(6,8100)
 8100 FORMAT(/,1X,'BOOK DEP. METHOD ; SUM OF THE YEARS DIGITS DEP.')
      GO TO 8301
  777 WRITE(6,8200)
 8200 FORMAT(/,1X,'BOOK DEP. METHOD ; DOUBLE DECLINING BALANCE DEP.')
 8301 GO TO(999,805,705),ITDEP
  999 WRITE(6,8400)
 8400 FORMAT(/,1X,'TAX DEP. METHOD ; STRAIGHT LINE DEP.')
      GO TO 8402
  805 WRITE(6,8500)
 8500 FORMAT(/,1X,'TAX DEP. METHOD ; SUM OF THE YEARS DIGITS DEP.')
      GO TO 8402
  705 WRITE(6,8600)
 8600 FORMAT(/,1X,'TAX DEP. METHOD ; DOUBLE DECLINING BALANCE DEP.')
 8402 WRITE(6,4500) (COST(I),I=1,N)
 4500 FORMAT(/,2X,'COST DATA;',/,(2X,5(F13.1),/))
      STOP
      END

      SUBROUTINE DDBDEP(DEP,INC,SALV,NX,NY)
      REAL INC,NX,NY
      DIMENSION DEP(50),BV(50),RBVSL(50),SDEP(50)
      IF(NX.GE.NY) GO TO 1561
      N=NY
      K=NX
      GO TO 116
 1561 N=NX
      K=NY
  116 D=2./NY
      BV(1)=INC*(1.-D)**1.
      DEP(1)=INC-BV(1)
      RBVSL(1)=(INC-SALV)/NY
      SDEP(1)=DEP(1)
```

```
      P=INC-SALV
      M=0
      DO 200 I=2,N
      IF(NY.GT.NX) K=N
      IF(I.LE.K) GO TO 220
      GO TO 210
  220 AI=I
      IF(M.EQ.1) GO TO 205
      BV(I)=INC*(1.-D)**AI
      ADEP=BV(I-1)
      DEP(I)=ADEP-BV(I)
      IF(I.EQ.K) GO TO 900
      RBVSL(I)=(BV(I)-SALV)/(NY-AI)
      IF(DEP(I).LE.RBVSL(I-1)) GO TO 600
      IF(M.EQ.0) DEP(I)=ADEP-BV(I)
      GO TO 250
  205 IF(M.EQ.1) DEP(I)=RBVSL(II)
      GO TO 250
  600 IF(M.EQ.1) GO TO 205
      M=1
      II=I-1
      GO TO 205
  900 SDEP(I)=DEP(I)+SDEP(I-1)
      IF(SDEP(I).LT.P) DEP(I)=P-(INC-SDEP(I))
      DEP(I)=INC-SDEP(I-1)-SALV
      GO TO 200
  210 DEP(I)=0.
      GO TO 200
  250 SDEP(I)=DEP(I)+SDEP(I-1)
      IF(SDEP(I).LT.P) GO TO 200
      DEP(I)=P-SDEP(I-1)
      JI=I+1
      DO 255 L=JI,N
  255 DEP(L)=0.
      GO TO 70
  200 CONTINUE
   70 RETURN
      END

      SUBROUTINE SLDEP(DEP,INC,SALV,NX,NY)
      REAL INC,NX,NY
      DIMENSION DEP(50)
      IF(NX.GE.NY) GO TO 1564
      N=NY
      K=NX
      GO TO 115
 1564 N=NX
      K=NY
  115 DO 200 I=1,N
      IF(I.GT.NX) GO TO 210
      DEP(I)=(INC-SALV)/NX
      GO TO 200
  210 DEP(I)=0.0
  200 CONTINUE
      RETURN
      END

      SUBROUTINE SYDDEP(DEP,INC,SALV,NX,NY)
      REAL INC,NX,NY
      DIMENSION DEP(50)
      IF(NX.GE.NY) GO TO 1562
```

```
      N=NY
      K=NX
      GO TO 117
 1562 N=NX
      K=NY
      DO 200 I=1,N
      IF(I.LE.K) GO TO 220
      GO TO 210
  220 DEP(I)=2.*(NY-I+1.)/NY/(NY+1.)*(INC-SALV)
      GO TO 200
  210 DEP(I)=0.
  200 CONTINUE
      GO TO 230
  117 DO 201 I=1,N
      DEP(I)=2.*(NY-I+1.)/NY/(NY+1.)*(INC-SALV)
  201 CONTINUE
  230 RETURN
      END

$ENTRY
```

YEAR	BOOK DEP	S.L.DEP	TAX DEP	DEFERRED TAX
1	142000.00	130000.00	247619.00	61161.88
2	142000.00	130000.00	235238.00	54723.76
3	142000.00	130000.00	222856.90	48285.61
4	142000.00	130000.00	210476.00	41847.55
5	142000.00	130000.00	198095.00	35409.43
6	142000.00	130000.00	185714.10	28971.38
7	142000.00	130000.00	173333.10	22533.22
8	142000.00	130000.00	160952.20	16095.17
9	142000.00	130000.00	148571.30	9657.11
10	142000.00	130000.00	136190.30	3218.99
11	142000.00	130000.00	123809.50	-3219.06
12	142000.00	130000.00	111428.50	-9657.14
13	142000.00	130000.00	99047.56	-16095.27
14	142000.00	130000.00	86666.63	-22533.35
15	142000.00	130000.00	74285.69	-28971.44
16	142000.00	130000.00	61904.76	-35409.50
17	142000.00	130000.00	49523.80	-41847.61
18	142000.00	130000.00	37142.84	-48285.70
19	142000.00	130000.00	24761.90	-54723.79
20	142000.00	130000.00	12380.94	-61161.88

YEAR	TAX CREDIT	AMT TAX CR	CAP INV	CAP INT
			3240000.00	
1	119999.90	6000.00	2922837.00	240000.00
2	0.00	6000.00	2732112.00	0.00
3	0.00	6000.00	2547825.00	0.00
4	0.00	6000.00	2369976.00	0.00
5	0.00	6000.00	2198565.00	0.00
6	0.00	6000.00	2033592.00	0.00
7	0.00	6000.00	1875057.00	0.00
8	0.00	6000.00	1722960.00	0.00
9	0.00	6000.00	1577301.00	0.00
10	0.00	6000.00	1438081.00	0.00
11	0.00	6000.00	1305299.00	0.00
12	0.00	6000.00	1178955.00	0.00
13	0.00	6000.00	1059049.00	0.00
14	0.00	6000.00	945582.20	0.00
15	0.00	6000.00	838553.60	0.00
16	0.00	6000.00	737963.00	0.00
17	0.00	6000.00	643810.50	0.00
18	0.00	6000.00	556096.10	0.00
19	0.00	6000.00	474819.80	0.00
20	0.00	6000.00	399981.60	0.00

YEAR	RET ON DEBT	RET ON EQTY	TAX	OPRT.COST	ANNUAL COST OR MIN ANN REV
1	194400.00	194399.80	-224061.90	100000.00	581899.60
2	175370.10	175370.00	141760.30	100000.00	783224.30
3	163926.60	163926.50	135801.30	100000.00	747940.10
4	152869.50	152869.40	130260.80	100000.00	713847.30
5	142198.50	142198.50	125138.80	100000.00	680945.10
6	131913.80	131913.80	120435.10	100000.00	649234.20
7	122015.50	122015.40	116150.00	100000.00	618714.10
8	112503.30	112503.30	112283.30	100000.00	589385.20
9	103377.50	103377.50	108835.10	100000.00	561247.30
10	94638.06	94638.00	105805.30	100000.00	534300.30
11	86284.88	86284.81	103194.10	100000.00	508544.80
12	78317.94	78317.88	101001.40	100000.00	483980.00
13	70737.31	70737.25	99227.19	100000.00	460606.40
14	63542.95	63542.92	97871.38	100000.00	438423.80
15	56734.94	56734.92	96934.13	100000.00	417432.40

16	50313.23	50313.20	96415.38	100000.00	397632.20
17	44277.79	44277.76	96315.13	100000.00	379023.00
18	38628.64	38628.61	96633.19	100000.00	361604.60
19	33365.77	33365.75	97369.94	100000.00	345377.60
20	28489.19	28489.18	98525.06	100000.00	330341.50

MIN. ANNUAL REV. REQMT BY NORMALIZING METHOD

INPUT DATA;

INITIAL INVESTMENT COST=3000000.00 SALVAGE VALUE= 400000.00
NO. OF USEFUL LIFE FOR BOOK PURPOSE= 20.
NO. OF USEFUL LIFE FOR TAX PURPOSE= 20.
RATE OF RETURN ON EQUITY=0.150 RATE OF RETURN ON DEBT=0.100
DEBT RATIO=0.600 TAX RATE=0.520
CAPITAL INTEREST RATE=0.080 INVESTMENT TAX CREDIT RATE=0.040
TAX SALVAGE VALUE= 400000.00

BOOK DEP. METHOD ; STRAIGHT LINE DEP.

TAX DEP. METHOD ; SUM OF THE YEARS DIGITS DEP.

COST DATA;

100000.0	100000.0	100000.0	100000.0	100000.0
100000.0	100000.0	100000.0	100000.0	100000.0
100000.0	100000.0	100000.0	100000.0	100000.0
100000.0	100000.0	100000.0	100000.0	100000.0

APPENDIX E
ECONOMIC USEFUL LIFE

This appendix provides a computer program for determining the economic useful life using the flow-through method. It allows the use of the three depreciation models given in Appendix C for tax purposes. An example of the output given by this program is also provided.

```
$JOB
C      DETERMINATION OF ECONOMIC USEFUL LIFE.(FLOW THROUGH METHOD)
       DIMENSION DEPT(50),ESTSAL(50),CAPREC(50),FE(50),FD(50),INVTXC(50),
      1TAX(50),COST(50),ANCOST(50),EQANC(50),AEQC(50),CAPINT(50),AMTXC(5
      20)
       REAL INITC,INVTCR,NB,NT,INVTXC
C      INPUT DATA ENTERED AS FOLLOW;
       READ(5,5000) INITC
       READ(5,5000) SALV
       READ(5,5000) NB
       READ(5,5000) NT
       READ(5,5000) RE
       READ(5,5000) RD
       READ(5,5000) DR
       READ(5,5000) T
       READ(5,5000) CAPIRA
       READ(5,5000) INVTCR
       N=NB
       READ(5,5000) (COST(I),I=1,N)
       READ(5,5000) (ESTSAL(I),I=1,N)
       CAP=INITC
C      CALL APPROPRIATE SUBROUTINE FOR TAX DEPRECIATION METHOD.
       CALL SYDDEP(DEPT,CAP,SALV,NB,NT)
       INVTXC(1)=INITC*INVTCR
       DO 200 I=2,N
       INVTXC(I)=0.
   200 CAPINT(I)=0.
       CAPINT(1)=INITC*CAPIRA
       FD(1)=CAP*DR*RD
       FE(1)=CAP*(1.-DR)*RE
       CAPREC(1)=CAP-ESTSAL(1)
       TAX(1)=(CAPREC(1)+FE(1)-DEPT(1))*T/(1.-T)-INVTXC(1)/(1.-T)
       ANCOST(1)=CAPREC(1)+FD(1)+FE(1)+TAX(1)+COST(1)
       DO 500 I=2,N
       AMTXC(I)=INVTXC(1)/NB
       CAPREC(I)=ESTSAL(I-1)-ESTSAL(I)
       FD(I)=ESTSAL(I-1)*DR*RD
       FE(I)=ESTSAL(I-1)*(1.-DR)*RE
       TAX(I)=(CAPREC(I)+FE(I)-DEPT(I))*T/(1.-T)-INVTXC(I)/(1.-T)
       ANCOST(I)=CAPREC(I)+FD(I)+FE(I)+TAX(I)+COST(I)
   500 CONTINUE
       B=RE*(1.-DR)+RD*DR
       AEQC(1)=ANCOST(1)*(1.+B)**(-1)
       EQANC(1)=ANCOST(1)
       DO 450 J=2,N
       AJ=J
       AEQC(J)=AEQC(J-1)+ANCOST(J)*(1.+B)**(-AJ)
       EQANC(J)=AEQC(J)*(B*(1.+B)**AJ)/((1.+B)**AJ-1.)
   450 CONTINUE
       LIFE=1
       SMALL=EQANC(1)
       DO 475 K=2,N
       IF(EQANC(K).LT.SMALL) LIFE=K
       IF(EQANC(K).LT.SMALL) SMALL=EQANC(K)
   475 CONTINUE
       WRITE(6,6200) CAP
       DO 700 I=1,N
       WRITE(6,6100) I,ESTSAL(I),CAPREC(I),DEPT(I),FE(I),FD(I)
   700 CONTINUE
  6200 FORMAT(1H1,5H YEAR,12H SALV VALUE ,12HCAP REC COST,12H   TAX DEP
      1,12H RET ON EQTY,12H  DEBT INT   ,/,6X,F12.2)
```

```
 6100 FORMAT(I5,5F12.2)
 5000 FORMAT(F10.0)
      WRITE(6,6201)
      DO 800 I=1,N
      WRITE(6,6100) I,INVTXC(I),TAX(I),COST(I),ANCOST(I),EQANC(I)
  800 CONTINUE
 6201 FORMAT(/,1X,5H YEAR,12HINV TAX CRED,12H      TAX     ,12H    COST
     1   ,12H  ANN COST   ,14HEQIV ANN COST,//)
      WRITE(6,6300) LIFE
 6300 FORMAT(///,1X,'ECONOMIC USEFUL LIFE = ',I5,'    YEARS')
      WRITE(6,4000) INITC,SALV,NB,NT,RE,RD,DR,T,CAPIRA,INVTCR
 4000 FORMAT(///,1X,'INPUT DATA;',//,1X,'INITIAL INVESTMENT COST=',F10.2
     1,5X,'SALVAGE VALUE=',F10.2,/,1X,'NB=',F5.0,5X,'NT=',F5.0,/,1X,
     2'RATE OF RETURN ON EQUITY=',F10.3,3X,'RATE OF RETURN ON DEBT=',
     3F10.3,/,1X,'DEBT RATIO=',F10.3,5X,'TAX RATE=',F10.3,5X,/,1X,
     4'CAPITAL INTEREST RATE=',F10.3,5X,'INVESTMENT TAX CREDIT RATE='
     5,F10.3)
      WRITE(6,4500) (COST(I),I=1,N)
 4500 FORMAT(/,2X,'COST DATA;',/,(2X,5(F13.1),/))
      WRITE(6,4750) (ESTSAL(I),I=1,N)
 4750 FORMAT(/,2X,'ESTIMATED SALVAGE VALUE AT THE END OF THE YEAR;',/,
     1(2X,5(F13.1),/))
      STOP
      END

      SUBROUTINE DDBDEP(DEP,INC,SALV,NX,NY)
      REAL INC,NX,NY
      DIMENSION DEP(50),BV(50),RBVSL(50),SDEP(50)
      IF(NX.GE.NY) GO TO 1561
      N=NY
      K=NX
      GO TO 116
 1561 N=NX
      K=NY
  116 D=2./NY
      BV(1)=INC*(1.-D)**1.
      DEP(1)=INC-BV(1)
      RBVSL(1)=(INC-SALV)/NY
      SDEP(1)=DEP(1)
      P=INC-SALV
      M=0
      DO 200 I=2,N
      IF(NY.GT.NX) K=N
      IF(I.LE.K) GO TO 220
      GO TO 210
  220 AI=I
      IF(M.EQ.1) GO TO 205
      BV(I)=INC*(1.-D)**AI
      ADEP=BV(I-1)
      DEP(I)=ADEP-BV(I)
      IF(I.EQ.K) GO TO 900
      RBVSL(I)=(BV(I)-SALV)/(NY-AI)
      IF(DEP(I).LE.RBVSL(I-1)) GO TO 600
      IF(M.EQ.0) DEP(I)=ADEP-BV(I)
      GO TO 250
  205 IF(M.EQ.1) DEP(I)=RBVSL(II)
      GO TO 250
  600 IF(M.EQ.1) GO TO 205
      M=1
      II=I-1
      GO TO 205
```

```
  900 SDEP(I)=DEP(I)+SDEP(I-1)
      IF(SDEP(I).LT.P) DEP(I)=P-(INC-SDEP(I))
      DEP(I)=INC-SDEP(I-1)-SALV
      GO TO 200
  210 DEP(I)=0.
      GO TO 200
  250 SDEP(I)=DEP(I)+SDEP(I-1)
      IF(SDEP(I).LT.P) GO TO 200
      DEP(I)=P-SDEP(I-1)
      JI=I+1
      DO 255 L=JI,N
  255 DEP(L)=0.
      GO TO 70
  200 CONTINUE
   70 RETURN
      END

      SUBROUTINE SLDEP(DEP,INC,SALV,NX,NY)
      REAL INC,NX,NY
      DIMENSION DEP(50)
      IF(NX.GE.NY) GO TO 1564
      N=NY
      K=NX
      GO TO 115
 1564 N=NX
      K=NY
  115 DO 200 I=1,N
      IF(I.GT.NX) GO TO 210
      DEP(I)=(INC-SALV)/NX
      GO TO 200
  210 DEP(I)=0.0
  200 CONTINUE
      RETURN
      END

      SUBROUTINE SYDDEP(DEP,INC,SALV,NX,NY)
      REAL INC,NX,NY
      DIMENSION DEP(50)
      IF(NX.GE.NY) GO TO 1562
      N=NY
      K=NX
      GO TO 117
 1562 N=NX
      K=NY
      DO 200 I=1,N
      IF(I.LE.K) GO TO 220
      GO TO 210
  220 DEP(I)=2.*(NY-I+1.)/NY/(NY+1.)*(INC-SALV)
      GO TO 200
  210 DEP(I)=0.
  200 CONTINUE
      GO TO 230
  117 DO 201 I=1,N
      DEP(I)=2.*(NY-I+1.)/NY/(NY+1.)*(INC-SALV)
  201 CONTINUE
  230 RETURN
      END

      SUBROUTINE PDBDEP(DEP,INC,SALV,NX,NY)
      REAL INC,NX,NY
      DIMENSION BV(50),DEP(50),SDEP(50),P(50)
```

```
      IF(NX.GE.NY) GO TO 1563
      N=NY
      K=NX
      GO TO 118
 1563 N=NX
      K=NY
  118 Q=2.*INC/NY
      IF(SALV.NE.0.) GO TO 35
      D=1./NY
      GO TO 55
   35 D=1.-(SALV/INC)**(1./NY)
   55 BV(1)=INC*(1.-D)
      DEP(1)=INC-BV(1)
      P(1)=(INC-DEP(1))/(NY-1.)
      SDEP(1)=DEP(1)
      DO 200 I=2,N
      IF(NY.GT.NX) K=N
      IF(I.LE.K) GO TO 220
      GO TO 210
  220 AI=I
      IF(I.EQ.K) GO TO 20
      DEP(I)=D*(INC-SDEP(I-1))
      SDEP(I)=DEP(I)+SDEP(I-1)
      IF(SALV.NE.0.) GO TO 200
      P(I)=(INC-SDEP(I))/(NY-AI)
      IF(P(I).GT.Q) GO TO 150
      GO TO 200
   20 DEP(I)=INC-SDEP(I-1)-SALV
      GO TO 200
  210 DEP(I)=0.
      GO TO 200
  150 DEP(I)=Q
      SDEP(I)=SDEP(I-1)+Q
  200 CONTINUE
      RETURN
      END

$ENTRY
```

YEAR	SALV VALUE	CAP REC COST	TAX DEP	RET ON EQTY	DEBT INT
	100000.00				
1	91000.00	9000.00	16363.63	18000.00	2000.00
2	82000.00	9000.00	14727.26	16380.00	1820.00
3	73000.00	9000.00	13090.90	14760.00	1640.00
4	64000.00	9000.00	11454.54	13140.00	1460.00
5	55000.00	9000.00	9818.18	11520.00	1280.00
6	46000.00	9000.00	8181.81	9900.00	1100.00
7	37000.00	9000.00	6545.45	8280.00	920.00
8	28000.00	9000.00	4909.09	6660.00	740.00
9	19000.00	9000.00	3272.73	5040.00	560.00
10	10000.00	9000.00	1636.36	3420.00	380.00

YEAR	INV TAX CRED	TAX	COST	ANN COST	EQIV ANN COST,
1	0.00	11522.73	10000.00	50522.73	50522.73
2	0.00	11540.46	10000.00	48740.46	49712.70
3	0.00	11558.18	10000.00	46958.18	48955.93
4	0.00	11575.91	10000.00	45175.91	48251.75
5	0.00	11593.63	10000.00	43393.63	47598.88
6	0.00	11611.37	15000.00	46611.36	47499.45
7	0.00	11629.08	20000.00	49829.08	47679.81
8	0.00	11646.82	25000.00	53046.81	48005.11
9	0.00	11664.53	30000.00	56264.53	48402.18
10	0.00	11682.27	40000.00	64482.27	49021.63

ECONOMIC USEFUL LIFE = 6 YEARS

INPUT DATA;

INITIAL INVESTMENT COST= 100000.00 SALVAGE VALUE= 10000.00
NB= 10. NT= 10.
RATE OF RETURN ON EQUITY= 0.225 RATE OF RETURN ON DEBT= 0.100
DEBT RATIO= 0.200 TAX RATE= 0.520
CAPITAL INTEREST RATE= 0.000 INVESTMENT TAX CREDIT RATE= 0.000

COST DATA;

10000.0	10000.0	10000.0	10000.0	10000.0
15000.0	20000.0	25000.0	30000.0	40000.0

ESTIMATED SALVAGE VALUE AT THE END OF THE YEAR;

91000.0	82000.0	73000.0	64000.0	55000.0
46000.0	37000.0	28000.0	19000.0	10000.0

APPENDIX F
AREAS UNDER THE NORMAL CURVE

Table F-1[a]

Proportion of total area under the curve that is under the portion of the curve from $-\infty$ to $\frac{X_i - \overline{X}'}{\sigma'}$. ($X_i$ represents any desired value of the variable X)

$\frac{X_i - \overline{X}'}{\sigma'}$	.09	.08	.07	.06	.05	.04	.03	.02	.01	.00
−3.5	.00017	.00017	.00018	.00019	.00019	.00020	.00021	.00022	.00022	.00023
−3.4	.00024	.00025	.00026	.00027	.00028	.00029	.00030	.00031	.00033	.00034
−3.3	.00035	.00036	.00038	.00039	.00040	.00042	.00043	.00045	.00047	.00048
−3.2	.00050	.00052	.00054	.00058	.00058	.00060	.00062	.00064	.00066	.00069
−3.1	.00071	.00074	.00076	.00079	.00082	.00085	.00087	.00090	.00094	.00097
−3.0	.00100	.00104	.00107	.00111	.00114	.00118	.00122	.00125	.00131	.00135
−2.9	.0014	.0014	.0015	.0015	.0016	.0016	.0017	.0017	.0018	.0019
−2.8	.0019	.0020	.0021	.0021	.0022	.0023	.0023	.0024	.0025	.0026
−2.7	.0026	.0027	.0028	.0029	.0030	.0031	.0032	.0033	.0034	.0035
−2.6	.0036	.0037	.0038	.0039	.0040	.0041	.0043	.0044	.0045	.0047
−2.5	.0048	.0049	.0051	.0052	.0054	.0055	.0057	.0059	.0060	.0062
−2.4	.0064	.0066	.0068	.0069	.0071	.0073	.0075	.0078	.0080	.0082
−2.3	.0084	.0087	.0089	.0091	.0094	.0096	.0099	.0102	.0104	.0107
−2.2	.0110	.0113	.0116	.0119	.0122	.0125	.0129	.0132	.0136	.0139
−2.1	.0143	.0146	.0150	.0154	.0158	.0162	.0166	.0170	.0174	.0179

−2.0	.0183	.0188	.0192	.0197	.0202	.0207	.0212	.0217	.0222	.0228
−1.9	.0233	.0239	.0244	.0250	.0256	.0262	.0268	.0274	.0281	.0287
−1.8	.0294	.0301	.0307	.0314	.0322	.0329	.0336	.0344	.0351	.0359
−1.7	.0367	.0375	.0384	.0392	.0401	.0409	.0418	.0427	.0436	.0446
−1.6	.0455	.0465	.0475	.0485	.0495	.0505	.0516	.0526	.0537	.0548
−1.5	0.559	.0571	.0582	.0594	.0606	.0818	.0630	.0643	.0655	.0668
−1.4	.0681	.0694	.0708	.0721	.0735	.0749	.0764	.0778	.0793	.0808
−1.3	.0823	.0838	.0853	.0889	.0885	.0901	.0918	.0934	.0951	.0968
−1.2	0.0983	.1003	.1020	.1380	.1057	.1075	.1093	.1112	.1131	.1151
−1.1	.1170	.1190	.1210	.1230	.1251	.1271	.1292	.1314	.1335	.1357
−1.0	.1379	.1401	.1423	.1448	.1469	.1492	.1515	.1539	.1582	.1587
−0.9	.1611	.1635	.1660	.1685	.1711	.1736	.1762	.1788	.1814	.1841
−0.8	.1867	.1894	.1922	.1949	.1977	.2005	.2033	.2061	.2090	.2119
−0.7	.2148	.2177	.2207	.2236	.2266	.2297	.2327	.2358	.2389	.2420
−0.6	.2451	.2483	.2514	.2548	.2573	.2611	.2643	.2676	.2709	.2743
−0.5	.2776	.2810	.2843	.2877	.2912	.2946	.2981	.3015	.3050	.3085
−0.4	0.3121	0.3158	.3192	.3228	.3264	.3300	.3336	.3372	.3409	.3448
−0.3	.3483	.3520	.3557	.3594	.3632	.3669	.3707	.3745	.3783	.3821
−0.2	.3859	.3897	.3938	.3974	.4013	.4052	.4090	.4129	.4168	.4207
−0.1	.4247	.4288	.4325	.4364	.4404	.4443	.4483	.4522	.4562	.4602
−0.0	.4641	.4881	.4721	.4761	.4801	.4840	.4880	.4920	.4960	.5000

[a]From E. L. Grant and R. S. Leavenworth, *Statistical Quality Control,* 4th Ed., McGraw-Hill Book Co., New York, 1972. By permission of the publishers.

Table F-2[a]

$\frac{X_i - \bar{X}'}{\sigma'}$	.00	.01	.02	.03	.04	.05	.06	.07	.08	.09
+0.0	.5000	.5040	.5080	.5120	.5160	.5199	.5239	.5279	.5319	.5359
+0.1	.5398	.5438	.5478	.5317	.5557	.5596	.5636	.5675	.5714	.5753
+0.2	.5793	.5832	.5871	.5910	.5948	.5087	.6026	6064	.6103	.6141
+0.3	.6179	.6217	.6255	.6293	.6331	.6388	.6406	.6443	.6480	.6517
+0.4	.6554	.6591	.6628	.6664	.6700	.6736	.6772	.6808.	.6844	.6879
+0.5	.6915	.6950	.6985	.7019	.7054	.7088	.7123	.7157	.7190	.7224
+0.6	.7257	.7291	.7324	.7357	.7389	.7422	.7454	.7486	.7517	.7549
+0.7	.7580	.7611	.7642	.7673	.7704	.7734	.7764	.7794	.7823	.7852
+0.8	.7881	.7910	.7939	.7967	.7995	.8023	.8051	.8079	.8106	.8133
+0.9	.8159	.8186	.8212	.8238	.8264	.8289	.8315	.8340	.8365	.8389
+1.0	.8413	.8438	.8481	.8485	.8508	.8508	.8554	.8577	.8599	.8621
+1.1	.8843	.8665	.8686	.8708	.8729	.8749	.8770	.8790	.8810	.8830
+1.2	.8849	.8869	.8888	.8907	.8925	.8944	.8962	.8980	.8997	.9015
+1.3	.9032	.9049	.9066	.9082	.9099	.9115	.1931	.9147	.9162	.9177
+1.4	.9192	.9207	.9222	.9236	.9251	.9265	.9279	.9292	.9306	.9319
+1.5	.9332	.9345	.9357	.9370	.9382	.9394	.9406	.9418	.9429	.9441
+1.6	.9452	.9483	.9474	.9484	.9495	.9505	.9515	.9525	.9535	.9545

+1.7	.9554	.9564	.9573	.9582	.9591	.9599	.9608	.9616	.9625	.9633
+1.8	.9641	.9649	.9658	.9664	.9671	.9678	.9686	.9693	.9699	.9706
+1.9	.9713	.9719	.9726	.9732	.9738	.9744	.9750	.9758	.9761	.9767
+2.0	.9773	.9778	.9783	.9788	.9793	.9798	.9803	.9808	.9812	.9817
+2.1	.9821	.9826	.9830	.9834	.9838	.9842	.9848	.9850	.9854	.9857
+2.2	.9861	.9884	.9868	.9871	.9875	.9878	.9881	.9884	.9887	.9890
+2.3	.9893	.9896	.9898	.9901	.9904	.9906	.9909	.9911	.9913	.9916
+2.4	.9918	.9920	.9922	.9925	.9927	.9929	.9931	.9932	.9934	.9936
+2.5	.9938	.9940	.9941	.9943	.9945	.9946	.9948	.9949	.9951	.9952
+2.6	.9953	.9955	.9956	.9957	.9959	.9960	.9961	.9962	.9963	.9964
+2.7	.9965	.9966	.9967	.9968	.9969	.9970	.9971	.9972	.9973	.9974
+2.8	.9974	.9975	.9976	.9977	.9977	.9978	.9979	.9979	.9980	.9981
+2.9	.9981	.9982	.9983	.9983	.9984	.9984	.9985	.9985	.9986	.9986
+3.0	.99865	.99869	.99874	.99878	.99882	.99886	.99889	.99893	.99896	.99900
+3.1	.99903	.99906	.99910	.99913	.99915	.99918	.99921	.99924	.99926	.99929
+3.2	.99931	.99934	.99936	.99938	.99940	.99942	.99944	.99946	.99948	.99950
+3.3	.99952	.99953	.99955	.99957	.99958	.99960	.99961	.99962	.99964	.99965
+3.4	.99966	.99967	.99969	.99970	.99971	.99972	.99973	.99974	.99975	.99976
+3.5	.99977	.99978	.99978	.99979	.99980	.99981	.99981	.99982	.99983	.99983

[a]Ibid.

APPENDIX G
UNIT NORMAL LOSS INTEGRAL, $L_N D$

Table G-1[a]

D	.00	.01	.02	.03	.04	.05	.06	.07	.08	.09
.0	.3989	.3940	.3890	.3841	.3793	.3744	.3697	.3649	.3602	.3556
.1	.3509	.3464	.3418	.3373	.3328	.3284	.3240	.3197	.3154	.3111
.2	.3069	.3027	.2986	.2944	.2904	.2863	.2824	.2784	.2745	.2706
.3	.2668	.2630	.2592	.2555	.2518	.2481	.2445	.2409	.2374	.2339
.4	.2304	.2270	.2236	.2203	.2169	.2137	.2104	.2072	.2040	.2009
.5	.1978	.1947	.1917	.1887	.1857	.1828	.1799	.1771	.1742	.1714
.6	.1687	.1659	.1633	.1606	.1580	.1554	.1528	.1503	.1478	.1453
.7	.1429	.1405	.1381	.1358	.1334	.1312	.1289	.1267	.1245	.1223
.8	.1202	.1181	.1160	.1140	.1120	.1100	.1080	.1061	.1042	.1023
.9	.1004	.09860	.09680	.09503	.09328	.09156	.08986	.08819	.08654	.08491
1.0	.08332	.08174	.08019	.07866	.07716	.07568	.07422	.07279	.07138	.06999
1.1	.06862	.06727	.06595	.06465	.06336	.06210	.06086	.05964	.05844	.05726
1.2	.05610	.05496	.05384	.05274	.05165	.05059	.04954	.04851	.04750	.04650
1.3	.04553	.04457	.04363	.04270	.04179	.04090	.04002	.03916	.03831	.03748
1.4	.03667	.03587	.03508	.03431	.03356	.03281	.03208	.03137	.03067	.02998
1.5	.02931	.02865	.02800	.02736	.02674	.02612	.02552	.02494	.02436	.02380
1.6	.02324	.02270	.02217	.02165	.02114	.02064	.02015	.01967	.01920	.01874
1.7	.01829	.01785	.01742	.01699	.01658	.01617	.01578	.01539	.01501	.01464
1.8	.01428	.01392	.01357	.01323	.01290	.01257	.01226	.01195	.01164	.01134
1.9	.01105	.01077	.01049	.01022	$.0^{2}9957$	$.0^{2}9698$	$.0^{2}9445$	$.0^{2}9198$	$.0^{2}8957$	$.0^{2}8721$
2.0	$.0^{2}8491$	$.0^{2}8266$	$.0^{2}8046$	$.0^{2}7832$	$.0^{2}7623$	$.0^{2}7418$	$.0^{2}7219$	$.0^{2}7024$	$.0^{2}6835$	$.0^{2}6649$
2.1	$.0^{2}6468$	$.0^{2}6292$	$.0^{2}6120$	$.0^{2}5952$	$.0^{2}5788$	$.0^{2}5628$	$.0^{2}5472$	$.0^{2}5320$	$.0^{2}5172$	$.0^{2}5028$
2.2	$.0^{2}4887$	$.0^{2}4750$	$.0^{2}4616$	$.0^{2}4486$	$.0^{2}4358$	$.0^{2}4235$	$.0^{2}4114$	$.0^{2}3996$	$.0^{2}3882$	$.0^{2}3770$
2.3	$.0^{2}3662$	$.0^{2}3556$	$.0^{2}3453$	$.0^{2}3352$	$.0^{2}3255$	$.0^{2}3159$	$.0^{2}3067$	$.0^{2}2977$	$.0^{2}2889$	$.0^{2}2804$
2.4	$.0^{2}2720$	$.0^{2}2640$	$.0^{2}2561$	$.0^{2}2484$	$.0^{2}2410$	$.0^{2}2337$	$.0^{2}2267$	$.0^{2}2199$	$.0^{2}2132$	$.0^{2}2067$
2.5	$.0^{2}2004$	$.0^{2}1943$	$.0^{2}1883$	$.0^{2}1826$	$.0^{2}1769$	$.0^{2}1715$	$.0^{2}1662$	$.0^{2}1610$	$.0^{2}1560$	$.0^{2}1511$
2.6	$.0^{2}1464$	$.0^{2}1418$	$.0^{2}1373$	$.0^{2}1330$	$.0^{2}1288$	$.0^{2}1247$	$.0^{2}1207$	$.0^{2}1169$	$.0^{2}1132$	$.0^{2}1095$
2.7	$.0^{2}1060$	$.0^{2}1026$	$.0^{3}9928$	$.0^{3}9607$	$.0^{3}9295$	$.0^{3}8992$	$.0^{3}8699$	$.0^{3}8414$	$.0^{3}8138$	$.0^{3}7870$
2.8	$.0^{3}7611$	$.0^{3}7359$	$.0^{3}7115$	$.0^{3}6879$	$.0^{3}6650$	$.0^{3}6428$	$.0^{3}6213$	$.0^{3}6004$	$.0^{3}5802$	$.0^{3}5606$

Table G-1[a]

D	.00	.01	.02	.03	.04	.05	.06	.07	.08	.09
2.9	$.0^{3}5417$	$.0^{3}5233$	$.0^{3}5055$	$.0^{3}4883$	$.0^{3}4716$	$.0^{3}4555$	$.0^{3}4398$	$.0^{3}4247$	$.0^{3}4101$	$.0^{3}3959$
3.0	$.0^{3}3822$	$.0^{3}3689$	$.0^{3}3560$	$.0^{3}3436$	$.0^{3}3316$	$.0^{3}3199$	$.0^{3}3087$	$.0^{3}2978$	$.0^{3}2873$	$.0^{3}2771$
3.1	$.0^{3}2673$	$.0^{3}2577$	$.0^{3}2485$	$.0^{3}2396$	$.0^{3}2311$	$.0^{3}2227$	$.0^{3}2147$	$.0^{3}2070$	$.0^{3}1995$	$.0^{3}1922$
3.2	$.0^{3}1852$	$.0^{3}1785$	$.0^{3}1720$	$.0^{3}1657$	$.0^{3}1596$	$.0^{3}1537$	$.0^{3}1480$	$.0^{3}1426$	$.0^{3}1373$	$.0^{3}1322$
3.3	$.0^{3}1273$	$.0^{3}1225$	$.0^{3}1179$	$.0^{3}1135$	$.0^{3}1093$	$.0^{3}1051$	$.0^{3}1012$	$.0^{4}9734$	$.0^{4}9365$	$.0^{4}9009$
3.4	$.0^{4}8666$	$.0^{4}8335$	$.0^{4}8016$	$.0^{4}7709$	$.0^{4}7413$	$.0^{4}7127$	$.0^{4}6852$	$.0^{4}6587$	$.0^{4}6331$	$.0^{4}6085$
3.5	$.0^{4}5848$	$.0^{4}5620$	$.0^{4}5400$	$.0^{4}5188$	$.0^{4}4984$	$.0^{4}4788$	$.0^{4}4599$	$.0^{4}4417$	$.0^{4}4242$	$.0^{4}4073$
3.6	$.0^{4}3911$	$.0^{4}3755$	$.0^{4}3605$	$.0^{4}3460$	$.0^{4}3321$	$.0^{4}3188$	$.0^{4}3059$	$.0^{4}2935$	$.0^{4}2816$	$.0^{4}2702$
3.7	$.0^{4}2592$	$.0^{4}2486$	$.0^{4}2385$	$.0^{4}2287$	$.0^{4}2193$	$.0^{4}2103$	$.0^{4}2016$	$.0^{4}1933$	$.0^{4}1853$	$.0^{4}1776$
3.8	$.0^{4}1702$	$.0^{4}1632$	$.0^{4}1563$	$.0^{4}1498$	$.0^{4}1435$	$.0^{4}1375$	$.0^{4}1317$	$.0^{4}1262$	$.0^{4}1208$	$.0^{4}1157$
3.9	$.0^{4}1108$	$.0^{4}1061$	$.0^{4}1016$	$.0^{5}9723$	$.0^{5}9307$	$.0^{5}8908$	$.0^{5}8525$	$.0^{5}8158$	$.0^{5}7806$	$.0^{5}7469$
4.0	$.0^{5}7145$	$.0^{5}6835$	$.0^{5}6538$	$.0^{5}6253$	$.0^{5}5980$	$.0^{5}5718$	$.0^{5}5468$	$.0^{5}5227$	$.0^{5}4997$	$.0^{5}4777$
4.1	$.0^{5}4566$	$.0^{5}4364$	$.0^{5}4170$	$.0^{5}3985$	$.0^{5}3807$	$.0^{5}3637$	$.0^{5}3475$	$.0^{5}3319$	$.0^{5}3170$	$.0^{5}3027$
4.2	$.0^{5}2891$	$.0^{5}2760$	$.0^{5}2635$	$.0^{5}2516$	$.0^{5}2402$	$.0^{5}2292$	$.0^{5}2188$	$.0^{5}2088$	$.0^{5}1992$	$.0^{5}1901$
4.3	$.0^{5}1814$	$.0^{5}1730$	$.0^{5}1650$	$.0^{5}1574$	$.0^{5}1501$	$.0^{5}1431$	$.0^{5}1365$	$.0^{5}1301$	$.0^{5}1241$	$.0^{5}1183$
4.4	$.0^{5}1127$	$.0^{5}1074$	$.0^{5}1024$	$.0^{6}9756$	$.0^{6}9296$	$.0^{6}8857$	$.0^{6}8437$	$.^{6}8037$	$.0^{6}7655$	$.0^{6}7290$

[a]These tables of Unit Normal Loss Integral appear in *Probability and Statistics for Business Decisions* by Robert Schlaifer, published by McGraw-Hill Book Company in 1959. They are reproduced here by specific permission of the copyright holder, the President and Fellows of Harvard College.

INDEX